Computational Molecular Science and Engineering Forum 2021

Held at the 2021 AIChE Annual Meeting

Boston, Massachusetts, USA and Online
7 - 11 November and 15 - 19 November 2021

ISBN: 978-1-7138-5714-3

Printed from e-media with permission by:

Curran Associates, Inc.
57 Morehouse Lane
Red Hook, NY 12571

AIChE
120 Wall Street, FL 23
New York, NY 10005-4020

Phone: (800) 242-4363
Fax: (203) 775-5177

www.aiche.org

Additional copies of this publication are available from:

Curran Associates, Inc.
57 Morehouse Lane
Red Hook, NY 12571 USA
Phone: 845-758-0400
Fax: 845-758-2633
Email: curran@proceedings.com
Web: www.proceedings.com

TABLE OF CONTENTS

632190 The Molecular Simulation Design Framework (MoSDeF) Project: A Collaboration Towards Transparent, Reproducible, Usable By Others, Extensible (TRUE) Simulations

Abstract:

The MoSDeF project is a multi-university collaboration focused on the development of software that enables more reproducible scientific workflows [1]. MoSDeF consists of a collection of Python libraries that facilitate the programmatic initialization of arbitrarily complex systems, force field parameterization, and creation of syntactically correct input files for a variety of molecular dynamics and Monte Carlo simulation engines. This project has enabled multiple large-scale computational screening studies of various soft-matter systems [2-5]. A key aspect of the MoSDeF project is the ability to use a single initialization workflow to support different simulation engines (i.e., GROMACS, LAMMPS, HOOMD-Blue, Cassandra, GOMC) and algorithms, enabling direct comparison and validation of results and software [4,6-8]. In recent years, the collaboration has been focusing on further improving the interoperability between tools within the MoSDeF ecosystem (e.g. mBuild, Foyer, Signac, Cassandra, GOMC), as well as with tools that are widely adopted by the simulation community (e.g. GROMACS, LAMMPS). These efforts result in more integration between different packages, and making them more accessible for user groups coming from different backgrounds. The MoSDeF collaboration also promotes a common set of practices to improve the reproducibility of computational simulation researches, namely the TRUE (Transparent, Reproducible, Usable by others, and Extensible) [9]. The TRUE standard essentially encourages better effort in documenting and reporting the simulation workflows, including all the softwares utilized in the process, in a way such that the results can be regenerated by the audience [9]. Libraries in the MoSDeF ecosystem provide utilities necessary to streamline simulation workflow and ease the reporting process, which have been demonstrated in several studies [4, 6-10]. Here, we provide an overview of the continued improvement, development, and updates to the MoSDeF software stack, highlighting recent case studies [4, 9].

References:

1. MoSDeF, https://github.com/mosdef-hub.
2. A. Z. Summers, J. B. Gilmer, C. R. Iacovella, P.T. Cummings, and C. McCabe. "MoSDeF, a Python Framework Enabling Large-Scale Computational Screening of Soft Matter: Application to Chemistry-Property Relationships in Lubricating Monolayer Films", Journal of Chemical Theory and Computation 2020 16 (3), 1779-1793, DOI: 10.1021/acs.jctc.9b01183. Chem. B 123, 1340–1347.
3. M. W. Thompson, R. Matsumoto, R. L. Sacci, N.C. Sanders, P. T. Cummings, 2019. Scalable Screening of Soft Matter: A Case Study of Mixtures of Ionic Liquids and Organic Solvents. J. Phys https://doi.org/10.1021/acs.jpcb.8b11527
4. P.T. Cummings, C. McCabe, C. R. Iacovella, et al. "Open-source molecular modeling software in chemical engineering focusing on the Molecular Simulation Design Framework". *AIChE J.* 2021; 67:e17206. https://doi.org/10.1002/aic.17206

5. A. Z. Summers, C. R. Iacovella, P. T. Cummings, and C. McCabe. "Investigating Alkylsilane Monolayer Tribology at a Single-Asperity Contact with Molecular Dynamics Simulation". Langmuir 2017 33 (42), 11270-11280. DOI: 10.1021/acs.langmuir.7b02479
6. C. Klein, A. Z. Summers, M. W. Thompson, J. B. Gilmer, C. McCabe, P. T. Cummings, J. Sallai, C. R. Iacovella. Formalizing atom-typing and the dissemination of force fields with foyer. Computational Materials Science 2019, 167, 215-227.
7. C. Klein, et al. A Hierarchical, Component Based Approach to Screening Properties of Soft Matter BT - Foundations of Molecular Modeling and Simulation: Select Papers from FOMMS 2015. Edited by Randall Q Snurr et al., Springer Singapore, 2016, pp. 79–92, doi:10.1007/978-981-10-1128-3_5.
8. GMSO: https://github.com/mosdef-hub/gmso.
9. M. W. Thompson, Justin B. Gilmer, Ray A. Matsumoto, Co D. Quach, Parashara Shamaprasad, Alexander H. Yang, Christopher R. Iacovella, Clare McCabe & Peter T. Cummings (2020) Towards molecular simulations that are transparent, reproducible, usable by others, and extensible (TRUE), Molecular Physics, 118:9-10.
10. B. Crawford, J. J. Potoff, M. Jansons. "Prediction of Physical Properties and Phase Behavior for Future Jet Fuel Selection Via Monte Carlo Simulations". November 17 2020, 2020 AIChE Annual Meeting, https://youtu.be/GPaP8tIUY88

632587 Testing and Validation of an Automated Iterative Boltzmann Inversion (IBI) Code

Abstract:

Iterative Boltzmann Inversion (IBI) is a systematic coarse-graining (CG) method in which tabular CG potentials are derived that reproduce target distributions generated from atomistic reference simulations. In IBI, an initial guess for the CG bonded and pair potential is iteratively refined by means of a correction proportional to the difference between the atomistic (targets) and coarse-grained potentials of the mean force. We report here on the release version of a software code which automates the development of multi-bead coarse-grained potentials using IBI. Two major problems make automation difficult: 1) noisy distributions derived from sampling; 2) low sampling regions which introduce discontinuities in the sampling. Both of these make IBI iteration to calculate energy and energy differentiation to calculate forces difficult and error prone. Our code addresses these problems by the use of an approach which combines data smoothing, fitting to functional expansions (bonded potentials), and flexible extrapolation schemes to handle low sampling regions without discontinuity or data distortion. The code is built on a series of reusable Python modules which are designed to enable data handling and data interoperability for applications involving multiscale molecular dynamics data. They also provide hierarchical data representation enabling automated conversion of molecular systems from all-atom (AA) to CG models according to user specified mapping directives. Several use cases will be discussed aimed at providing guidance on usage and comparison with previous work.

630655 Screening of Organic Photovoltaic Morphologies Enabled By Mosdef Tools, Continuous Integration, and Test-Driven Development.

Abstract:

Sustainable power generation is a key part of solving climate change, and organic solar cells may cost-effectively meet this need if their morphologies can be reliably optimized. High performance computing workflows with modeling techniques linking millimeter morphologies to femtosecond charge dynamics are informing organic solar cell formulations and processing, but it is challenging to keep such complex software pipelines Transferable, Reproducible, Usable by others, and Extensible (TRUE). Here we develop open-source containerized software stacks for OPV simulation workflows that are transferable between XSEDE resources, GPU clusters, and developer machines. We develop unit tests and use continuous integration to manage collaborative workflow updates, making extensive use of MoSDeF.org libraries. Using this development framework we test two major component updates: A different forcefield (GAFF vs OPLS) for molecular dynamics simulation, and a different quantum chemical library (pyscf vs ORCA) for calculating charge hopping rates within predicted morphologies. We validate morphology predictions for three previously studied electron donating materials across 450 thermodynamic state points. We validate charge transfer integral distributions and predicted mobilities are insensitive to the pyscf update, which also enables container deployment that was hindered by ORCA's license. This work shows how a prioritization on transferability can guide the modularization of a software stack that both enhances its extensibility and maintains its correctness.

622443 How a Quantum Computer Could Solve a Microkinetic Model

Abstract:

Most advances in quantum computing applied to physical chemistry revolve around solving the Schrödinger equation for atoms and molecules. Little attention has been paid to using quantum computers for solving chemical kinetics problems. Microkinetic models, a widespread approach to solving chemical kinetics problems, are a communicating channel between the molecular mechanism of a reaction and the reactor behavior. Microkinetic models often take the form of a set of nonlinear, coupled ordinary differential equations and/or Bayesian updates. In this oral presentation, a CO oxidation microkinetic model is set up for a quantum circuit. Current limitations in executing the quantum circuit to obtain the solution will be discussed.[1]

1. Walker, E. A.; Pallathadka, S. A., How a Quantum Computer Could Solve a Microkinetic Model. *J. Phys. Chem. Lett.* **2021**, *12*, 592-597.

Graphical abstract

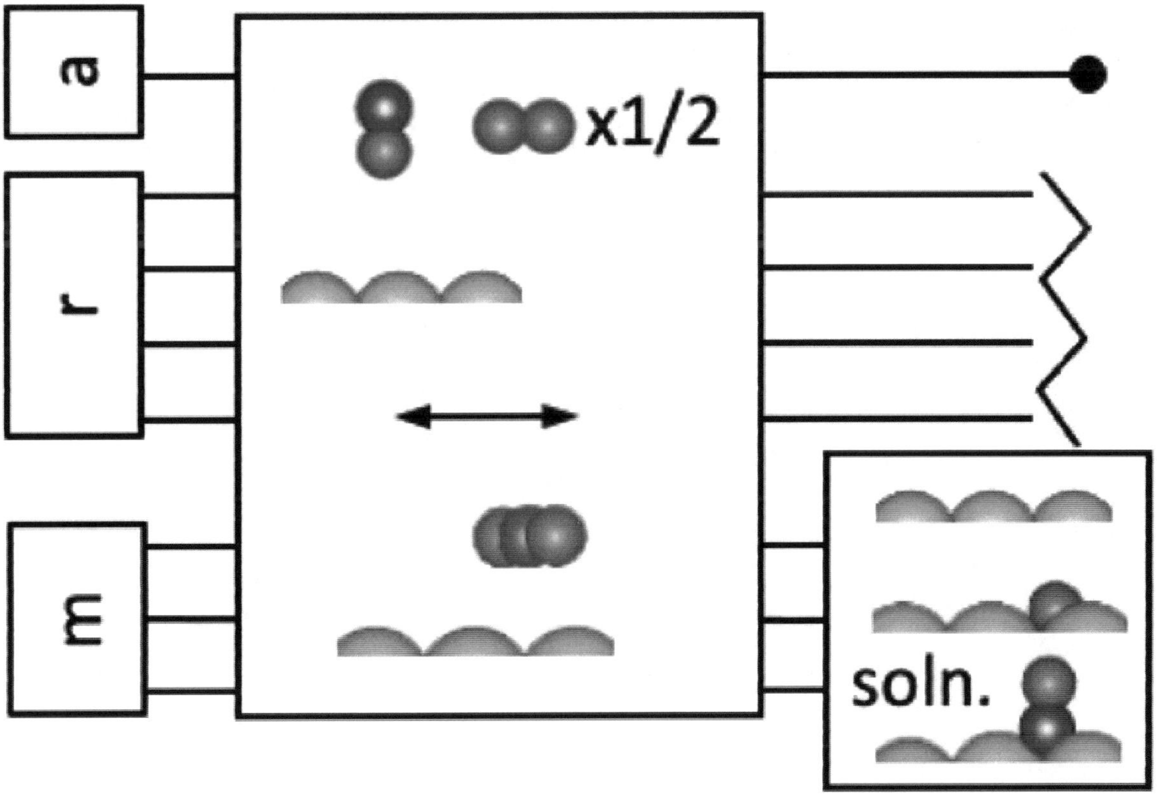

631511 Performant Python Extensions in HOOMD-Blue Version 3.0

Abstract:

In this talk, we describe how we enable performant Python extensions in the core of HOOMD-blue version 3.0, and showcase examples highlighting these features. HOOMD-blue is a molecular simulation engine for molecular dynamics and hard particle Monte Carlo simulations. Our new major release, version 3.0, increases the extensibility, flexibility, and usability of the package by enabling extensions to HOOMD-blue's run loop in Python. With this flexibility, users and developers can build plug-ins and packages written in Python for use with the main package. Python plug-ins provide researchers with quicker prototyping, easier compatibility (no linker or compiler issues), and simpler distribution (conda-forge, PyPI) than compiled C++ plug-ins. This extensibility is accomplished, in part, through custom actions. Custom actions are Python classes that can perform simulation protocols such as writing out system data, resizing the simulation box, or tuning a neighbor list's hyper-parameters for performance. However, pure Python code can be much slower than compiled languages like C++. For this reason, when implementing custom actions, we worked to minimize any performance penalties associated with using the feature. One way we accomplish this is through the implementation of zero-copy access to HOOMD-blue's main data buffers (e.g. particle positions, bonded particles, etc.) both on the CPU and GPU. This allows packages like NumPy, SciPy, and others to efficiently compute on the CPU and packages like PyTorch, CuPy, and others to efficiently compute on the GPU. Via Python's extensions system and packages like Numba and Cython, the custom action's computationally expensive tasks can even be compiled for further performance gains. In many cases, pure Python plug-ins can be made to be sufficiently efficient for research-quality simulations. To further increase extensibility, performance, and facile plug-in creation, we are enabling just-in-time compilation (JIT) for new use cases in HOOMD-blue version 3.0 that will continue to drive and facilitate new research.

630102 GOMC-MoSDeF: A Module for Generating the Required Files for Conducting Monte Carlo Simulations via the GOMC and MoSDeF Software

Abstract:

One of the most critical, time-consuming, and significant bottlenecks in molecular simulations can be generating the initial molecular configurations and the simulation engine's control files. Traditionally, constructing these GPU Optimized Monte Carlo (GOMC) files requires in-depth knowledge of molecular simulations [1, 2], which presents a significant learning curve for new users. The newly added GOMC module expands the capabilities of the existing Molecular Simulation Design Framework (MoSDeF) software [1, 2, 3, 4, 5, 6], allowing molecular configuration, connectivity, force field, and GOMC control files to be created easily via straightforward Python code. The new GOMC module's functionality drastically reduces the entry barrier for new users [1, 2], minimizes the upfront training, and human error in the file generation.

Molecular simulations can help identify and optimize new materials or chemical systems via large-scale screening efforts, in advance of experimental testing, provided that minimal user interaction is needed to generate the required simulation files [1, 2]. In many cases, these large-scale screenings are not possible or prohibitively expensive with current experimental methods. Conventionally constructing these GOMC simulations was time and labor-intensive [1, 2], and required users to write their own detailed workflows. The recent extension of MoSDeF to support GOMC enables easy integration with the Signac software [1, 2, 3, 4, 5, 6, 7, 8, 9], allowing for a significant chemical space to be explored with minimal user interaction. A few examples using this new GOMC-MoSDeF module will highlight the new capabilities, flexibility, and end simulation results using the GOMC simulation engine [1, 2, 3, 4, 5, 6].

References

[1] A. Summers, J. Gilmer, C. Iacovella, P. Cummings and C. McCabe, "MoSDeF, a Python Framework Enabling Large-Scale Computational Screening of Soft Matter: Application to Chemistry-Property Relationships in Lubricating Monolayer Films," J. Chem. Theory. Comput., vol. 16, no. 3, pp. 1779-1793, 2020.

[2] M. Thompson, R. Matsumoto, R. Sacci, N. Sanders and P. Cummings, "Scalable Screening of Soft Matter: A Case Study of Mixtures of Ionic Liquids and Organic Solvents," J. Phys. Chem. B, vol. 123, no. 6, p. 1340–1347, 2019.

[3] Y. Nejahi, M. Barhaghi, J. Mick, B. Jackman, K. Rushaidat, Y. Li, L. Schwiebert and J. Potoff, "GOMC: GPU Optimized Monte Carlo for the simulation of phase equilibria and physical properties of complex fluids," SoftwareX, vol. 9, p. 20–27, 2019.

[4] Y. Nejahi, M. Barhaghi, G. Schwing, L. Schwiebert and J. Potoff, "Update 2.70 to GOMC: GPU Optimized Monte Carlo for the simulation of phase equilibria and physical properties of complex fluids," SoftwareX, vol. 13, p. 100627, 2021.

[5] MoSDeF - the Molecular Simulation Design Framework, "https://github.com/mosdef-hub," 2019. [Online]. [Accessed March 2021].

[6] C. Klein, A. Summers, M. Thompson, J. Gilmer, C. McCabe, P. Cummings, J. Sallai and C. Iacovella, "Formalizing atom-typing and the dissemination of force fields with foyer," Computational Materials Science, vol. 167, pp. 215-227, 2019.

[7] M. Thompson, J. Gilmer, R. Matsumoto, C. Quach, P. Shamaprasad and A. Yang, "Towards molecular simulations that are transparent, reproducible, usable by others, and extensible (TRUE)," Mol. Phys., vol. 118, p. e1742938, 2020.

[8] V. Ramasubramani, C. Adorf, P. Dodd, B. Dice and S. Glotzer, "signac: A Python framework for data and workflow management," in Proceedings of the 17th Python in Science Conference, 152-159, 2018.

[9] C. Adorf, P. Dodd, V. Ramasubramani and S. Glotzer, "Simple data and workflow management with the signac framework," Computational Materials Science , vol. 146, no. C, pp. 220-229, 2018.

629422 Aiidalab, an Ecosystem for Developing, Executing, and Sharing Scientific Workflows

Abstract:

Cloud platforms allow users to execute tasks directly from their web browser and are a key enabling technology not only for commerce but also for computational science. Research software is often developed by scientists with limited experience in (and time for) user interface design, which can make research software difficult to install and use for novices. When combined with the increasing complexity of scientific workflows (involving many steps and software packages), setting up a computational research environment becomes a major entry barrier. AiiDAlab [1,2] is a web platform that enables computational scientists to package scientific workflows and computational environments and share them with their collaborators and peers. By leveraging the AiiDA workflow manager [3,4] and its plugin ecosystem, developers get access to a growing range of simulation codes through a python API, coupled with automatic provenance tracking of simulations for full reproducibility. Computational workflows can be bundled together with user-friendly graphical interfaces and made available through the AiiDAlab app store. Being fully compatible with open-science principles, AiiDAlab provides a complete infrastructure for automated workflows and provenance tracking, where incorporating new capabilities becomes intuitive, requiring only Python knowledge.

[1] A. V. Yakutovich et al., Comp. Mat. Sci. 188, 110165 (2021). DOI:10.1016/j.commatsci.2020.110165

[2] https://www.materialscloud.org/work/aiidalab (accessed on 2020-04-12)

[2] S.P. Huber et al., Scientific Data 7, 300 (2020). DOI:10.1038/s41597-020-00638-4

[3] M. Uhrin et al., Comp. Mat. Sci. 187 (2021). DOI:10.1016/j.commatsci.2020.110086

632666 Signac: Data Management and Workflows for the Molecular Sciences

Abstract:

The signac data management framework (https://signac.io) helps researchers execute reproducible computational studies, scaling from laptops to supercomputers and emphasizing portability and fast prototyping. With signac, users can track, search, and archive data and metadata for file-based workflows and automate workflow submission on high performance computing (HPC) clusters. We will discuss recent improvements to the software's feature set, scalability, scientific applications, usability, and community. Newly implemented synced data structures, workflow subgraph execution, and performance optimizations will be covered, as well as recent research using the framework and the project's efforts on improving documentation, contributor onboarding, and governance. Motivated by the needs of computational molecular scientists for heterogeneous data processing with bespoke parallel execution patterns on widely varying hardware, we will demonstrate how these new capabilities enhance signac's integrated approach to molecular simulations and facilitate the high level of flexibility required by this domain of research.

623495 Delineating and Controlling the Electrode-Electrolyte Interfacial Reactions in High Energy Density Batteries

Abstract:

Rapid increase in global energy use and growing environmental concerns have prompted the development of clean, sustainable, alternative energy technologies. Renewable energy sources like solar and wind are a promising solution, but electrical energy storage (EES) is critical to efficiently utilize electricity produced from renewable sources as they are intermittent. EES is also the only viable near-term option for electrification of transportation sector. Rechargeable batteries are prime candidates for EES, but their widespread adoption for electric vehicles and grid electricity storage requires optimization of cost, cycle life, safety, energy density, power density, and environmental impact, all of which are directly linked to severe component materials challenges and the interfacial reactions between them. This presentation will focus on delineating and controlling the electrode-electrolyte interfacial reactions and consequent crossover of chemical species in high energy density batteries.

With an aim to develop sustainable battery technologies at an affordable cost with high energy density and free from supply chain issues, the presentation will focus on interfacial reactions in two battery systems: (i) lithium-ion batteries with low- or cobalt-free, high-nickel cathodes and graphite or lithium-metal anodes and (ii) lithium-sulfur batteries with sulfur cathode and lithium-metal anode or anode-free cells with Li_2S cathode. The high-nickel cathodes are synthesized with a coprecipitation of transition-metal hydroxide precursors followed by calcining with LiOH under controlled temperature and oxygen flow in order to obtain commercially-relevant particle size and morphologies. Practically-relevant pouch lithium-ion cells or lithium-sulfur cells fabricated with the cathodes and anodes are disassembled and characterized after hundreds or thousands of cycles with advanced characterization techniques, such as X-ray photoelectron spectroscopy, time-of-flight secondary ion mass spectrometry, and high resolution transmission electron microscopy, to track the anode-electrolyte interfacial reactions, cathode-electrolyte interfacial reactions, as well as the crossover of chemical species between the two electrodes through the electrolyte/separator during cycling. Based on the data, the impact of interfacial reactions and crossover of chemical species between the two electrodes on cell performance is established. The understanding gained is utilized to condition the electrode surfaces or modify electrolyte compositions in order to minimize undesirable interfacial reactions and enhance the cycle life of lithium-ion and lithium-sulfur cells.

623899 Investigating Li-Ion Behavior in ALD Coated NMC Cathode Materials Via Molecular Dynamics

Abstract:

High nickel content $LiNi_{1-y-z}Mn_yCo_zO_2$ (NMC) has emerged as a cathode capable of delivering high capacities in Li-ion batteries. In particular, high-nickel NMCs are being used commercially in electric vehicles (EVs) and are projected to comprise over a third of EV cathode chemistries by 2026.[1] However, high-nickel cathodes experience structural instability with the charge/discharge cycling of the battery, which leads to large capacity fading, voltage decay, and low rate capability of the overall battery. Atomic layer deposited (ALD) lithium-containing nano-coatings of NMC cathodes have been shown to help stabilize the transition metal sites while still allowing the movement of Li-ions across the coating, leading to greater capacity retention of the Li-ion battery.[2]

Our prior experimental work has demonstrated that alumina particle ALD produces a less than two nanometer, mixed Al_2O_3 – $LiAlO_2$ film that preferentially covers the transition metals of the cathode powder while leaving lithium relatively uncovered and presumably available for transport. Expanding upon this work, molecular dynamics simulations were used to investigate the behavior of surface lithium of uncoated cathodes and ALD coated cathodes. The Interface Force Field in the 12-6 Lennard Jones form was used to model the pairwise interactions.[3] Both amorphous and crystalline films were simulated at 393 K, with the crystalline film being used as a comparison to the amorphous, ALD-like one. The thickness and uniformity of the ALD coating affects the surface lithium movement. For example, lithium is able to completely travel through one simulated Al_2O_3 layer, while it cannot for two or more simulated Al_2O_3 layers. Further, the energy of the interface between the cathode and coating continues to decrease (i.e., become more favorable) as the simulated coating thickness increases until the energy plateaus off and is about the same for all later coating thicknesses. At this energy plateau, the representative energy of deposition for Al_2O_3 on $LiCoO_2$ is -1760 mJ/m^2 for the amorphous film and -2180 mJ/m^2 for the crystalline film.

By 2030, electric vehicles are projected to comprise almost 30% of new car sales globally. A protective nano-coating on the cathode that preserves battery capacity should help hasten the global transition from combustion vehicles to EVs.

References:

(1) Frith, James. Battery Prices and Manufactured Costs (Highlights from Bloomberg NEF EVO Report 2020), 2021.

(2) Hoskins, A. L.; McNeary, W. W.; Millican, S. L.; Gossett, T. A.; Lai, A.; Gao, Y.; Liang, X.; Musgrave, C. B.; Weimer, A. W. Nonuniform Growth of Sub-2 Nanometer Atomic Layer Deposited Alumina Films on Lithium Nickel Manganese Cobalt Oxide Cathode Battery Materials. *ACS Appl. Nano Mater.* **2019**, *2* (11), 6989–6997. https://doi.org/10.1021/acsanm.9b01490.

(3) Dhakane, A.; Varshney, V.; Liu, J.; Heinz, H.; Jain, A. Molecular Dynamics Simulations of Separator-Cathode Interfacial Thermal Transport in a Li-Ion Cell. *Surf. Interfaces* **2020**, *21*, 100674. https://doi.org/10.1016/j.surfin.2020.100674.

632863 Reserve Lithium-Ion Batteries for Lithium-Ion Free Cathodes

Abstract:

Lithium-ion Batteries' (LIBs) influence on every aspect of human lives has been unimaginable. One sector, which has received a tremendous boost because of LIBs is transportation. Electric vehicle (EV) sales have increased almost 20 folds from 2011 until now. This has caused a huge increase in the demand for LIBs across the world. Presently, there is heavy dependence on cathodes like $LiFePO_4$, $LiNiO_2$, $LiCoO_2$, and $LiNi_xMn_yCo_zO_2$ to meet energy expectations. Compounds without lithium in their lattice rarely have been used as cathodes either due to inability for reverse reactions or require lithium electrode, which causes lithium dendrites on continuous cycling. This has resulted in fewer efforts towards the development of such lithium-free cathodes. Some of these materials are ZnO, TiO_2, Cr_3O_4, MnO_2, V_2O_5, etc. Although these materials possess high theoretical capacities, many of them have a low potential for faradic reactions and have the risk of reducing to metal states during the electrochemical reactions. Hence, for such materials to commercialize, few measures have been taken namely, pre-lithiation of the cathode, film-forming additives, lithium additives, blended cathodes, etc. Some strategies like pre-lithiation are not practical and feasible, even though they help in overcoming solid electrolyte layer interface (SEI). Other strategies alter the cell voltage characteristics due to the presence of voids in the electrode.

We present a pioneering configuration called reserve lithium-ion batteries for in situ lithiation of the lithium-free electrodes with the use of lithium-electrode as a reservoir. For the initial formation of the SEI and lithiation of the MCMB electrode, a reservoir lithium electrode was brought into the cell circuitry. V_2O_5 cathode with meso carbon microbeads (MCMB) was used to fabricate high specific capacity and energy-dense reserved lithium-ion batteries (RLIBs). The RLIB system delivered a high charge specific capacity of 245 mAh g^{-1} at 0.2C rate. The tri-electrode system provided an initial high energy density of >880 Wh kg^{-1}. A parallel comparison of electrochemical performance between V_2O_5 half-cell, V_2O_5-MCMB full-cell, and V_2O_5-MCMB RLIB was conducted and observed. There is a distinct resemblance in the performance of RLIB performance with half-cell profile and in terms of capacity versus cycle number. The pouch-cell configurations of this work yielded a high energy-dense system, which cycled stably over more than 200 cycles at 0.5C rate. Using this innovative concept of reservoir charging would be of great importance for applications in various sectors of the society viz., space exploration, UPS, healthcare, EVs, grid storage, etc.

625813 Model for Disordered Proteins with Strongly Sequence-Dependent Liquid Phase Behavior

Abstract:

Phase separation of intrinsically disordered proteins is important for the formation of membraneless organelles, or biomolecular condensates, which play key roles in the regulation of biochemical processes within the cells. In this talk, we present the phase separation behavior of different sequences of a coarse-grained model for intrinsically disordered proteins and show that they exhibit a surprisingly rich phase behavior. Both the fraction of total hydrophobic parts and the distribution hydrophobic parts are investigated. We observed not only conventional liquid-liquid phase separation, but also reentrant phase behavior, in which the liquid phase density decreases at lower temperatures. For some sequences, we also observe formation of open phases consisting of aggregates, rather than a normal liquid. These aggregates had overall lower densities than the conventional liquid phases and complex geometries with large interconnected fibril-like or membrane-like clusters. Minor alterations in the ordering of a given set of amino acids may lead to large changes in the overall phase behavior, a fact of significant potential relevance for biology.

633387 A Touch of Non-Linearity: Mesoscale Swimmers and Active Matter in Fluids

Abstract:

Living matter, such as biological tissue, can be seen as a nonequilibrium hierarchical assembly of assemblies of smaller and smaller active components, where energy is consumed at many scales. The functionality and versatility of such living or "active-matter" systems render it a promising candidate to study and to synthetically design. While many active-matter systems reside in fluids (solution, blood, ocean, air), so far, studies that include hydrodynamic interactions have focussed on microscopic scales in Stokes flows, where the active particles are <100um and the Reynolds number, Re <<1. At those microscopic scales viscosity dominates and inertia can be neglected. However, what happens as swimmers slightly increase in size (say ~0.1mm-100cm) or as they form larger aggregates and swarms? The system then enters the intermediate Reynolds regime where both inertia and viscosity play a role, and where nonlinearities in the fluid are introduced. In this talk, I will present a simple model swimmer used to understand the transition from Stokes to intermediate Reynolds numbers, first for a single swimmer, then for pairwise interactions and finally for collective behavior. We show that, even for a simple model, inertia can induce hydrodynamic interactions that generate novel phase behavior, steady states and transitions.

624231 Binary Solute Adsorption in Surfactant Bilayers with Gibbs Ensemble Monte Carlo Simulations

Abstract:

Solute uptake into microstructured soft materials is a complex phenomenon with applications spanning the delivery of therapeutics to the release of personal care products. Previous attempts to simulate this process have successfully estimated the critical density of solute that could load into a surfactant system, but determining structural details at the molecular level has proved challenging. Using configurational-bias Monte Carlo simulations in the osmotic Gibbs ensemble, we have been able to determine the loading of unary n-nonane and 1-hexanol into a triethylene glycol mono-n-decyl ether (C10E3) surfactant bilayer, as well as the resulting structural effects caused by each solute. However, due to solute interactions, adsorption becomes more complex in multicomponent systems. Here, we apply a similar method to the binary adsorption of both n-nonane and 1-hexanol into the same surfactant bilayer system, allowing us to determine the partial molar properties of each solute and compare the favorability of molecular interactions within this bilayer system. We further investigate these solute interactions by comparing simulation to the behavior predicted by Ideal Adsorbed Solution Theory.

622070 Chemistry Under Extreme Conditions: Accelerating Research through Artificial Intelligence

Abstract:

Many of the most challenging problems in chemical engineering involve reactivity in complicated systems. First principles methods like density functional theory (DFT) and density functional tight binding (DFTB) are attractive for studying these problems due to their predictive power, however, both methods are confined to systems of a few 100s of atoms, and in the case of DFT, 10s of ps. These limitations have motivated extensive work towards development of accurate, scalable, and reliable reactive interatomic potentials but these models generally rely on complicated functional forms requiring time consuming parameterization schemes. As a result, the physiochemical space over which parameters are available is limited.

We have developed ChIMES, an artificial intelligence-based reactive interatomic potential generation and simulation approach, to overcome these challenges. ChIMES is particularly well suited for studies that would otherwise rely on costly quantum simulations, and to date, has been successfully applied to problems spanning chemical reactivity under extreme conditions to hydriding in solid state materials (i.e., problems with characteristic time and lengths many orders of magnitude greater than is accessible to standard DFT). Ultimately, ChIMES has enabled quantum accurate atomistic simulations on scales overlapping with experiments, in many cases, for the first time. Details of the ChIMES framework will be presented along with its application towards achieved an improved understanding of shockwave-driven chemistry in organic molecular materials.

This work was performed under the auspices of the U.S. Department of Energy by Lawrence Livermore National Laboratory under Contract DE-AC52-07NA27344.

632118 Modeling Diffusion Processes in the Zeolite Nanopores

Abstract:

Transport of reactants and products in the nanopores of zeolites is of fundamental importance in field of catalysis. Although there has been selected investigation to determine transport of gas molecules in the zeolitic pores, such investigations are rarely undertaken in the condensed phase. In this talk, we will share transport properties of aqueous solution of β-glucose in the nanopores of selected zeolitic frameworks.

623078 Discovery of Novel Compounds and Pathways through Identification of Bioprivileged Molecules

Abstract:

Bioprivileged molecules are biology-derived chemical intermediates that can be efficiently converted to a diversity of chemical products including both novel molecules and drop-in replacements. Bridging chemical and biological catalysis by bioprivileged molecules provides a useful and flexible new paradigm for producing biobased chemicals. However, the discovery of bioprivileged molecules has been demonstrated to require extensive experimental effort over a long period of time. To meet this need, we developed a computational framework for identification of candidate bioprivileged molecules from $C_6H_xO_y$ (C6), $C_4H_xO_y$ (C4), $C_5H_xO_y$ (C5), and $C_7H_xO_y$ (C7) molecule subspaces that ranks the molecules according to a number of diverse criteria and generates products emanating from them using automated reaction network generation. All top candidates were analyzed for their key functional moieties using a random forest model, and this algorithm was applied to compare the functional group space occupied by bioprivileged molecules of various databases of molecules with a focus on evaluating how closely the molecules were aligned with those known to biology. The framework is sufficiently automated and flexible that it can be easily expanded to include other chemical formulae to screen for bioprivileged candidates. This in turn facilitates the retrosynthesis process inherent in the framework to identify those bioprivileged intermediates in other subspaces that lead to target molecules. The application to discovery of known and novel monomers for poly(hydroxyurethanes) that are derived from biobased molecules and lead to recyclable materials will be discussed.

623936 The Machine Learning Route to Accelerated Discovery and Inverse Design of Chemical and Materials Systems

Abstract:

The use of modern machine learning, informatics, and data mining approaches is a relatively new development in the chemical and materials domain. These techniques have been exceedingly successful in other application fields, and since there is no fundamental reason why they should not have a similarly transformative impact on chemical and materials research, there is now a concerted effort by the community to introduce data science in this new context. They hold tremendous promise for the practical realization of accelerated discovery and inverse design. However, adapting techniques from other application domains for the study of chemical and materials systems requires a substantial rethinking and redevelopment of the existing methods.

In this presentation, we will discuss our work on designing advanced, physics-infused neural network architectures, the fusion of unsupervised clustering with supervised regression for local ensemble models, active and transfer learning techniques, bootstrapping approaches to minimize our training data footprint, methods to increase the applicability domain of data-derived models, and automated hyperparameter optimization.

623770 A Universal Framework for Featurization of Atomistic Systems

Abstract:

A major challenge in the development of machine-learning force fields is the fact that most featurization schemes are element-specific, causing them to scale poorly with the number of elements in the system, and requiring the development of specialized machine-learned force fields for systems containing different combinations of elements. This prevents the construction of general-purpose machine-learned force fields, and makes datasets containing many (5+) elements impractical to treat with feature-based machine-learning models. In this talk we introduce the Gaussian multi-pole (GMP) featurization scheme (https://arxiv.org/abs/2102.02390) that utilizes physically-relevant multipole expansions of the electron density around atoms to yield feature vectors that interpolate between element types and have a fixed dimension regardless of the number of elements present. We combine GMP with neural networks to directly compare it to the widely-used Behler-Parinello symmetry functions for the MD17 dataset, revealing that it exhibits improved accuracy and computational efficiency. Further, we demonstrate that GMP-based models can achieve chemical accuracy for the QM9 dataset, and their accuracy remains reasonable even when extrapolating to new elements. Finally, we test GMP-based models for the Open Catalysis Project (OCP) dataset, revealing comparable performance and improved learning rates when compared to graph convolutional deep learning models. The results indicate that this featurization scheme fills a critical gap in the construction of efficient and transferable reactive force fields.

631803 Transfer Learning Using Large-Scale ML Models for Catalyst/Molecular Datasets

Abstract:

The development of renewable energy technologies relies heavily on efficient and cost-effective catalysts. Unfortunately, screening millions-billions of unique catalysts is infeasible with traditional ab-initio techniques. Model development has been limited by the availability of large, diverse datasets spanning across composition and chemical space. To address this, we present a transfer learning approach to use learned representations from a large pre-trained model to aid in training alternative, smaller datasets. Pretrained models from the Open Catalyst 2020 Dataset, trained on 130M+ calculations spanning 55 elements and 80 adsorbates, were used to retrain smaller, densely sampled datasets. Initial layers of ML models tend to learn basic representations of input features, atomic embeddings in this case. These representations and embeddings can be assumed to not change much across different catalytic systems and therefore a pre-trained model helps in providing a good prior for the task. We demonstrate this approach for datasets from 2 different domains. One in the catalysis domain using a literature CO dataset and the other in the domain of small molecules on the MD17 dataset. We show that the pre-trained model gives a 15% improvement for in-domain CO dataset and gives better results when tested on unseen molecules for MD17. This work aims to be the benchmark for how pre-trained ML models can be used for predictions of various small-scale catalyst or molecular datasets.

625886 Semi-Supervised Learning Detects Safe Islands Where Density Functional Theory Is Applicable for Chemical Discovery

Abstract:

Density functional theory (DFT) plays an essential role in the discovery of new molecules and materials because of its optimal balance between accuracy and computational cost. Despite this balance, DFT can be inaccurate for strongly correlated systems (e.g. containing transition metals or stretched bonds) and chemical reactions that involve bond breaking and formation. To address this challenge, numerous multi-reference (MR) diagnostics have been developed over the years. These diagnostics aim to detect strong correlation and identify when single-reference methods (such as DFT) are insufficient for obtaining molecular properties. Despite the emergence of dozens of diagnostics, no single diagnostic has been universally predictive of MR character. Additionally, many such diagnostics are based on computationally demanding calculations (i.e., wave function theory, or WFT) that may also be intractable. Therefore, we aim to generate a low-cost and universal metric that measures whether a system is suitable for DFT evaluation. We generate a dataset that contains 15 MR diagnostics from multiple levels of theory on equilibrium and distorted geometries of organic molecules. We observe poor linear correlation across 15 MR diagnostics, with more expensive WFT based diagnostics outperforming the DFT based diagnostics in predicting a figure of merit for MR character. We train supervised learning models to predict more demanding WFT diagnostics using the DFT diagnostics along with the molecular geometries as inputs, to eliminate the cost of performing the WFT calculations. We show how a semi-supervised learning classifier unravels the hidden consensus of the 15 MR diagnostics by utilizing the distribution of the diagnostics for all data points. We demonstrate that this classifier is robust to noisy inputs and thus can achieve the same accuracy even it only sees the ML-predicted WFT diagnostics as the model inputs. Since our workflow only requires DFT calculations, we expect this data-driven decision engine to be useful in detecting "safe islands" for DFT based high-throughput computation and guiding the selection of the optimal electronic structure methods for the evaluation of new chemical systems in chemical discovery.

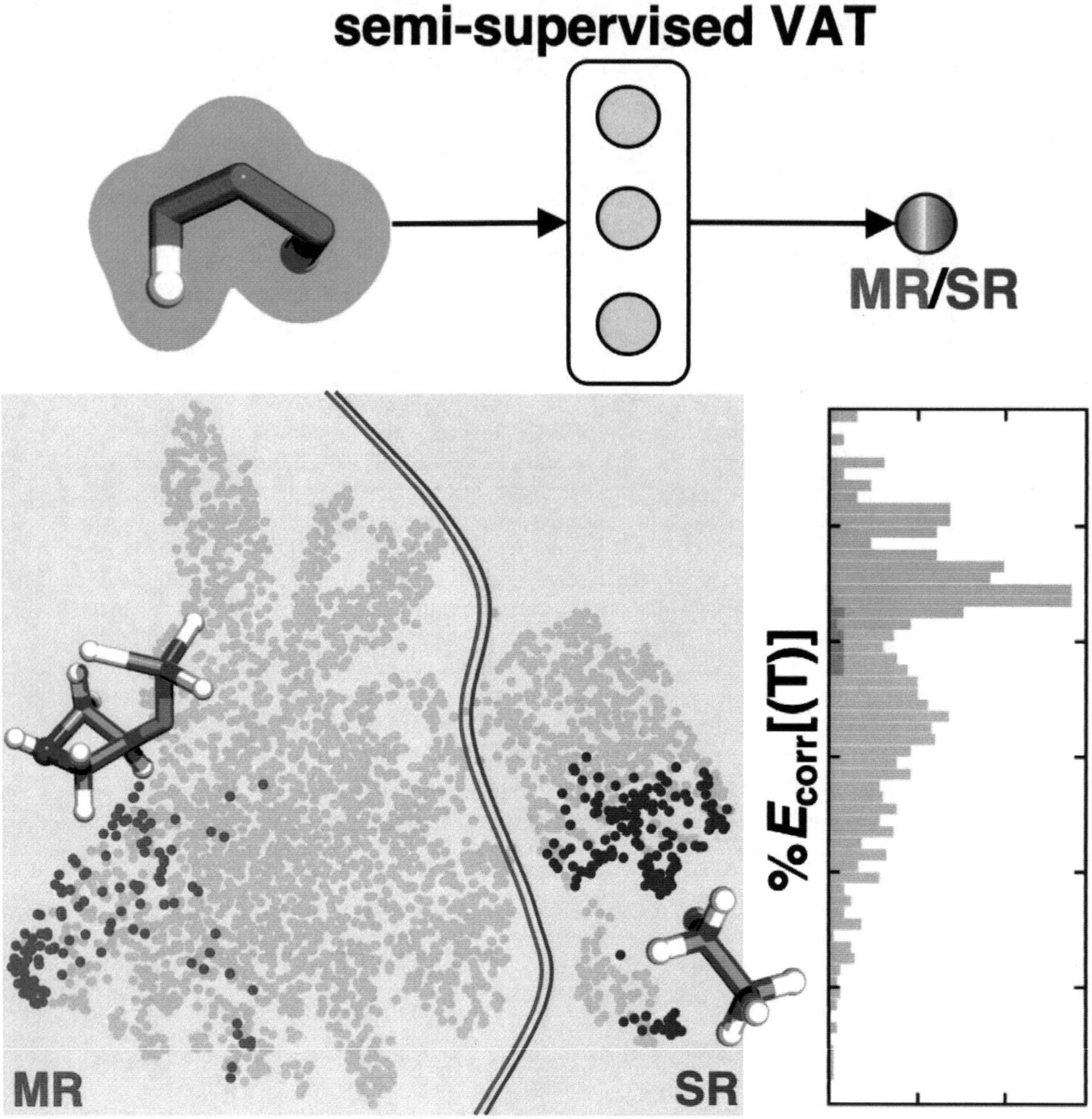
semi-supervised VAT
MR/SR
MR
SR
%E_{corr}[(T)]

622036 *In silico* discovery of Target Peptide Inhibitors for *C. Diff* toxins a and B

Abstract:

Clostridium difficile infection is the leading cause of diarrhea and colitis (inflammation of the colon) in North America and Europe. The fraction of the population infected by the disease is increasing as new strains associated with significant morbidity and mortality have appeared. There is a growing concern about the failure of the first line of treatment for *C. diff* infection (metronidazole and vancomycin) and thus, current attention is focused on the need for alternative treatment options like biologic drugs which are safer and more efficacious. A viable and cost-effective strategy is the development of targeted peptide-based therapeutics that prevent and treat *C. diff* infections by deactivating the pathogen while leaving the resident gut microbiota unharmed.

During infection, the *C. diff* pathogen produces two large virulent toxins (Toxin A and B) that share 71% sequence homology. The glucosyltransferase domain (GTD) of these toxins acts by binding UDP (Uridine diphosphate)-glucose, hydrolyzing it into glucose and UDP, and attaching the glucose monomer to the human Rho family of GTPases. Glycosylation of the GTPases by TcdA and TcdB GTDs leads to disruption of the cytoskeleton, apoptosis and cell death.

The objective of this project is to computationally design peptide inhibitors that bind with high affinity and specificity to the glucosyltransferase domain of *C. diff* toxins A and B to inhibit its activity. We have been successful in developing a proof-of-concept disruptive computational approach to design peptides that bind to a biomolecular target with high affinity. Our peptide search algorithm employs Monte Carlo methods, self-consistent mean-field theory, and the concerted rotation (CONROT) technique to search for peptides in sequence and conformation space that bind to a target protein. The starting input structure for the algorithm requires a reference ligand which is usually an experimentally determined peptide sequence that has a good binding affinity to the target protein.

We have discovered two peptides inhibitors, NPA and NPB, using EGWHAHTGGG (K_d = 100 nM) as the reference ligand, a peptide identified by the Feig lab using phage display. Simulations of NPA and NPB at the binding site of Toxin A show that they have a high affinity for the toxin. We estimated the binding free energies for these peptides and found that they were better binders than the reference ligand. The peptides have shown good neutralization against Toxin A when tested *in vitro* in jejunum (intestinal) cells but showed no effect in colon cells. Current work is focused on furthering our understanding of the interaction of these peptides with the toxins, redesigning shorter peptide inhibitors for Toxin A, and discovering peptides that can bind to Toxin B.

627711 You May Not Have Noticed, but Your Neural Network Did: Machine Learning from Simulated Enzyme Variants

Abstract:

Molecular simulation of enzymes has long been established as a valuable tool for making predictions about and developing rationalizations for observable behaviors such as reaction efficiency or thermostability. These (relatively) macroscopic observables are emergent properties of highly intricate interatomic relationships, but the exact nature of the relationships between particular subsets of the whole system and the observables of interest are most often too complex for humans to interpret. Nevertheless, it must be the case that given the appropriate framing and enough data, a suitable model could be trained to accurately quantify the relationship between each component of the system and the whole. Here, we train a machine learning model on molecular features from a large set of simulation data for variants of a particular engineered enzyme. We investigate the ability of the model to identify non-obvious relationships between components of the system, and to make testable predictions that help guide directed evolution experiments. Eventually, models that encode molecular-level understanding of enzyme structure-function relationships can be used to intelligently select desirable multiple mutants, even when the component single mutations are undesirable, thereby addressing a major shortcoming of experimental directed evolution.

627825 Deep Reinforcement Machine Learning Driven Vaccine Design Against Highly Mutable Pathogens

Abstract:

Made abundantly clear by the current COVID-19 pandemic, there is a pressing need for new approaches to rapidly develop robust vaccines against highly mutable pathogens. SARS-CoV-2 has demonstrated the worst-case scenario in this regard, however, HIV and influenza continue to pose serious threats annually because of the lack of universal vaccines against these pathogens. To this end, computational methods have the potential to revolutionize the development of such vaccines. Recent advances in agent-based models of affinity maturation (AM) – the Darwinian process by which antibodies evolve against a pathogen – have enabled the use of real nucleotide and amino acid sequences of B cell receptors and pathogenic proteins (antigens; Ags), respectively, in rational vaccine design. However, the vast sequence space of the vaccine-candidate Ags does not feasibly allow for iterative Ag sequence design due to the high computational demands. Here, we present the development of an efficient and robust computational pipeline for designing vaccines against highly mutable pathogens. This is facilitated by coupling deep reinforcement learning (DRL) algorithms to agent-based models of AM to effectively 'steer' the AM process towards an optimal solution, namely, towards evolving large quantities of cross-reactive antibodies against the administered Ags. We first describe coupling off-the-shelf DRL algorithms to a previously published coarse-grained model of AM to determine optimal coupling performance parameters, considering both speed and accuracy against previously published results. We then describe coupling DRL to the more realistic AM model described earlier, exploring the optimal way in which to administer four HIV-based Ags with fixed sequences across three sequential immunizations. Our results serve as a proof-of-concept for efficient and biologically-relevant Ag sequence design using next generation computational and data-driven methods. Future work will include development of a computational pipeline to explore the Ag sequence space and determine the optimal amino acid sequences of HIV-based Ags for evolving cross-reactive antibodies across various numbers of immunizations. Furthermore, we plan to optimize this pipeline to encompass a larger repertoire of pathogens, such as influenza and malaria. Using this approach, we hope to accelerate the development of universal vaccines against a variety of highly mutable pathogens that remain significant burdens on global health today.

622757 Data-Driven Reconstruction of Molecular Folding Trajectories from Single-Molecule Experimental Measurements

Abstract:

Single molecule fluorescence resonance energy transfer (smFRET) measures intramolecular distances in real time, allowing for experimental tracking of dynamics of molecules. These distances serve as low-dimensional observables of the molecular trajectory but do not directly furnish the high-dimensional atomistic trajectory. Takens' Delay Embedding Theorem asserts that under sufficiently long and frequent sampling a time series of generic scalar observables contain the same dynamical systems information as the all-atom trajectory up to symmetries in the observable. By combining Takens' Theorem with manifold learning, deep learning, and molecular simulations we present an approach termed Single-molecule TAkens Reconstruction (STAR) to reconstruct atomistic trajectories from experimentally measurable observables. We demonstrate the approach in applications to long all-atom simulations the proteins Chignolin, BBL, Villin and NTL9. In addition, we investigate the role noise, data and time resolution to show that the robust denoising and filtering properties of the manifold learning techniques enable molecular trajectories can be reconstructed to within 0.2 nm accuracy in the mean atomic positions under realistic conditions of experimental smFRET data collection. This synthesis of dynamical systems theory, data-driven learning, and statistical thermodynamics establishes STAR as a useful new tool to analyze and process single molecule measurements allowing for robust extraction of atomistic molecular trajectories.

632098 High-Throughput Screening of Tribological Properties of Monolayer Films Using Molecular Dynamics and Machine Learning

Abstract:

Monolayer films are promising candidates as lubricants for nanoscale devices in which sliding surfaces are in contact. In such systems traditional lubricants have been shown to have limited effectiveness as they can undergo a phase transition (from liquid to solid) when confined at the nanoscale [1, 2]. Monolayer films consisting of chains chemisorbed to a surface have been shown to provide protection against wear and reduce frictional forces in the event of contact. Various tunable properties of the monolayer including terminal group chemistry, backbone chain composition as well as monolayer composition have an impact on the lubricating ability of these films [2]. However, the parameter design-space is vast and requires significant computational resources to study. To address this, we have conducted a high-throughput screening study using molecular dynamics (MD) simulations and machine learning (ML) to determine optimal monolayer film designs, focusing on terminal group chemistry and monolayer composition. Approximately 120,000 simulations have been performed, representing nearly 10,000 unique monolayer film combinations. The cheminformatics lib RDKit, which is used to determine the "molecular fingerprint" of each system [7], and random forest regressor machine learning algorithm was used to create predictive ML models from the simulation data [8]. The effectiveness of the ML models and the optimal amount of training data is assessed. Several promising systems to be further investigated were identified highlighting how ML models can be used in lieu of simulation data to direct screening studies [3]. Such early evaluations of thin film designs can speed up the discovery process by minimizing the number of compute-intensive simulations that need to be performed. The MD simulations and high-throughput screening workflow was set up using the MoSDeF framework, ensuring the simulations are TRUE (transferable, reproducible, usable by others and extensible) [9]. The Signac framework was used to create and manage the parameter space and control the simulation workflow [5].

References

1. R. Maboudian, W. R. Ashurst, and C. Carraro. Tribological Challenges in Micromechanical Systems. *Tribology Letters* 12, 95–100 (2002). https://doi.org/10.1023/A:1014044207344
2. P. T. Cummings, H. Dochert, C. R. Iacovella, and J. K. Singh. (2010), Phase transitions in nanoconfined fluids: The evidence from simulation and theory. AIChE J., 56: 842-848. https://doi.org/10.1002/aic.12226
3. Y. L. Pavlov. (2019) 'Random forests', Random Forests, pp. 1–122. doi: 10.1201/9780429469275-8.
4. "MoSDeF" [Online]. Available: https://mosdef.org.
5. "Signac Framework" [Online]. Available: https://signac.io.
6. M. J. Abraham, T. Murtola, R. Schulz, S. Páll, J. C. Smith, B. Hess, E. Lindahl. "GROMACS: High performance molecular simulations through multi-level parallelism from laptops to supercomputers". SoftwareX, Volumes 1–2, 2015, Pages 19-25. https://doi.org/10.1016/j.softx.2015.06.001.
7. "RDKit Library" [Online]. Available: https://github.com/rdkit/rdkit.
8. "Scikit-learn" [Online]. Available: https://github.com/scikit-learn/scikit-learn
9. M. W. Thompson, J. B. Gilmer, R. A. Matsumoto, C. D. Quach, P. Shamaprasad, A. H. Yang,

C. R. Iacovella, C. McCabe and P. T. Cummings (2020) Towards molecular simulations that are transparent, reproducible, usable by others, and extensible (TRUE), Molecular Physics, 118:9-10, DOI: 10.1080/00268976.2020.1742938

633516 Opening Remarks By the Chair/Co-Chair

Abstract:

The Chair/Co-Chair will introduce the session on Plasma Catalysis

628891 DFT-Informed Energetics of Plasma-Enabled Reactions Pathways and Microkinetic Modeling for Ammonia Synthesis on Transition and Low-Melting Point Metals

Abstract:

Ammonia is a crucial chemical (> 200 millions tons a year) due to its widespread use in fertilizer production, and its potential as a hydrogen carrier in a future "hydrogen economy." Currently synthesized using thermal catalysis, ammonia synthesis is currently a fossil energy-powered, high temperature-high pressure, energy-intensive process that is responsible for ~2% of the world's CO2 emissions. Ideally, as the world transitions to cleaner, renewable energy, the net CO2 emissions with ammonia synthesis would decrease. However, the harsh conditions required for thermally catalyzed ammonia synthesis demands a highly centralized, continuously-on process that is incompatible with renewable energy. Thus, currently, alternative methods to synthesize ammonia under milder conditions are actively sought, including plasma catalysis in plasma reactors.

Optimizing plasma-assisted ammonia synthesis, including optimizing the catalyst to be loaded on the reactor, is currently challenging due to an incomplete understanding of the most relevant reaction mechanisms involved in the process. One snippet of this debate include the role of vibrational versus radical species on ammonia formation. Some first principles-informed models have attempted to shed light on the relevant mechanism, but centered on vibrational species and only the typical Langmuir-Hinshelwood reactions associated with thermal catalysis were considered [1]. More comprehensive models including plasma radicals also attempted similar insights, but used generic, metal-agnostic parameters, and while including Eley-Rideal reactions, it only considered NH_X species [2]. These models thus may be at odds with experiments reporting the appearance of N2Hx species on the metal [3].

To enable first principles-informed comprehensive modeling of all plausible Langmuir-Hinshelwood (LH) and Eley-Rideal (ER) reactions involving N_YH_X surface species and plasma radicals we performed density functional theory (DFT) to obtain reaction energies for 50+ reaction events across 9 metal surfaces of different degrees of nitrophilicity (Fe, Co, Ni, Pd, Cu, Au, Ag, Ga, Sn), and used scaling relationships (both from the literature and new ones generated from our own select DFT calculations) to provide corresponding activation energies. Hydrogen dissolution events were also included in our calculations, motivated by previous observation of a hydrogen sink effect in experiments by some of us [4]. Ultimately, the obtained data was input into a microkinetic model, which we used to explore the effect of plasma composition on catalyst state and dominant reaction pathways. Our results indicate that plasma radical species such as N(g) and H(g) only need minimal concentration in the plasma phase (even several orders of magnitude lower than that of vibrational species) to make ER pathways dominant and the formation of N_YH_X species plausible. However, N_YH_X seem more likely to hydrogenate until saturation and spontaneous dissociation via ER reactions with H(g) radicals than to dissociate early via LH reactions. The dominant role of ER reactions makes for a crucial factor blurring differences in activities across different metals. Finally, we present relationships between energetic descriptors and experimentally measured catalyst activities in DBD and RF reactors.

1. Metha et al. Nature Catal. 2018, 1, 269-265

2. Van't Veer et al. *J. Phys. Chem. C* 2020, 124, 42, 22871–22883

3. Lea et al. *ACS Catal.* 2020, 10, 24, 14763–14774

4. Shah et al. ChemCatChem 2020, 12, 1200-1211

626387 Plasma-Assisted Catalytic Approaches for the Production of Chemicals

Abstract:

Plasma-assisted catalysis has grown rapidly in the past few years as a nonthermal, ambient pressure approach to activate stable molecules (e.g., N2). Utilizing renewable energy, plasma processes have the potential to reduce the CO2 footprint compared to thermal processes, and convert reactants at point sources owing to the modular nature of these systems. Our research in the area of plasma-assisted catalysis is focused on the ability to control the catalyst performance by tuning the plasma environment or plasma-catalyst interactions. We have been particularly interested in the use of nonthermal plasmas for C-H and N2 activation via dry reforming of methane and ammonia synthesis reactions, respectively. In recent publications from our groups, we observed remarkable synergy between the plasma and catalyst for NH3 synthesis from N2/H2: 1) the rate enhancements with plasma stimulation were more sensitive to N2 concentration, 2) the ideal catalyst was not the same in a plasma environment as in traditional thermal catalysis, and 3) NH3 yields above the thermal equilibrium limits were observed and sensitive to the catalyst composition. Further advances in plasma-assisted catalysis requires the fundamental knowledge of plasma interactions with catalyst surfaces. However, the consequences of plasma activation for the interaction of gas species with a catalyst surface are currently speculative, and observation of these species on the surface is very challenging. This presentation will focus on exploiting catalyst-plasma interactions for nitrogen and hydrocarbon activation, and the role of the plasma in enhancing catalytic performance.

627791 Models and Observations of Plasma-Catalytic Nitrogen Oxidation

Abstract:

Coupled plasma and catalyst enables reaction yields that exceed the contributions of the individual components. Reliable kinetic models are essential to rationalize observations and to guide the identification of appropriate plasma-catalyst combinations. Kinetic models of plasma- and of heterogeneous-catalytic-chemistries are individually well developed. Models to couple the two are still primitive. Here, we demonstrate DFT-parametrized microkinetic models to couple plasma and surface reactions. The models are applied to N_2 oxidation, an alternative nitrogen fixation process, enabled only with coupled non-thermal plasma and catalysts. We compare the intrinsic activity of Pt and Au under plasma conditions. We build integral reactor models coupling simplified plasma reaction and surface reaction networks to predict product yield. We first show plasma excited species can enhance turnover frequencies of both Pt and Au. The enhancing potential of catalysts to chemical yield from plasma, however, are condition- and material- dependent. Plasma/Pt generates more or comparable nitric oxide than plasma/Au under most thermal and plasma parametric conditions. Plasma/Au outperforms plasma/Pt under low thermal but high plasma excitations. Results are compared against experimental observations, highlighting that material selection must consider plasma properties, thermal conditions and reactor configurations.

630666 Methane-Products Process Intensification through a Nanosecond Plasma Discharge

Abstract:

The energy industry, such as other sections, moves toward more efficient systems and newer technologies hires by companies. Due to excellent energy-engineering setups that we have developed in many years, we could revolute the energy industry in the 20th century. The usages of natural gas, a clean feedstock, increase every year, and the majority of people in civilized countries have decided to move toward less polluted earth. Among all products that are produced at different plants, hydrogen plays a vital role. Producing hydrogen more efficiently at the existing plants is crucial. Moreover, hydrogen is a clean energy source and can help move toward a sustainable energy economy.

Most of (about 95%) hydrogen is produced from fossil fuels by steam reforming of natural gas, partial oxidation of methane, and coal gasification as of 2020. All these methods of hydrogen production require a considerable amount of energy due to high temperate operations. Therefore, the first question is how to produce hydrogen and light hydrocarbons cheaper with more eco-friendly technologies.

In this work, a nanosecond pulsed plasma reactor was designed to produce hydrogen and liquid hydrocarbons. The conversion rate, efficiency, and cost of the designed-made reactor show a promising method for converting hydrogen. Currently, hydrogen is priced between $3 and $8 per kilogram. We are presently producing hydrogen below $1.

Most of the researches that have been done on converting methane to hydrogen or other products have focused on the reactor's efficiency. This study attempts to present a novel reactor design and perform process intensifications for future industrial uses. Moving toward hydrogen as a source of energy can diversify our energy sources and enhance energy resiliency and security. Moreover, to avoid constructing gigantic plants to make them profitable, new-more efficient technologies are required to make the production in smaller-scale plants profitable. Also, having more distributed-smaller production plants would create opportunities to reach stranded reservoirs.

631226 Plasma-Assisted Upgrading of Methane: Mechanistic Insights from in-Situ PM-IRAS and OES Spectroscopy

Abstract:

Non-thermal plasmas (NTPs) produce reactive chemical environments including electrons, ions, radicals, and vibrationally-excited molecules and drive chemistry at temperatures at which such species are thermally inaccessible. The integration of a catalyst with reactive NTPs can promote novel chemical transformations. In particular, we are interested in upgrading of light hydrocarbons (e.g. methane) by direct coupling with nitrogen to produce value-added liquid chemicals (e.g. pyrrole and pyridine) in a plasma-assisted catalytic process. To design effective catalysts and plasma-catalytic processes, we need comprehensive understanding of the plasma catalytic processes including characterization of plasma-catalyst interactions. However, to date, the fundamental understanding of plasma-catalyst interactions is still in its infancy. In this work, we use a newly developed multi-modal polarization-modulation infrared-reflection absorption spectroscopy (PM-IRAS) tool with optical emission spectroscopy (OES) to probe plasma-catalyst interactions of CH_4/N_2 plasma species with different model catalytic surfaces (Ni, Pd, and Au). CH_4/N_2 feed gas mixture is introduced in three different ways: (1) CH_4-N_2 in sequence, (2) N_2-CH_4 in sequence, and (3) simultaneous with varying CH_4/N_2 ratio. We characterize excited species in the plasma-phase using OES, detect surface-adsorbed species using PM-IRAS, and detect gas-phase products using mass spectrometry. We find that different sequences in the feed gas introduction result in different surface-adsorbed species, and the presence of different surfaces dictates the surface-adsorbed species present. We also find that varying CH_4/N_2 feed ratio affects the macroscopic properties of the plasma and the production of intermediates such as cyanide (CN).

628510 Computational Evaluation of Plasmonic Photocatalysis in Au and Ag Nanoparticles

Abstract:

Many industrial chemical processes require high operating temperatures to drive catalytic reactions which are crucial to human development. These high temperatures are both energy demanding and detrimental to the long term life of the catalyst. Plasmonic metal nanoparticles offer an interesting alternative to traditional heterogenous catalytic processes due to their ability to be activated by light. One potential mechanism of action is the localized surface plasmon resonance, which involves oscillations in surface electron density that decay into so called "hot" electrons. These hot electrons can then transfer to unoccupied orbitals of surface intermediates to break chemical bonds or accelerate reactions. While plasmonic photocatalysis is a well-known event, the exact mechanism of these reactions is still debated. Understanding the precise workings of plasmon transfer is important for rational design of novel catalytic structures. Here, we utilize real-time, time-dependent density functional theory (RT-TD-DFT) to excite a system with oscillating electric fields and track the subsequent excited state dynamics in real time. This method models light of varying frequency in order to observe the mechanism of plasmon-mediated reactions in Au and Ag nanoparticles. Our results indicate that adsorbates can have qualitatively different charge transfer behavior. Specifically, O_2 loses electrons while N_2 gains electrons when exposed to identical electric fields on both Au and Ag nanoparticles. Additionally, charge density calculations indicate that these excitations are $\pi \rightarrow \pi^*$ on short time scales and a mixture of π, $\sigma \rightarrow \pi^*$, σ^* as time evolves. We intend to further explain the differences between O_2 and N_2 via careful inspection of the electronic structure.

623502 Deep Learning for Morphology Detection of Self-Assembled Oligomers in Atomistic Simulations: Pointnet

Abstract:

Amphiphiles including block polymers and block oligomers can self-assemble into numerous ordered morphologies spanning length scales from a few to hundreds of nanometers. Molecular simulations using atomistic or coarse-grained force fields are widely used as a powerful tool to understand and predict the self-assembly phase behavior of these complex systems. In molecular simulations, screening block oligomer structures to search for promising candidates for functional materials is facilitated by straightforward but effective structure identification techniques. However, capturing ordered network structures with periodicity in all three dimensions is challenging because the size of the periodic simulation box and the number of molecules inside the box must be commensurate with the unit cell of the resulting network geometry. Simulations with incorrect system sizes can yield imperfect and distorted structures, and common morphology identification strategies including structure factors and order parameters may fail in assigning the morphology. Here we present a new structure detection approach based on PointNet, a neural network designed for computer vision applications using point clouds. Our PointNet was trained using atomic coordinates from molecular dynamics simulation trajectories and synthetic point clouds for ordered network morphologies which were absent from previous simulations. The trained PointNet model achieves an accuracy as high as 0.996 with high robustness against missing data for globally ordered morphologies formed by block oligomers of various architectures and their mixtures, and it also allows for the discovery of emerging ordered patterns from non-equilibrium systems.

625581 Automated Solid State Electrolyte Conductivity Predictions Via Probability Density Analysis

Abstract:

The ionic conductivity is the paramount property of the solid state electrolyte (SSE) to enable solid state lithium batteries, but is often multiple orders of magnitude lower than in liquid electrolytes. Current machine learning (ML) models predicting ionic conductivity directly from the crystal structure are limited by the low number of training points and offer little mechanistic insight into diffusion mechanisms for a given crystal. Here, we train a ML model to predict the lithium ion probability density from crystal structure (**Figure 1**). The probability density qualitatively predicts the ionic conduction pathways and topological analysis provides quantitative predictions of diffusion activation energy barriers and ionic conductivity. We train a convolutional neural network with a Gaussian representation of the crystal structure electron density as the input, and a binarized probability density as the output. The newly proposed model is trained on a dataset of 100 materials with a final training accuracy of 94% and a test accuracy of 91%. By screening the ICSD, we rediscover Li_6NBr_3 which has an experimentally measured ionic conductivity of 0.0035 mS/cm at 150°C and propose a series of conduction pathways which would not have been feasible through density functional theory due to the large number of atoms. Additional crystalline families were found such as $LiGaS_2$, and $Li_2NaB(PO_4)_2$ whose ionic conductivities are further optimized through aliovalent substitution and are being verified with *ab initio* molecular dynamics. Our resulting ML algorithm is applicable over a wide range of compositions, scales efficiently to hundreds of atoms, and can analyze disorder systems making it a powerful new tool to analyze, optimize, and screen for new SSEs.

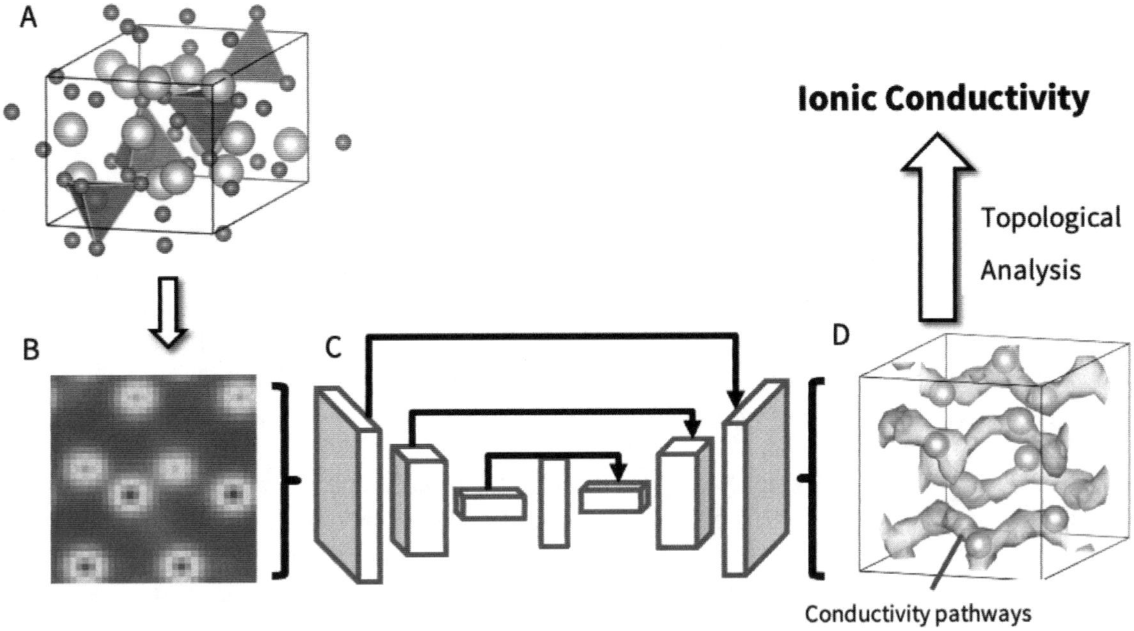

Figure 1: General flowchart of model used to predict probability density and calculate ionic conductivity. Using the crystal structure as an input (A), the 3-dimensional electron density is approximated as a sum of Gaussians centered at each atom (B). A 3-dimensional convolution neural network (C) is fit to predict the 3-dimensional probability density (D). Topological descriptors of the probability density are calculated which are used to calculate ionic conductivity.

632687 Bayesian Model Selection: Applying Parsimony to Build Better Molecular Models

Abstract:

Non-covalent interactions play critical roles in molecular biology. Molecular dynamics is the most common *in silico* method to probe these interactions, as simple force fields enable the larger system sizes and longer timescales required to capture the larger-scale interactions that drive many biological processes. The speed and power of these models comes at the cost of explicit electronic interactions, which are replaced with simpler approximations; this forces the modeler to make many decisions while selecting the appropriate force field, including choices of models for atom typing, combination rules, polarizability, and charge modeling.

In the Open Force Field Initiative, we aim to develop force fields and their non-covalent interactions, using data-driven techniques. To this end, we explore the use of Bayesian inference to make data-driven choices between non-covalent interaction parameters and functional forms, by calculating Bayes factors, a quantitative measure of the relative evidence between models. Bayes factors incorporate parsimony, penalizing unnecessary complexity naturally through the Bayesian paradigm. More complex models generally have more parameters with less certain values, putting them at a disadvantage against simpler models which are able to reproduce the same quantities of interest. To justify its additional complexity, a model must demonstrate a significant improvement in its accuracy.

We test this strategy on the 2-center Lennard-Jones plus Quadrupole (2CLJQ) model for simple fluids, leveraging analytical surrogate models from literature to rapidly evaluate the large number of parameter sets required to compute Bayes factors. With this framework we are able to evaluate whether including a quadrupole is necessary when designing a model to reproduce temperature-dependent density, saturation pressure, and surface tension data for several small molecules. In most cases, we find that the inclusion of a quadrupole parameter is not justified, indicating that a simpler representation is sufficient to reproduce the quantities of interest. This process also produces parameter probability distributions for each compound, which provides valuable information about the parameter uncertainty and sensitivity. This work demonstrates the utility of Bayesian inference as a tool for model selection and informs our future application of this technique to more complex decisions required in fitting biomolecular force fields.

632269 Evaluating Polymer Stabilizer Performance Using Molecular Descriptors and Machine Learning on a Small Dataset

Abstract:

Mining experimental data from the literature is an important exercise for informing future experimental studies, even if available data is sparse. However, extracting insights from small materials datasets, such as those within research papers and patents, is inherently challenging due to high complexity, high dimensionality, and heterogeneous reporting across sources. This case study presents a situation where judicious molecular representation, feature importance, and physicochemical interpretation were integrated to extract machine learning insights on a small dataset. Here, experimental data from a single patent was analyzed to learn from the small molecule additives that were most effective in mitigating the degradation of poly(ethylene terephthalate) (PET). MACCS-166 and alvaDesc molecular descriptors were calculated for the dataset of 39 additive candidates to yield two sets of 166 and 1875 different features, respectively. Performing *k*-means clustering using these molecular descriptors revealed evidence that performance differences were sensitive to variations in molecular structure. To pinpoint the features responsible for improved performance, a supervised reduced design region approach was applied to analyze descriptors both individually and in multiple dimensions to determine effectiveness in a binary classification of high and low performance. Not only were the most influential descriptors justifiable with respect to degradation chemistry, but also the selected features successfully trained random forest models with good cross validated performance. In comparing molecular descriptor approaches, we find that judicious interpretation of underlying physicochemical behavior is indispensable in validating the effectiveness of small data machine learning, especially for prioritizing experimental work toward a richer dataset.

630943 Machine-Learning-Guided Discovery of New Electrochemical Reactions

Abstract:

Machine-Learning-Guided Discovery of New Electrochemical Reactions

*Andrew Zahrt, Yiming Mo, Yanfei Guan, Esther Heid, Kakasaheb Nandiwale, Klavs Jensen**

Significance

The discovery of new transformations constitutes a significant portion of research effort in organic chemistry, with hundreds of researchers globally working to develop new synthetic methods. The impact of field is exemplified by the incorporation of modern methods into medicinal chemistry programs is exemplified by the types of reactions used in discovery campaigns. Between 1976 and 2015, the types of synthetic transformations employed by medicinal chemists has increased by 27 %, with 123 different reaction types accounting for 95 % of reactions used in 1976 to 159 reactions used in 2015.[1] Further, a comparison between methods employed in both 1984 and 2014 reveals a number of synthetic method developments have changed the landscape of medicinal chemistry, including the Suzuki-Miyaura coupling, the Buchwald-Hartwig coupling, urea formation reactions, the Mitsunobu reaction, and others.[2] However, despite these developments, the chemical space explored by most discovery campaigns remains limited. Five types of reactions make up over 60 % of reactions used in drug candidate surveys.2,[3] Further, a survey of drug molecules published in 2010 indicated that the top 50 molecular frameworks covered ~50% of experimentally approved drugs,[4] a number not significantly different than a similar study using data from 1996.[5] In addition, a study of drugs developed before 2013 indicated that only 2% of known monocyclic or bicyclic rings systems are present in drug compounds.[6],[7] Although one might argue that this relative structural invariability is a consequence simply of the privileged bioactivity of these select scaffolds, the commonly cited "escape from flatland" concept indicates that this is likely not the case and that exploring 3D-chemical space is essential to further the field of drug development.[8] Clearly, the toolbox of the medicinal chemist must expand to increase the rate of drug discovery.

Background

Applications of computer guided approaches to new reaction discovery are rare, with most early examples never extending beyond proof-of-principle.[9]-[10][11][12] Recently, Cronin and coworkers have re-invigorated the field of computer-guided reaction discovery, demonstrating the ability to use classification models to predict productive reactant combinations.18 However, *the synthetic chemistry community still has no computer-guided, generalizable method capable of discovering new reactions.* This work seeks to rectify this limitation by combining high-throughput experimentation with a machine-learning-guided protocol for reaction discovery capable of computationally evaluating vast quantities of virtual reactions, including those with reactants not included in the training data. This workflow will increase the success rate of experimental reaction discovery campaigns. In this case, we explore the chemical space of convergent-paired electrolytic reactions.

Results

To achieve this goal, we postulated that advanced molecular representations will be required to produce generalizable models with sparse datasets. To develop this representation, a literature dataset of 370 electrochemical reactions was gathered with the objective of predicting the site selectivity of electrochemical oxidation reactions. Different molecular representations were evaluated and their performance compared in generating models classifying atomic centers as reactive (i.e. that atom is oxidized in the reaction) or unreactive. In the design of this study, we chose to divide the dataset on the basis of the overall transformation, in which one category (or "reaction template") is left out as the test set. In this data partitioning design, fingerprint-based models struggle to predict into the "left out" reaction class. However, augmenting the DFT features with additional electronic structure information derived from NBO analysis of the neutral, oxidized, and reduced molecules improved the model significantly (Precision/Recall with Morgan Fingerprints vs. Engineered Features is 0.18/0.31 and 0.51/0.68, respectively). This finding is significant, suggesting that engineered features outperform structure-based fingerprints when generalization/extrapolation is the goal.

The engineered DFT-level features were atom-level features used to classify individual atoms and reactive and unreactive. In order to combine these atom-level features to create a molecular representation, the graph2vec method[13] was used to embed the molecular graph (with atom-level features as node attributes) to a fixed-length vector. To validate this molecular representation, models were produced to predict molecular properties (e.g. ionization potential, HOMO energy, etc.) or molecules using a data-partitioning scheme designed to force extrapolative predictions. In these cases, the engineered vector significantly outperformed Morgan fingerprints. With a validated representation identified, DFT calculations were performed for ~40K molecules containing atoms H, C, N, O, F, Si, P, S, Cl, Br, I, Li, Na, and K and this data used learn the desired representation, enabling instantaneous generation of the desired vectors.

With this computational groundwork established, an automated microfluidic platform previously published by our laboratory[14] was used to test ~150 electrochemical reactions. In the reactions, the cathodic reactions was intended to be held constant as the 1-electron reduction of 1,4-dicyanobenzene and the anodic reaction varied with a series of different partners to discover new convergent-paired electrolytic reactions. Using this dataset, a model was generated capable of classifying whether a reactant would be a competent reactive partner at 74 % accuracy. To date, this workflow has resulted in the discovery of 4 new electrochemical reactions. We now seek to expand this dataset and couple the reaction discovery model and site-selectivity model to produce a catalog of hypothetical reactions to be evaluated and (if successful) optimized on the automated high-throughput platform.

References

([1]) Schneider, N.; Lowe, D. M.; Sayle, R. A.; Tarselli, M. A.; Landrum, G. A. Big data from pharmaceutical patents: a computational analysis of medicinal chemists' bread and butter. *J. Med. Chem.* **2016**, *59*, 9, 4385-4402.

([2]) Brown, D. G.; Boström, J. Analysis of past and present synthetic methodologies on medicinal chemistry: where have all the new reactions gone? *J. Med. Chem.* **2016**, *59*, 4443–4458.

([3]) Roughley, S. D.; Jordan, A. M. The medicinal chemist's toolbox: an analysis of reactions used in the pursuit of drug candidates. *J. Med. Chem.* **2011**, *54*, 3451–3479.

([4]) Wang, J.; Hou, T. Drug and drug candidate building block analysis. *J. Chem. Inf. Model.* **2010**, *50*, 55–67.

([5]) Bemis, G. W.; Murcko, M. A. The properties of known drugs. 1. Molecular frameworks. *J. Med. Chem.* **1996**, *39*, 2887–2893.

([6]) Taylor, R. D., MacCoss, M.; Lawson, A. D. Rings in drugs. *J. Med. Chem.* **2014**, *57*, 5845–5859.

([7]) Taylor, R. D., MacCoss, M.; Lawson, A. D. Combining molecular scaffolds from FDA approved drugs: application to drug discovery. *J. Med. Chem.* **2017**, *60*, 1638–1647.

([8]) Lovering, F.; Bikker, J.; Humblet, C. Escape from Flatland: Increasing Saturation as an Approach to Improving Clinical Success. *J. Med. Chem.* **2009**, *52*, 6752-6756.

([9]) Balaban, A.T. Chemical graphs. 3. Reactions with Cyclic 6-Membered Transition States. *Rev. Roum. Chem.* **1967**, *12*, 875-902.

([10]) a) Hendrickson, J.B. The Variety of Thermal Pericyclic Reactions. *Angew. Chem, Int. Ed.* **1974**, *13*, 47-76. b) Arens, J.F. A Formalism for the Classification and Design of Organic Reactions. I. The Class of (-+)n+ and (-+)n – reactions. *Recl. Des Trav. Chim. Des Pays-Bas*. **1979**, *98*, 395-399. c) Arens, J.F. A Formalism for the Classification and Design of Organic Reactions II. The Classes of (+-)nC reactions. *Recl. Des Trav. Chim. Des Pays-Bas.* **1979**, *98*, 471-483. d) Arens, J.F. A Formalism for the Classification and Design of Organic Reactions III. The Class of (+-)nC Reactions. *Recl. Des Trav. Chim. Des Pays-Bas*. **1979**, *98*, 471-483. e) Zefirov, N.S.; Tratch, S.S. Formal-Logical Approach to Multicentered Processes with Cyclic Electron Transfer. *Match*. **1977**, 263-264. f) Zefirov, N.S.; Tratch, S.S.; Tratch, S.S. Systematization of Tautomeric Processes and Formal-Logical Approach to the Search for New Topological and Reaction Types of Tautomerism. *Chem. Scr.* **1980**, *15*, 4-12.

([11]) a) Bauer, J.; Herges, R.; Fontain, E.; Ugi, I. IGOR and Computer Assisted Innovation in Chemistry. *Chimia (Aarau).* **1985**, *39*, 43-53. b) Bauer, J. IGOR2: A PC-program for Generating New Reactions and Molecular Structures. *Tetrahedron. Comput. Methodol.* **1989**, *2*, 269-280.

([12]) a) Herges, R.; Reaction Planning: Prediction of New Organic Reactions. *J. Chem. Inf. Comput. Sci.* **1990**, *30*, 377-383. B) Herges, R.; Hoock, C. Reaction Planning: Computer-Aided Discovery of a Novel Elimination Reaction. *Science*. **1992**, *255*, 711-713. c) Zefirov, N.S.; Baskin, I.I.; Palyulin, V.A. SYMBEQ Program and Its Application in Computer-Assisted Reaction Design. *J. Chem. Inf. Comput. Sci.* **1994**, *34*, 994-999. d) Zefirov, N.S.; Tratch, S.; Molchanova, M. The Argent Program System: A Second-Generation Tool Aimed at Combinatorial Search for New Types of Organic Reactions. *1 Main Concepts Potentialities MatchCommunication Math. Comput. Chem. 46* **2002 SRC,** 253-273. e) Molchanova, M.S.; Tratch, S.S.; Zefirov, N.S. Computer-Aided Design of New Organic Transformations: Exposition of the ARGENT-1 Program. *J. Phys. Org. Chem.* **2003**, *16*, 463-474.

([13]) Narayanan, A.; Chandramohan, M.; Venkatesan, R.; Chen, L.; Liu, Y.; Jaiswal, S. graph2vec: Learning Distributed Representations of Graphs. arXiv:1707.05005.

([14]) (A) Mo, Y.; Lu, Z.; Rughoobur, G.; Patil, P.; Gershenfeld, N.; Akinwande, A.I.; Buchwald, S.L.; Jensen, K.F. Microfluidic Electrochemistry for Single-Electron Transfer Redox-Neutral Reactions. *Science*, **2020**, *368*, 1352-1357. (B) Mo, Y.; Rughoobur, G.; Nambiar, A.M.K.; Zhang, K.; Jensen, K.F. A Multifunctional Microfluidic Platform for High-Throughput Experimentation of Electroorganic Chemistry. *Angew. Chem. Int. Ed.* **2020**, *59*, 20890-20894.

625957 A Deep Learning Potential to Study Large-Scale Anhydrous Proton Transport Systems

- Download/View Extended Abstract

Abstract:

Designing and studying new proton conducting materials are essential to operate proton exchange membranes fuel cells at intermediate temperatures and conditions of low humidity. We use graphane functionalized with hydroxyl groups (graphanol) as the material of interest. Previous methods to study the occurrences of proton conduction of graphanol involved performing highly expensive density functional theory (DFT) calculations[1]. To overcome this computational barrier, we made use of dense neural networks. Our previous work with using deep-learning potentials (DP) for the precursor material of graphanol, graphane, showed tremendous capabilities of capturing molecular physics with small training data sets. We have developed a DP for graphanol that is capable of reproducing DFT forces to very high accuracies, within meV/Å accuracy. This DP is also able to simulate proton hopping, aided by the rotation of hydroxyl groups. We used the DeePMD[2] formalism along with an active learning scheme (DP-GEN)[3] to develop a DP for graphanol. We show that our DP is capable of accurately predicting phonon, dynamic properties, and proton self-diffusion coefficients of graphanol for larger system sizes and varying operating temperatures. We also test this DP by varying the number of added protons to the graphanol system. Future work will involve testing this DP for larger multi-layered graphanol systems and to understand charge build up and polarization at interfaces of the multi-layered systems.

References

1. Bagusetty, A.; Johnson, J. K., Unraveling anhydrous proton conduction in hydroxygraphane. *The journal of physical chemistry letters* **2019,** *10* (3), 518-523.
2. Wang, H.; Zhang, L.; Han, J.; Weinan, E., DeePMD-kit: A deep learning package for many-body potential energy representation and molecular dynamics. *Computer Physics Communications* **2018,** *228*, 178-184.
3. Zhang, Y.; Wang, H.; Chen, W.; Zeng, J.; Zhang, L.; Wang, H.; Weinan, E., DP-GEN: A concurrent learning platform for the generation of reliable deep learning based potential energy models. *Computer Physics Communications* **2020,** *253*, 107206.

630820 Leveraging Machine Learning Techniques to Identify Ionic Liquids Possessing High Ionic Conductivity

Abstract:

Ionic liquids are a class of salts that are liquid at room temperature with several unique and desirable properties such as negligible vapor pressure, precise designer tunability, high thermal and electrochemical stability. Because of these favorable properties, ionic liquids are currently being considered as a potential replacement for conventional organic electrolytes in Li-ion batteries. However, a major limitation of ionic liquids is their sluggish transport properties that are limiting their widespread application in battery technology. This drawback can be tackled by choosing the right cation-anion combination based on chemical intuition that could improve transport properties. The vast chemical space of available cations and anions for such manipulation could be challenging and unfeasible. The increase in the availability of experimental data in the open literature for ionic liquids provides an excellent opportunity to apply machine learning methods for correlating the ionic liquid properties and using the knowledge to discover new ionic liquids with desired properties.

Much of the data found in the literature is primarily focused on imidazolium-based cations because of their low viscosity and high ionic conductivity characteristics. However, imidazolium-based cations are associated with low electrochemical stability because of the presence of an acidic proton in the imidazolium ring. Alternates to the imidazolium-based ionic liquid family are classes of cations derived from groups such as phosphonium, piperidinium, and pyrrolidinium, which are not known to have high electrochemical stability, but are often characterized by very low ionic conductivity and high viscosity.

In this presentation, we will discuss our efforts developing a machine learning model based on the ionic conductivity data covering several different cation families obtained from the NIST ILThermo Database. A feed-forward artificial neural network (FFAAN) comprised of an input layer, a hidden layer, and an output layer is obtained from the data on all ionic liquid types. We will demonstrate that the model can capture ionic conductivity with very high accuracy for both the training and test cases. We will present our approach on using data mining techniques to identify patterns governing ionic conductivity trends within the data. This insight is fed into a classification model that is used to identify high ionic conductivity ionic liquids with implications in advancing the field of ionic liquids as next generation electrolytes for Li-ion, Na-ion, and K-ion batteries.

627650 High-Throughput Electrochemical Screening of Deep Eutectic Solvents for Use in Redox Flow Batteries

Abstract:

Deep eutectic solvents (DESs) are a class of materials with varied applications including catalysis and synthesis, extraction processes, drug solubilization, and battery electrolytes. Their appeal stems from their broad electrochemical stability window, high electrical conductivity, low vapor-pressure, and low-flammability. These solvents present a depression in the melting point at specific molar ratios of organic components that can result in a liquid solution at moderate temperatures. A breadth of candidates with varying concentrations can be used to form DESs, leading to a vast design space. High throughput experiments and data-driven design strategies are key to accelerate the optimization of materials based on their physicochemical and electrochemical properties as well as engineering criteria (e.g. cost, safety, toxicity) for candidate DESs. The implementation of high-throughput tools allows for a rapid evaluation and screening based on metrics such as the melting point, potential stability window and ionic conductivity. Several high-throughput protocols have been designed for identifying the design space of molecules under investigation, their formulation and material characterization. Using data science principles, the molecules composing our basis set were identified using scoring based on metrics such as cost, melting point, toxicity and molecular weight. The formulation of deep eutectic solvents was automated through the use of a pipetting robot, combinatorial techniques and well-plates. Next, the melting point of the proposed mixtures was detected using an IR camera and hot-plate set-up. To screen for their electrochemical properties, 96-well plates with screen printed electrodes in combination with measurement techniques such as Cyclic Voltammetry (CV) and Electrochemical Impedance Spectroscopy (EIS) were implemented. Finally, to adapt high-throughput principles to the analysis of the data collection obtained through the aforementioned protocols, machine learning was leveraged for the data classification and data modeling. Furthermore, open-source Python-based packages have been developed and made available on GitHub. The combination of high-throughput experimentation and data analysis greatly accelerated the design and screening candidate DES systems. This overall workflow could be easily adapted to other design spaces and applications.

631708 Bridging Homogenous and Heterogenous Catalysts: Computational Design of Catalysts at Atomic Scale

Abstract:

Achieving low cost, high activity and selectivity at the same time is highly critical and desirable but challenging in developing innovative catalysts. Learning from enzymes, homogenous and heterogenous catalysts to design new catalysts that can integrate their advantages is a promising solution. In this talk, I will discuss my past work on leveraging computational tools to accelerate the design process towards bridging homogenous and heterogenous catalysts. Take Mo/W containing catalysts for example, I will show examples from high-throughput screening on Mo/W molecular complexes for bio-inspired formate oxidation catalysts design, to modeling transition metal doped MoS_2 for catalyzing hydrogen evolution reaction (HER) based on first-principles simulations. From those examples, I will elaborate the challenges in bridging molecular complexes and heterogenous 2D materials for catalysis based on physical chemistry principles. The knowledge will provide new principles for future rational catalysts design.

631653 First-Principles Design of Heterogeneous Catalysts with Automatic Simulations

Abstract:

Producing high-valued commodities and energy-storing molecules from non-petroleum sources is among the challenges of decarbonizing the economy by 2050. However, known catalysts are often expensive, display poor reaction kinetics, or are limited in synthetic accessibility. Discovering cost-effective materials for clean energy reactions is decisive to overcome pressing climate goals. In this talk, I will discuss how a wide range of computational methods can accelerate catalyst discovery. In particular, I will show how combining graph theory, high-throughput simulations, density functional theory (DFT), and machine learning (ML) can elucidate different aspects of materials discovery in thermal and electrocatalysis.

I first illustrate with the case of zeolite catalysts. Due to their topological diversity, zeolites suffer from phase competition and transformations, which directly impact their stability. Using graphs to model atomic rearrangements, I demonstrate which topologies are related by diffusionless transformations and intergrowths, reproducing results from the literature and proposing new synthesis pathways for zeolites. Then, to quantify phase competition in zeolite synthesis, I developed a computational platform to predict host-guest interactions two orders of magnitude faster than the existing methods. By simulating millions of pairings between frameworks and organic structure-directing agents (OSDAs), binding metrics that explain phase selectivity in zeolites are proposed. The theory reproduces outcomes from more than one thousand papers extracted from the literature, and enables the design of selective, chemically simple OSDAs through a web-based design platform. Experiments show that the computer-designed OSDAs significantly improve catalytic properties of zeolite frameworks and produce previously inaccessible structures. The kinetic mechanisms of confined reactions are further studied using DFT, elucidating the origins of such enhancement in catalytic properties and introducing new design principles for OSDA selection. Finally, machine learning approaches are used to explore the space of OSDAs towards millions of possible molecules. These works can enable yet unrealized zeolites to be accessed, and reactions previously unfavorable to be catalyzed by novel nanoporous materials.

A second example is shown with multicomponent transition metal oxides. In contrast with zeolites, these materials are limited in structural diversity, but display a combinatorial number of realizable compositions. To simultaneously optimize their activity and stability, automated DFT simulations are used to map composition to electronic structure descriptors, which are then interpreted using atomic features. Literature data shows that DFT-based stability correlates with synthesis outcomes of these materials, thus validating the design strategy. The data is then used to train ML models, providing faster evaluation of electronic properties and reduction of prediction uncertainties in an active learning loop. Through an interplay between experiments and theory, materials selection strategies are successively refined, leading to experimental realization of potentially superior structures in terms of activity and stability. This iterative approach provides a roadmap to improving electrocatalysts from theory-driven research.

624338 Two-Dimensional Energy Histograms As Features for Machine Learning to Predict Adsorption in Metal-Organic Frameworks

Abstract:

Feature engineering is one of the most important steps in a machine learning (ML) workflow. Features of a data set that are constructed in a systematic and physically informative way can facilitate the convergence of ML algorithms and enhance the interpretability of a ML model. In this work, we propose novel features for predicting adsorption in metal-organic frameworks (MOFs) and other nanoporous materials. In particular, we use two-dimensional (2D) histogram features derived from the adsorbate-adsorbent energy and energy gradient at grid points located throughout the adsorbent. The inclusion of the energy gradient encodes some spatial information of the adsorbent while maintaining a relatively low computational expense. The training and testing data for the ML model are obtained from grand canonical Monte Carlo simulations for different adsorbates, including Kr, Xe, ethane, propane, n-butane and n-hexane, at different pressures and temperatures. With these 2D histogram features, the ML model provides an overall improvement of the prediction quality (with the coefficient of determination R^2~0.95-0.99) over a model that uses 1D energy histogram features. We identify the restrictions of the 2D histogram features applied to n-butane and n-hexane with the help of dimensional reduction methods. The possibility of combining 2D histogram features with some other features such as those derived from persistent homology will also be discussed. The physical understanding of the adsorption behavior decoded from the ML model may help develop more advanced MOFs for gas storage and separation in the future.

623164 Disordered Hyperuniform Networks and Their Application in Atomic-Scale Low-Dimensional Materials

Abstract:

Disordered hyperuniformity is a recently discovered novel state of many-body systems that possesses vanishing normalized infinite-wavelength density fluctuations and a hidden long-range order similar to a perfect crystal, and yet is statistically isotropic with no Bragg peaks like a liquid or glass. In this work I will present a series of research centered around a new concept called "disordered hyperuniform quantum materials" that my collaborators and I have introduced, i.e., disordered hyperuniform atomic-scale low-dimensional materials with nontrivial quantum characteristics. In particular, we discover a hyperuniformity-preserving topological transformation in low-dimensional networks that involves continuous introduction of Stone-Wales (SW) defects. Our network models reveal unique electronic transport mechanisms and mechanical behaviors associated with distinct classes of disorder in low-dimensional materials. In particular, we find that when adding disorder in a hyperuniform manner, silica and pyrite systems exhibit a transition from insulating/semiconducting to metallic behavior, which is in contrast to the conventional wisdom that disorder generally diminishes electronic transport.

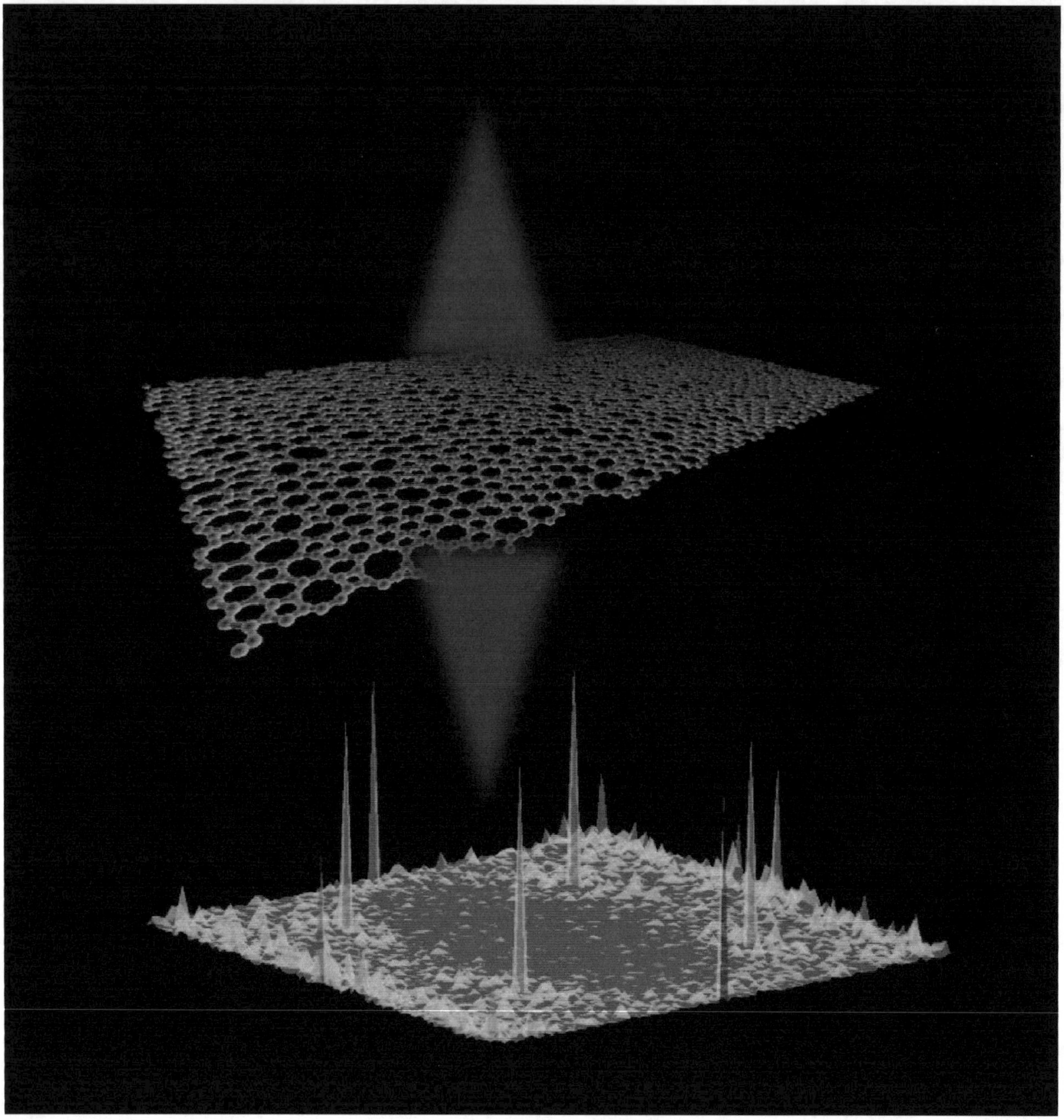

629129 Machine Learning-Directed Advanced Sampling Simulations of Reactions in Condensed Phases

Abstract:

Reactions in condensed phases are at the heart of many physical and chemical processes in molecular synthesis, heterogeneous catalysis, and material processing. The mechanism, rates, and chemical equilibria can be understood from free energy profiles of reactions. Predicting the energetics has involved a trade-off in accuracy between the enthalpic (i.e., interatomic potential) and the entropic (i.e., degrees of freedom) contributions to the free-energy. Traditionally, computational efforts in molecular reactions have been largely focused on the former. The latter, however, becomes increasingly important as the environmental effect and the size of the reactive species grow.

Here, we introduce two advanced sampling approaches that combine machine-learning algorithms and electronic structure calculations, to bridge the gap between accuracy and system size for reactions in condensed phases. First, we present an adaptive-biasing technique using neural networks, CFF-AIMD, for computing free energy surface with DFT-MD accuracy. We demonstrate how the neural network sampling method allows efficient sampling and direct computation of free energies through examples in molecular reactions on metallic surfaces and in liquid phases. Second, we develop a hierarchical resolution scheme for force field optimization with advanced sampling simulations. The method is applied to probe the formation, migration, and annihilation mechanisms of spin defects in wide-band gap semiconductors, which are relevant for understanding and designing material platforms for quantum technologies.

631622 Development and Implementation of Enhanced Sampling Approaches: Applications to Ion-Pairing in Battery Electrolytes and Nucleation of Nano-Porous Materials

Abstract:

Fast and accurate simulations of rare phenomena and their energetics are limited by the timescale accessible to conventional molecular dynamics (MD). The development of novel enhanced sampling approaches have provided significant thrusts to overcome those limitations. In this presentation, I will discuss two such examples from my research: one that utilizes an unconventional implementation of an existing method and another that develops a new statistical mechanical method for analyzing complex free energy landscapes.

In the first work, we utilize a biased-sampling (metadynamics) approach on a cluster model consisting of an ion-pair with a single shell of solvent. By sampling transitions between the ion-paired state and solvent separated ion-pair state, we demonstrate that quick estimates of the ion-pairing free energy can be obtained from cluster models that agree well with bulk solvent calculations [1]. However, the cluster model is 1 to 2 orders of magnitude faster than bulk solvent calculations. This has implications for the design of battery electrolytes because ion-pairing can significantly impact conductivity and performance of batteries. Our method provides a novel basis for the screening and assessment of candidates for electrolyte designs.

In the second work, we developed a novel graph theory-based sampling approach [2] for modeling the nucleation of weak electrolytes that overcomes limitations of existing approaches for bulk solvent systems. Our method seeks to exploit the property of materials whose crystal structure exhibit directional bonding and thus can be described as a "graph" of connected monomers. By utilizing a rigorous statistical-mechanics approach, we generate an ensemble of representative nuclei and their corresponding free energies via a "bootstrapping" approach in the nucleus size. Starting with a simple system of aqueous lithium fluoride (LiF), we elucidate the factors governing the nucleation and growth of LiF clusters. Subsequently, this work will be extended to complex materials such as zeolites and metal-organic frameworks.

---

[1] Ajay Muralidharan, Tyler Lytle, and Arun Yethiraj. "Why Lithium Ions Stick to Some Anions and not Others". The Journal of Physical Chemistry B (Accepted), 2021.

[2] Ajay Muralidharan, Xinyi Li, and J.R. Schmidt. "A Hierarchical Graph Theory-Based Sampling Approach to Study the Nucleation of Weak Electrolytes". (In preparation), 2021

625183 What Is a “Hydrophobic” Solute? a Detailed Examination of Driving Forces for Adsorption of Small, Neutral Solutes at Chemically Patterned Interfaces

Abstract:

While essential to life, water is also a valuable, critical component in the production of both energy and man-made materials. With global population growth, water demands will increase significantly, making recovery of water from waste-streams an essential to human life and the global economy. An efficient, cost-effective route towards meeting this demand is through membrane-based purification strategies. However, significant improvements will be necessary to meet projected global growth, in particular in terms of developing improved membrane materials for preventing fouling and allowing for selective recovery of potentially valuable contaminants. We begin to address this goal through a computational design strategy that couples detailed molecular simulations with genetic algorithm optimization. Specifically, we optimize the affinity of small, neutral solutes for model interfaces by adjusting chemical patterning of exposed surfaces. Affinities of solutes are decomposed in a detailed manner to pick apart specific contributions to binding, which we can show is driven largely by an increased ability of solvent to accommodate a solute at an interface compared to in bulk solution. Surprisingly, polar solutes adsorb to both predominantly methylated and hydroxylated surfaces, which precludes the use of common metrics, such as solvation or octanol-water transfer free energies, to predict adsorption. We directly show that chemical repatterning of a surface to reduce affinity is tied to reducing the “space” available to accommodate a solute at an interface. Finally, we demonstrate that specific patterns of charged surface moieties can flip this contribution to binding from highly favorable to unfavorable, allowing for large shifts in binding affinity at fixed material composition, which further emphasizes the central role of chemical patterning in designing membrane materials.

625593 Data-Driven Design of Solid-State Electrolytes

Abstract:

Solid state batteries provide many safety and performance advantages by replacing the combustible organic liquid electrolyte in a lithium-ion battery with a ceramic solid-state electrolyte (SSE). However, the ionic conductivity in these SSEs is often several orders of magnitude lower than in their liquid counterparts. Predicting a material's ionic conductivity directly from its crystal structure is limited by data availability and incomplete material descriptors. First-principles (i.e., with density functional theory, or DFT) molecular dynamics (MD) is an established approach to calculate and study ionic conductivity but is limited in the number and type of materials that can be simulated due to the high computational cost. To this end, we leverage advanced machine learning (ML) algorithms to more efficiently calculate ionic conductivity and optimize material composition.

We accelerated the calculation of forces and energies in MD with an artificial neural network force field, which scales linearly and enables the calculation of ionic conductivity at experimentally relevant compositions. While these force-fields are system specific, we show the valence electron density is an effective universal descriptor, allowing direct calculation of ionic conductivity from the crystal structure with a partial least squares algorithm. To augment the limited number of training samples, we trained additional ML models to predict the lithium probability density. Our novel 3D material segmentation network provides both quantitative and qualitative insight into the topology of diffusion pathways to accelerate SSE design. From these models, we predict diffusion activation barriers in materials which are inaccessible with traditional DFT methods and identified several new promising classes of solid-state electrolyte candidates to have conductivities greater than 16 mS/cm as verified by DFT-MD simulations.

The proposed combinations of first principles simulation data with ML models will greatly accelerate the rate of materials design/discovery and can automate calculating structure-property relationships for other solid-state systems with high accuracy and without sacrificing interpretability.

624414 Generating Molecules with Optimized Solubility Using Iterative Graph Translation

Abstract:

Molecular discovery is key to solving many problems in chemical engineering, ranging from identifying cures for infectious diseases to developing technologies that address climate change. While there exists a large number of theoretically accessible molecules, the time and resource costs of experiments make it intractable to fully explore chemical space. To this end, we present a generative modeling framework that is capable of producing novel molecules that are optimized with respect to multiple objectives or constraints. Our approach involves training a Hierarchical Graph Neural Network (HGNN) to translate a given molecule into an improved one by providing training pairs of "less optimal" and "more optimal" molecules. Once the model is trained, it can then be used to translate all molecules in the dataset into improved molecules. These improved molecules are then added to the training data and used to retrain the translation model. In this way, we can iteratively train and translate in order to 1) build a robust translator that can be applied to a given set of candidate molecules and 2) discover new, highly optimized molecules for a given application.

In this talk, we will present our work applying this framework to the specific problem of designing molecules to optimize aqueous solubility. Aqueous solubility is an important property for many chemical applications ranging from drug design to climate prediction. We successfully trained a translator to improve aqueous solubility and found that the model was also capable of discovering molecules that were more soluble than any of the training examples. When we applied synthetic feasibility as a secondary optimization constraint, the resulting model generated synthetically feasible molecules 93.2% of the time. Additionally, we investigated the role that training dataset size plays in model performance and found that reasonable models could be trained on datasets containing only 10^2-10^3 molecules. This workflow serves as a general approach for generating molecules that are both optimized and synthetically feasible. These promising results have led us to explore how this framework can be applied to solve a variety of molecular design problems including designing novel dyes, enantioselective catalysts, and soluble drugs.

627808 Leveraging Thermodynamic Calculations Towards Predictive Solid-State Materials Synthesis

Abstract:

Computational approaches, such as density functional theory, play a larger and larger role in materials design with each passing year. However, a major gap in the existing computational pipeline is the ability to predict synthesizability or synthesis pathways, precluding truly theory-driven discovery of novel materials. Solid-state synthesis is the bedrock of inorganic materials chemistry, yet this approach is typically a "black box", where the only observation is what formed from the precursor materials after a reaction at a predetermined set of conditions (typically chosen by intuition). In order to rationally synthesize new inorganic materials, it is critical to not only understand but predict what intermediates form and how they influence the reaction towards the synthesized phase(s). In this talk, I'll provide two examples [1,2] where we use readily computable thermodynamic quantities as inputs to conceptual models to understand phase evolution during synthesis. The models we develop are validated through experimental synthesis experiments along with *in situ* X-ray diffraction and *in situ* electron microscopy measurements. Finally, I will discuss the limitations of our models and what steps are needed to truly reach a paradigm of predictive and prescriptive materials synthesis.

[1] A. Miura, H. Ito, C. Bartel, W. Sun, N.C. Rosero-Navarro, K. Tadanaga, H. Nakata, K. Maeda, G. Ceder, *Materials Horizons*, 2020, 7, 1310-1316, DOI: 10.1039/C9MH01999E

[2] A. Miura^, C. Bartel^, Y. Goto, Y. Mizuguchi, C. Moriyoshi, Y. Kuroiwa, Y. Wang, T. Yaguchi, M. Shirai, M. Nagao, N.C. Rosero-Navarro, K. Tadanaga, G. Ceder, W. Sun, *Advanced Materials*, 2021, Accepted, DOI: 10.1002/adma.202100312

633459 First-Principles Understanding and Design of Metal Oxides for Photocatalytic Water Splitting

Abstract:

Renewable sources of energy are increasingly needed and solar production of hydrogen fuel from water offers significant potential to contribute to these needs if new photocatalyst materials and/or more efficient photocatalytic systems can be identified for visible-light-driven water photoelectrolysis. A variety of materials have been synthesized and tested for use as electrocatalysts. Among others, transition metal oxides are considered as one of the most promising electrocatalysts as they are earth abundant, eco-friendly, and stable in aqueous solution. Particularly, Bismuth Vanadate (BiVO4) has recently been identified as an inexpensive, promising photoanode for use in water-splitting photoelectrochemical cells (PECs). However, pristine BiVO4 exhibits poor electron-hole separation and sluggish water-oxidation kinetics, limiting its solar energy conversion efficiency. Significant research efforts have been made to enhance the PEC performance of BiVO4 through doping with metal/nonmetal elements and loading of oxygen evolution co-catalysts. However, the design and successful realization of commercially viable BiVO4-based photoanodes are still rather elusive. This talk will highlight the strong effects of crystal phase and doping on the photocatalytic performance of BiVO4 that have been found from combined first-principles and experimental studies. The improved understanding offers important guidance for the rational design of metal oxide-based photocatalysts.

632359 Structural and Stability Trends in Single (ABO_3) Perovskite Oxides from DFT-Optimized Bond Valence Structures

Abstract:

Ternary perovskite oxides (ABO_3) are a diverse class of technically important materials that exhibit exceptional properties for a variety of applications; as such, ternary perovskite material properties have been characterized in several high-throughput density functional theory (DFT) investigations. To ensure computational tractability, however, these previous investigations employ structural and stability approximations that are not rigorously supported. Neglecting these assumptions, we report a comprehensive investigation of 200 experimental ABO_3 compositions generated in all possible ternary perovskite geometries— as dictated by the 11 unique Glazer tilts— and optimized with DFT. Glazer tilt enumeration was performed with the Structure Prediction and Diagnostic Software (SPuDS) optimization package— which uses a bond valence method (BVM) based global instability index (GII) minimization procedure to predict octahedral (BO_6) tilting— and relaxed with GGA+U parameters compatible with the Materials Project (MP) database. We report a general linear correlation between GII and DFT energy that enables GII minimized structures to predict DFT optimized structures, particularly those that exhibit significant octahedral tilting. ~95% of DFT predicted ground states arise from the heavily distorted a-b+a- (orthorhombic) Glazer tilt, verifying that BO_6 distortions are the primary mode of stabilization in these compositions. Furthermore, we define the "energetic well depth" to be the difference between the undistorted a0a0a0 (cubic) geometry and the DFT predicted ground state, and show that linear scaling of the well depth energy predicts the decomposition enthalpy of the ground state— computed relative to MP tabulated competing phases— with RMSE = 40 meV/atom. By leveraging this relationship and slightly modifying the GII minimization procedure used by SPuDS to improve cubic lattice predictions, stability can be rapidly and accurately predicted from a single DFT optimization of the 5-atom cubic geometry. This study thereby constitutes the first time an integrated SPuDS+DFT scheme has been used to examine perovskite oxides, and represents a valuable addition to the growing arsenal of high-throughput investigative tools for targeted ABO_3 applications.

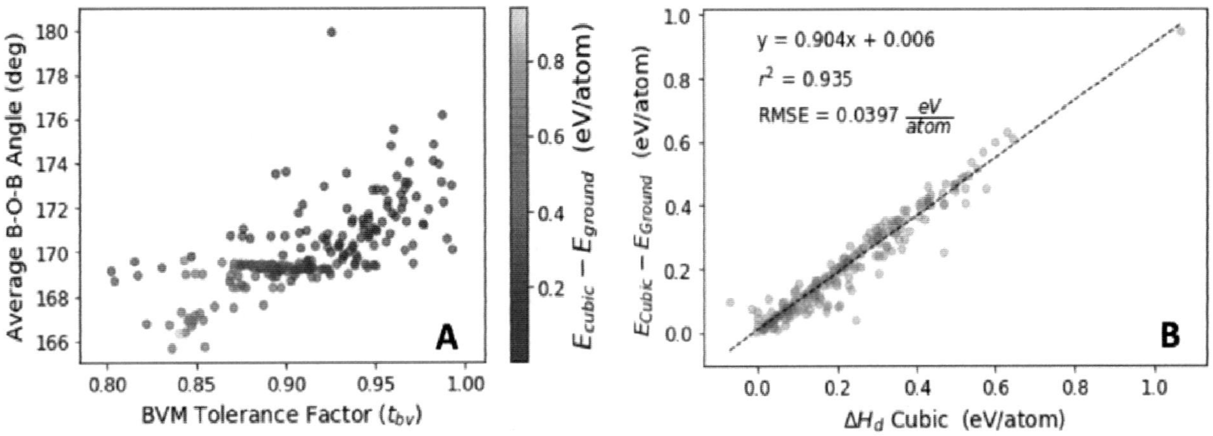

Shown for the 200 experimental compositions considered in our dataset: **A)** the relationship between the average B-O-B angle of the minimum DFT energy polymorph (i.e., the DFT predicted ground state, y_1-axis), the DFT energy difference between the undistorted cubic polymorph and the minimum DFT energy (distorted ground state, y_2-axis) polymorph, and the bond valence Goldschmidt tolerance factor (t_{bv}, x-axis). In the region $t_{bv} < 1.0$, the average B-O-B angle of the DFT ground state polymorph and the depth of the DFT predicted energetic well both increase as t_{bv} decreases, in agreement with SPuDS. In the region $t_{bv} > 1.0$, A and B site displacements supersede BO_6 tilting as the dominant form of distortion; this region is therefore beyond the scope of the current investigation. **B)** the stability of the cubic polymorph with respect to the convex hull (ΔH_d Cubic, x-axis) is linearly related to the depth of the energetic well (E_cubic - E_ground, y-axis). Linear regression shows that the magnitude of the energetic well depth is ~90% that of the cubic phase stability ($r^2 = 0.935$). Therefore, the DFT predicted ground state stability with respect to the Materials Project (MP) computed convex hull can be estimated as $\Delta H_{d,\,ground} = 0.1 \cdot \Delta H_{d,\,cubic}$ with reasonable accuracy (RMSE = 0.0397 eV/atom).

633322 Extracting the RedOx Thermodynamics of Perovskites via Combined Experiment and Theory

Abstract:

The analysis and design of thermochemical looping systems for renewable fuel production and energy storage requires detailed knowledge of the partial molar enthalpies and entropies of reduction of the active metal oxides. Experimental data normally consists of measured non-stoichiometries at a series of temperatures and oxygen partial pressures. Thermodynamic properties are then extracted via the van't Hoft or other analysis, such as the compound energy formula. These techniques often allow for trade off of enthalpy or entropy, providing non-unique solutions. In contrast computational methods such as density functional theory struggles to reproduce the complex temperature dependent behavior due to the large number of potential configurations. Here we present a new method for extracting thermodynamic properties which leverages a combination of DFT and experiment. This method removes the ambiguity between enthalpy and entropy, while providing temperature dependent behavior. The solution does not require pre-made assumptions.

633448 Materials Discovery and Development for Lower Temperature and Near Isothermal Thermochemical H2 Production

Abstract:

Materials discovery and development related to thermochemical materials for H2 production has historically been focused on materials that have large entropy changes and moderate enthalpies, in order that temperature swings between reduction and oxidation steps are minimized and thermal reduction temperatures are reduced below 1400 °C. Isothermal or near-isothermal cycling has been proposed as a means to decrease sensible heating requirements to the solid phase, but theoretical and experimental work has yielded mixed predictions and results. Based on our recent theoretical and experimental work, we postulate that the prior mixed results reported in literature stem from a combination of materials and conditions for isothermal operation (e.g., CeO2-δ at pO2 ~ 10-4 atm and 1400°C-1500 °C), that has been far from those desired for optimal performance. In this work, we implement a materials design concept that identifies and investigates new material classes, compositions, and operating conditions that are ideal for isothermal thermochemical H2 production. Starting from La-Sr-Mn perovskites, which have been theoretically and experimentally identified as having more favorable properties than ceria, we are utilizing high throughput screening coupled to direct DFT computations of down-selected materials for key performance properties such as oxide formation energies and oxygen vacancy formation and migration energies. DFT computation and screening is coupled closely with experimental synthesis of powdered, replica and direct foam compositions that afford enhanced heat and mass transfer properties. Compositions are experimentally characterized through thermogravimetric analysis and relaxation experiments at a larger scale to test the stability and performance of the replica and foamed compositions.

629134 Thermodynamic Characterization of Doped Spinels for Thermochemical Fuel Production

Abstract:

Technologies such as solar thermochemical hydrogen (STCH) production, which leverage the oxygen-exchange capacity of metal oxides to split water, are thermodynamically attractive, as the entire solar spectrum can be exploited to supply process heat. Here, a metal oxide is subjected to cyclic reduction-oxidation reactions that facilitate either oxygen or hydrogen production, depending on the operating conditions. Chemical thermodynamics suggest that the forward direction of each redox half-reaction is favorable in different temperature and/or oxygen partial pressure regions, and as a result, in practice, large irreversibilities generally arise from sensible heating between redox regimes. To address this issue, recent research has primarily focused on: (1) investigating novel perovskite materials with tunable thermodynamic properties that enable greater reduction extents at more moderate temperatures; and (2) integrating auxiliary techniques, such as mechanical or thermochemical pumping, that reduce the operating oxygen partial pressure (and thus temperature) in the endothermic reduction step. However, until highly effective and reliable solid-solid heat recovery methods are devised, inefficiencies inherent to temperature-swing operation will continue to negatively impact performance. Alternatively, at the expense of lower hydrogen productivity, the water-splitting (oxidation) step can be operated at temperatures where reduction is favorable, thereby minimizing thermal stresses and avoiding complexities associated with solid-solid heat recuperation. Herein, motivated by the immediate practicality of isothermal operation, the behavior of low-cost cobalt-iron aluminate spinels was characterized under conditions relevant to isothermal STCH. Using thermogravimetry, equilibrated changes in oxygen content were quantified at different oxygen partial pressures and temperatures to extract thermodynamic properties, namely the standard molar enthalpy and entropy of relevant defect reactions, and thus enable performance predictions. Results indicate that this class of redox intermediates is capable of higher hydrogen productivities under isothermal operation than previously established benchmark materials (e.g., ceria and lanthanum manganite-based perovskites) evaluated under temperature-swing. The favorable thermodynamic properties of spinels at conditions better suited for industrialization present a step forward towards sustainable hydrogen production with solar thermal energy.

622622 Sustainable High-Purity Nitrogen Production Via a Coupled PSA-Thermochemical Process for the Ammonia Industry

Abstract:

In order to achieve the Paris agreement's target of maximum global warming of 1.5 °C, a sustainable and CO_2 lean production, not only of fuels, but also of key commodities like ammonia, is necessary. Ammonia is a fundamental chemical commodity for fertilizer production, and becomes increasingly important as an energy vector, especially considered as fuel for maritime applications. Since not only the production of H_2, but also the production of the feedstock N_2 itself causes high CO_2 emissions, its sustainable production is mandatory. Economic analysis reveals that the combination of a solar-energy-powered pressure-swing absorption (PSA) and a redox-oxide-based, solar thermochemical cycle decreases the thermal energy input of high-purity N_2-production significantly. The viability of thermochemical air separation and the benefit of using pre-purified air has already been proven in laboratory scale experiments, but the approach of coupling PSA with such a thermochemical cycle has not yet been demonstrated.

In this contribution we present an overall approach including both the redox oxide material selection as well as the solar reactor design. The characteristics of the selected redox materials are analyzed by in-situ high-temperature x-ray diffraction (XRD) in a temperature range of 500-1100°C and thermogravimetric analysis (TGA). Various oxygen partial pressures (0-83 % O_2) are applied in both characterization methods.

The proposed air separation process is demonstrated with a PSA unit coupled to an indirectly heated, pressurized packed bed reactor (Fig. 1). The influence of the process parameters such as pressure, reduction and oxidation temperatures on the performance criteria are shown and compared to results gained by simulations. Moreover, we illustrate how the process parameters can be influenced by tuning the composition of a representative redox material $Ca_{1-x}Sr_xMnO_3$. Therefore, the impact of the Sr-amount on the structure, structural changes and the enthalpy are presented. The performance of the overall system is evaluated on the basis of the achieved purity of the outlet gases, the volume flow rates and the efficiency.

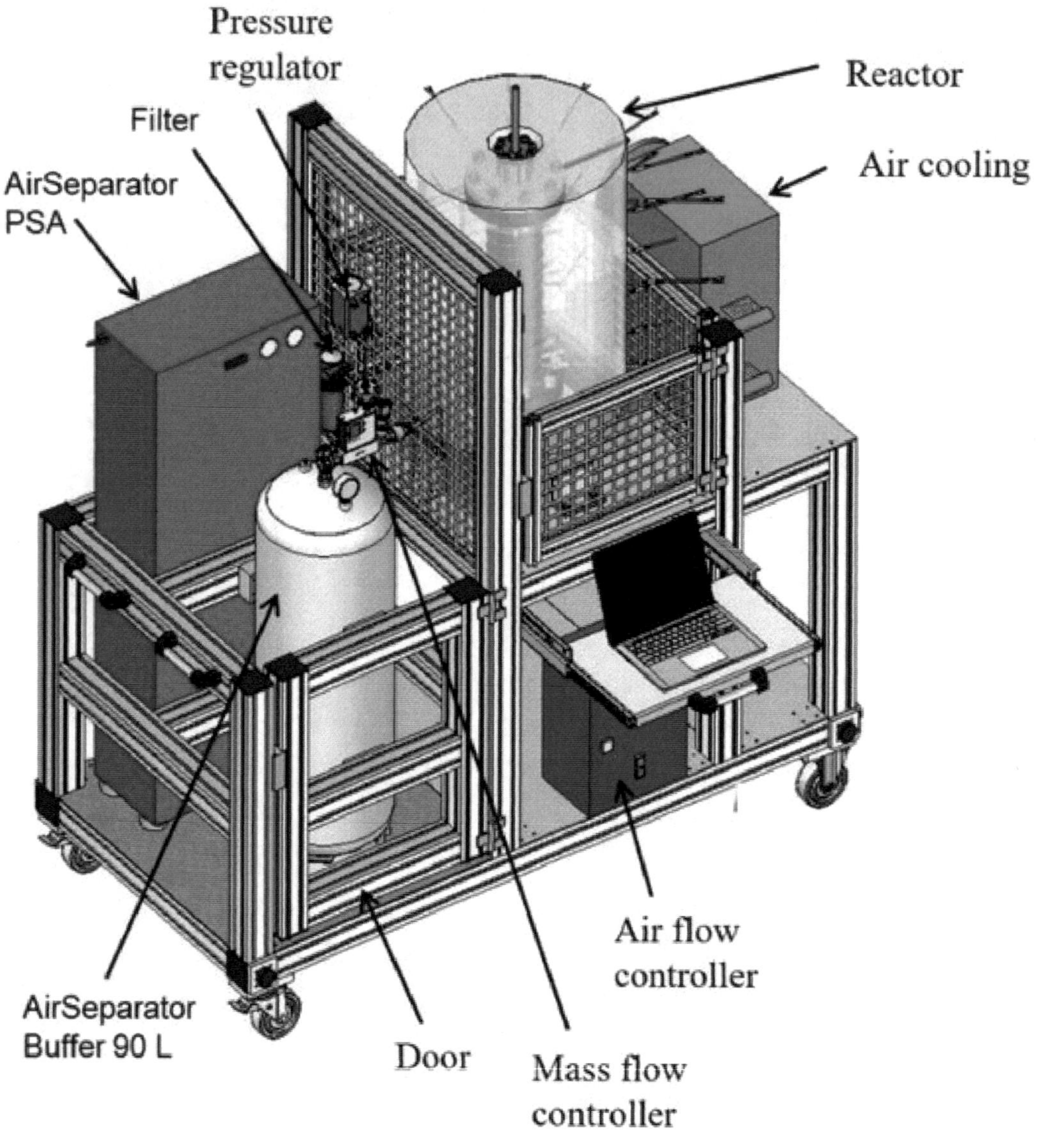

Fig. 1. Reactor design for a two-step air separation process

633022 High-Purity Nitrogen Production from Air By Pressure Swing Adsorption Combined with $SrFeO_3$ Redox Chemical Looping for Trace Oxygen Removal

Abstract:

This work focuses on the production of high purity nitrogen from air via a combination of pressure swing adsorption (PSA) with a downstream redox chemical looping cycle used as a deoxygenation unit. The non-stoichiometric perovskite $SrFeO_{3-\delta}$ was selected for the redox chemical looping cycle because of its favorable thermodynamics, rapid oxidation kinetics and intermediate reduction temperatures. A first principles analysis of the system showed that this combined process could improve upon both the energy balance and production capacity of a PSA system. The chemical stability of the material was demonstrated over 250 redox cycles via thermogravimetry with less than 1 % degradation in the chemical storage capacity. The oxidation kinetics were also measured at 523 K over a range of oxygen concentrations, and fit with a suitable kinetic model. The kinetics were then incorporated into a 1D convection-diffusion model of a packed bed reactor. The model indicates that, for a targeted oxygen impurity of $x_{O2} < 3 \times 10^{-6}$, a chemical looping unit added downstream to a PSA system could approximately triple the capacity and reduce the energy demand to 14 kJ/mol of N_2 [Bulfin *et al.* 2020]. The results are promising for further study, with the goal of bridging the gap between PSA and cryogenic systems in terms of scale, purity and cost.

Bulfin, B., Buttsworth, L., Lidor, A., & Steinfeld, A. (2020). High-purity nitrogen production from air by pressure swing adsorption combined with SrFeO3 redox chemical looping. Chemical Engineering Journal, 127734.

.

632564 Interlayer Vibrational and Thermal Transport Phenomena in 2D Layered Perovskites

Abstract:

2D lead halide perovskites offer a versatile platform for the creation of atomically precise interfaces between organic and inorganic layers. In this talk, I will describe the vibrational, thermal, and phase change properties of 2D alkylammonium lead halide perovskites. Highlights will include 1) selective order/disorder transitions of the organic subphase, 2) observation of zone-folded longitudinal acoustic (zf-LA) phonons in 2D superlattices, and 3) evidence for ballistic phonon transport in interlayer heat transfer.

634247 Boosting Hybrid 2D Perovskite Stability through Computational Screening of Ligand Chemistries

Abstract:

2D hybrid perovskite materials are attractive from the perspective of providing a much larger design space of organic ligand chemistries in comparison with bulk 3D materials. Likewise, many recent reports have highlighted how tuning the ligand chemistry can impact the opto-electronic properties and stability profiles of the resulting 2D materials. In this talk, we'll describe how we have implemented a high-throughput molecular simulation framework to screen prospective ligand chemistries for synthesizability and stabilization of the perovskite lattice. Using this framework, we have been able to characterize tens of thousands of perovskite materials and establish some novel design rules for ligand chemistries. We will also present a few case studies of materials that have been synthesized and tested in response to this effort.

625598 Engineering Entangled Photon Pairs with Metal–Organic Frameworks

Abstract:

The discovery and design of new materials with competitive optical frequency conversion efficiencies can accelerate the development of scalable photonic quantum technologies. Metal–organic framework (MOF) crystals without inversion symmetry have shown potential for these applications, given their nonlinear optical properties and the combinatorial number of possibilities for MOF self-assembly. In order to accelerate the discovery of MOF materials for quantum optical technologies, scalable computational assessment tools are needed. We develop a multi-scale methodology to study the wavefunction of entangled photon pairs generated by selected non-centrosymmetric MOF crystals *via* spontaneous parametric down-conversion (SPDC). Starting from an optimized crystal structure, we predict the shape of the $G^{(2)}$ intensity correlation function for coincidence detection of the entangled pairs, produced under conditions of collinear type-I phase matching. The effective nonlinearities and photon pair correlation times obtained are comparable to those available with inorganic crystal standards. Our work thus provides fundamental insights into the structure–property relationships for entangled photon generation with metal–organic frameworks, paving the way for the automated discovery of molecular materials for optical quantum technology.

634315 Invited Talk: A Charge Transfer Framework That Describes Supramolecular Interactions Governing Structure and Properties of 2D Perovskites

Abstract:

Organometal halide perovskites can efficiently absorb light; perovskite solar cells can generate electricity with power conversion efficiencies now exceeding 25%. However, halide perovskites decompose when exposed to light, heat, or moisture, making them unsuitable for use as active components in solar cells. Two-dimensional perovskites with organic ammonium spacer ligands offer improved environmental stability, but the power conversion efficiencies of the solar cells containing these two-dimensional perovskites are low, reflecting inefficient charge transport. As the interactions between organic ligands and the perovskite octahedra affect the octahedral tilts, directly affecting charge transport, there is an opportunity for a comprehensive molecular design strategy that can manipulate organic-inorganic interactions, control octahedral tilt, enhance charge transfer, and impact device performance. In this talk, I will discuss our approach to molecular design, and using both theoretical and experimental support, elucidate the role of the organic cation in controlling the macroscopic properties of the resulting two-dimensional perovskites. Importantly, we learned that attractive point interactions within the organic bilayer result in a withdrawal of the organic cation from the interstitial site, and less penetration of the organic cation correlates to better performing solar devices. This design paradigm can guide future innovations in solution processable perovskite optoelectronics, with the goal of creating both high performing and stable next generation solar cells.

627921 Molecular Modeling of Halide Diffusion in 2D Organic-Inorganic Hybrid Perovskites

Abstract:

Organic-inorganic hybrid perovskites are promising semiconductor materials for a wide range of optoelectronic applications due to their unique properties, including high carrier mobility, solution processibility, and strong light absorption. However, facile halide diffusion is a source of instability for 3D perovskites that leads to rapid degradation of device performance. In contrast, 2D perovskites exhibit relatively suppressed halide diffusion that is hypothetically further improvable via surface cation design. Nonetheless, the halide diffusion mechanism in 2D perovskites remains largely unknown and is challenging to experimentally characterize. To understand the microscopic origin of the diffusion and the role of organics cations, we have employed the molecular dynamics (MD) simulation to explore different halide migration paths at the atomic scale and the role of surface cation in suppressing diffusion. Steered MD and thermodynamic integration were used to characterize the free energy profiles of each migration path at surface/bulk layers for various systems. In comparison to 3D perovskites, we observe that 2D perovskites exhibit relatively high diffusion activation energies and a large the asymmetry across the accessible channels, which accounts for the suppression of halide diffusion. In addition, the change in the energy barrier when applying different organic cations and the number of inorganic layers offers guidelines for molecular engineering of perovskite materials toward long-term stability.

627960 Understanding the Impact of Pore-Polymer Interactions on the Mobility of Poly(ethyleneimine) Confined in Mesoporous SBA-15: Quasi-Elastic Neutron Scattering Studies

Abstract:

Supported amines are considered a promising class of CO_2 sorbents based on their capability of capturing CO_2 from various sources such as flue gas stream from power plants (5-15% CO_2) as well as ambient air (~400 ppm). Among supported amines, poly(amines) physically impregnated within the mesopores of solid oxide supports have been extensively used for achieving high amine contents (i.e., high capacity) and physical/chemical stability (i.e., limited loss via evaporation/oxidative degradation of amine compounds).[1] Within supported amine systems, the physical properties of the poly(amines) play key roles in determining CO_2 uptake performance. In particular, the chain dynamics of poly(amines) govern diffusion rates of CO_2, which ultimately determine CO_2 sorption/desorption dynamics.[2] Therefore, understanding the dynamics of poly(amine) species confined in solid supports will promote designing optimal supported amines for CO_2 capture.

Herein, we study the impact of the interaction modes between the pore surface and poly(amines) with a model system comprised of branched-poly(ethyleneimine) (BPEI) supported in mesoporous SBA-15 silica. Chain mobilities of BPEI are quantitatively analyzed using quasi-elastic neutron scattering (QENS) in tandem with coarse-grained molecular dynamics (MD) simulations. The pore-polymer interactions are tuned by grafting organosilanes having different end groups to the silica surface, leading to different levels of attraction between BPEI and the pore walls of silica. Chain mobilities of BPEI undergoing four different interaction modes with pore walls are investigated. The first mode brings covalent tethering of amine sites of BPEI to pore surface via halide-amine reaction, where alkyl halides are grafted onto pore surfaces. The second interaction is acid-base and hydrogen bonding interactions based on silanols decorating the native SBA-15 silica, which strongly interact with amine groups of BPEI. The third interaction considers relatively weak interaction with pore surface and BPEI by substituting acidic silanols into basic aminoalkyl groups on the pore walls, so that pore walls interact with BPEI mainly by inter-amino hydrogen bonding. Lastly, repulsive, van der Waals interaction (via grafting aliphatic chains on pore walls) yields the weakest interaction between pore walls and BPEI. QENS (High-Flux Backscattering Spectrometer; HFBS at NIST) and MD simulation revealed that the chain mobilities of BPEI are significantly influenced by different surface groups appearing on the pore walls. The strong/transient interaction (acid-base, hydrogen bonding) due to silanol-BPEI interaction yielded the slowest motions, covalent bonding and repulsive interactions resulted in slightly increased mobility, and hydrogen bonding (aminoalkyl-BPEI; without acid-base interaction) brought significantly recovered mobility as compared to bulk PEI. We hypothesize that PEI's mobility depends not only on the attraction level with the pore walls but also on its chain conformation in the presence of surface capping groups on the pore walls. To understand further, these results are currently being coupled with additional QENS experiments, alongside insights developed via other techniques.

References:

(1) Choi, S.; Drese, J. H.; Jones, C. W. Adsorbent Materials for Carbon Dioxide Capture from Large Anthropogenic Point Sources. *ChemSusChem* **2009**, *2* (9), 796–854.

(2) Holewinski, A.; Sakwa-Novak, M. A.; Carrillo, J.-M. Y.; Potter, M. E.; Ellebracht, N.; Rother, G.; Sumpter, B. G.; Jones, C. W. Aminopolymer Mobility and Support Interactions in Silica-PEI Composites for CO2 Capture Applications: A Quasielastic Neutron Scattering Study. *J. Phys. Chem. B* **2017**, *121* (27), 6721–6731.

622996 Overcoming the Permeability-Selectivity Tradeoff Via the Design of Heterogeneous Pore Wall Chemistries

Abstract:

Developing next-generation membranes for water treatment demands system-specific materials with precisely tuned functionality. In particular, the purification and reuse of highly contaminated water calls for solute-specific designs where a key challenge is the separation of small neutral solutes. One such solute is boric acid, which is strictly regulated in water reuse for agricultural irrigation. Precise tuning of membrane pore chemistry requires a fundamental understanding of the mechanism by which functional moieties at the pore wall modulate the transport of water and solutes. Simulations enable the nanoscopic investigation of the mechanism by which a chemically heterogeneous pore modulates hydration water and thus its water-mediated interactions with solute molecules. Here, we use molecular dynamics simulations to investigate water and boric acid structure, dynamics, and transport in a cylindrical nanopore patterned with nonpolar methyl and polar hydroxyl groups. We find that changes in the arrangement of the methyl and hydroxyl groups, in addition to the density of each group, affect the transport of both water and boric acid in the pore. These changes not only affect the selectivity of water over boric acid, but also the permeability of water through the pore, leading to improvements in the permeability-selectivity behavior, a pernicious trade-off in membrane technology. We thus apply an optimization procedure based on a genetic algorithm to design the chemical patterning of the pore wall to simultaneously optimize water selectivity and permeability. The optimization identifies non-intuitive patterns that significantly hinder the transport of boric acid through the pore, simply by altering the functional group patterning. This inverse design procedure and the accompanying molecular-level understanding of the effect of chemical heterogeneity on solute selectivity in a pore will inform the design of more efficient water purification membranes, as well as the modulation of water-mediated surface-solute interactions in other systems.

624568 Hydrogen Bond Patterns of Deep Eutectic Solvents and Non-Deep Eutectic Solvents

Abstract:

Deep eutectic solvents (DESs) have shown promise as sustainable solvents to extract substrates from aqueous solutions as they are composed of natural compounds. Currently, a trial-and-error method is used to find new DESs experimentally. This process is inefficient and time consuming as all permutations of binary systems must be tested and then ruled out if no DES is formed. Hydrogen bond interactions between the binary compounds are considered the prevailing reason for the eutectic point. We investigated the relationship between hydrogen bond features and the potential of forming DESs for a binary system using a computational approach. We conducted molecular dynamics simulations to analyze the structural and dynamic properties of hydrogen bonds in 50 DESs and 100 non-DES systems and investigated their pattern using a statistical approach. Our results could be used to develop molecular principles for designing new DES based on natural hydrogen bond donors and acceptors.

620632 Silica-Kaolinite Interface for Shale Gas Application: A Molecular Modeling Study

Abstract:

Methane mostly makes up the constituent of shale gas and it is currently being exploited in fulfilling the world's energy demands. Molecular modeling techniques including Density Functional Theory (DFT) and Molecular Dynamics (MD) techniques are employed to understand methane transport in the pores at typical reservoir conditions. Herein, we study the interfacial phenomenon of methane adsorption on Silica-Kaolinite interface. This interface is created as it simulates a shale model which in reality is made up of clay (kaolinite) and quartz-like (silica) materials. The simulations revealed that the interface is formed by a chemical bond between a silicon atom from the silica surface and two oxygen atoms from the kaolinite surface. The adsorption of methane was studied at three different positions on the Silica-Kaolinite interface namely, silica-dominated region, interface region, and kaolinite-dominated region. The mode of adsorption of methane irrespective of its position on the interface was found to be physisorption with the kaolinite region having a greater adsorption. This is in agreement to previous studies which used experimental techniques for study methane adsorption. The results found herein which will be useful in shale estimating the original gas in place for shale reservoir application.

This work is funded by NPRP grant 12S-0130-190023 from from the Qatar National Research Fund (a member of Qatar Foundation). The statements made herein are solely the responsibility of the authors.

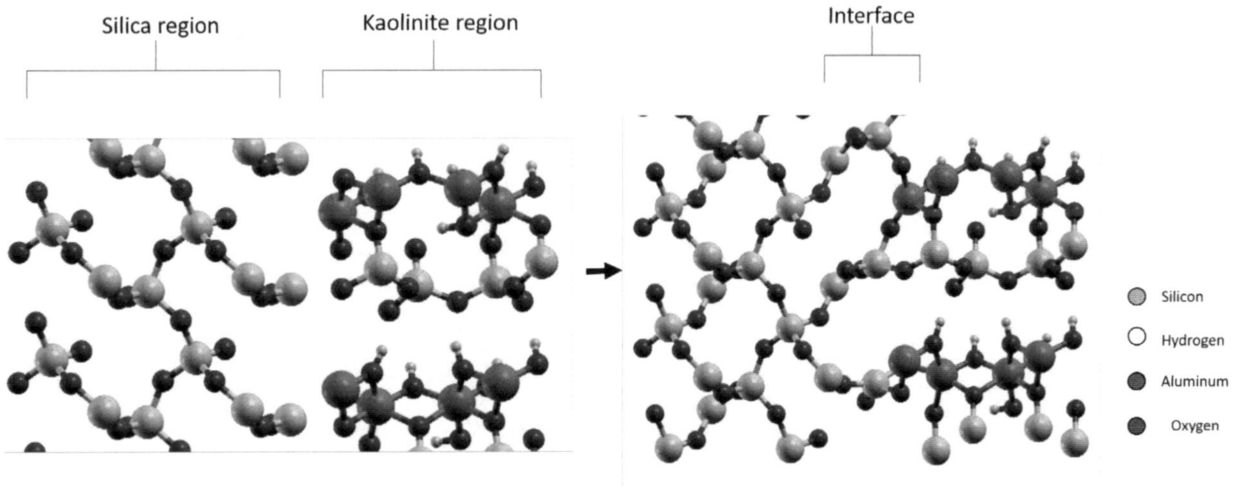

629486 Investigation on the Sodiation of Selenium-Graphene for Na-Ion Batteries: A First-Principles Study

Abstract:

Sodium–selenium (Na–Se) batteries have recently attracted attention as an emerging alternative for the traditional Li-ion batteries owing to their high theoretical volumetric capacity (~ 3250 mAh/cm^3) and electronic conductivity of Se (~ 10^{-3} S/m). Carbon composites are proposed as promising host materials to buffer the volume fluctuation of Se during charge/discharge, provide conductive channels, and physically trap the high-order polyselenides. In this talk, we evaluate the role of carbon host in establishing the reaction kinetics as the fundamental understanding of the sodiation of the Se-C composite is still limited. Based on first-principles simulations, we first present interfacial structure, charge transfer, and diffusion characteristics of Na atoms at the interface between Se and graphene (C). Our calculations demonstrate that there is a substantial enrichment of Na atoms at the interface and corresponding electron transfer to both Se and C nearby. The interactive strength of Se with graphene is found to increase with sodiation, which could lead to mechanical failures as compressive stress build-up at the interface. The presence of C host results in a more populated distribution of longer Se chains, as less reduced Se (due to electron transfer toward graphene) tends to be clustered. We observed a faster mobility of interfacial Na atoms along the graphene surface when compared to the diffusion of Na atoms in Se. Our calculations also show that the Na–Se/C system exhibits n-type metallic characteristic, which could provide fast electron flux for the sodiation reaction.

630087 Molecular Insights into the Electrowetting Behavior of Aqueous Ionic Liquid

Abstract:

Electro-wetting is the modulation in the wettability of a surface prompted by an external voltage [1, 2] across the solid/liquid interface. Ionic liquids (ILs) which have unique physicochemical properties,[3] gain industrial significance, due to its development as a promising alternative medium for electrowetting. This phenomenon is exploited in various applications such as variable fluid lens systems, electrowetting displays, programmable optical filters, paint drying, micro motors, electronic microreactors, liquid optics, or lab-on-a-chip devices.[4-7]

Molecular dynamics simulations were applied to study the wetting behavior of nanoscale aqueous hydrophilic and hydrophobic Imidazolium-based ionic liquid (IL) droplets on a solid graphite substrate subjected to the perpendicular electric field. Imminent transformation in the droplet configuration was observed at E = 0.08 V/Å both for hydrophobic ILs [EMIM][NTF_2] and SPC/E water droplets. However, for hydrophilic, [EMIM][BF_4], the droplet was completely stretched to be column-shaped under the electric field E = 0.09 V/Å for lower weight percentages of ILs and $E \geq 0.1$ V/Å at a higher weight percentage of ILs (> 30 wt%), because of the hydrophilic nature of [EMIM][BF_4]. We explore the impact of electric-field through various parameters such as mass density distribution, contact angle, orientation, and hydrogen bonds.

[1] A. Quinn, R. Sedev, J. Ralston, Contact Angle Saturation in Electrowetting, The Journal of Physical Chemistry B, 109 (2005) 6268-6275.

[2] J. Restolho, J.L. Mata, B. Saramago, Electrowetting of ionic liquids: Contact angle saturation and irreversibility, The Journal of Physical Chemistry C, 113 (2009) 9321-9327.

[3] J.T. Hirvi, T.A. Pakkanen, Wetting of Nanogrooved Polymer Surfaces, Langmuir, 23 (2007) 7724-7729.

[4] D. Bratko, C.D. Daub, K. Leung, A. Luzar, Effect of field direction on electrowetting in a nanopore, Journal of the American Chemical Society, 129 (2007) 2504-2510.

[5] Y.S. Nanayakkara, H. Moon, D.W. Armstrong, A tunable ionic liquid based rc filter using electrowetting: A new concept, ACS Applied Materials & Interfaces, 2 (2010) 1785-1787.

[6] Y.S. Nanayakkara, H. Moon, T. Payagala, A.B. Wijeratne, J.A. Crank, P.S. Sharma, D.W. Armstrong, A fundamental study on electrowetting by traditional and multifunctional ionic liquids: possible use in electrowetting on dielectric-based microfluidic applications, Analytical chemistry, 80 (2008) 7690-7698.

[7] Y. Wang, G.A. Voth, Unique spatial heterogeneity in ionic liquids, Journal of the American Chemical Society, 127 (2005) 12192-12193.

621833 Data-Driven Protein Engineering Using Deep Representational Active Learning

Abstract:

The design of synthetic proteins with desired function is a long-standing goal in biomolecular science with broad applications in biochemical engineering, agriculture, medicine, and public health. Rational *de novo* design and experimental directed evolution have achieved remarkable successes but are challenged by the requirement to find functional needles in the vast haystack of protein sequence space. We have developed data-driven models based on semi-supervised deep representational learning and Bayesian optimization to provide a predictive genotype to phenotype map that can prospectively identify functional sequences. Designs are realized by high-throughput gene synthesis and experimentally assayed along one or more dimensions of function. Experimental measurements passed back to retrain the models within a virtuous feedback loop that rationally traverses protein sequence space to optimize protein function. We will discuss applications of this data-driven design platform to Sho1, S1A, and avGFP proteins.

A.L.F. is a co-founder and consultant of Evozyne, LLC and a co-author of US Provisional Patents 62/853,919 and 62/900,420 and International Patent Applications PCT/US2020/035206 and PCT/US20/50466. R.R. is a co-founder and consultant of Evozyne, LLC and a co-author of US Provisional Patent 62/900,420 and International Patent Application PCT/US20/50466.

625489 Conformational Landscape of Amyloidal Polypeptides

Abstract:

One of the hallmarks of neurodegenerative diseases is amyloid fibril deposits. These are large, structured β-sheet aggregates that are insoluble and thermodynamically very stable. Amyloid formation begins with the assembly of often unfolded or misfolded relatively short polypeptide segments. The propensity of a polypeptide chain to aggregate into amyloids is dependent on several factors including charge, sequence, and hydrophobicity. The secondary structure of the polypeptide segment also plays an important role, and many amyloidal polypeptides have been found to originate from misfolded a-helical structures. It is widely believed that amyloid aggregates initiate from locally unfolded secondary structures. It has been recently suggested that these unfolded states are accessible under physiological conditions, for example, via thermal fluctuations. We performed extensive analysis on protein datasets and found that polypeptide sequences classified as amyloidal have geometric signatures that differ from non-amyloidal sequences. With the new geometric interpretation, we map the energy landscape of the polypeptide backbone on the Ramachandran dihedral coordinates and suggest low energy barrier pathways for misfolding from a-helix to β-sheet structures.

628717 Simulating the Flow of Concentrated Suspensions

Abstract:

Concentrated suspensions are ubiquitous in technologies ranging from personal care products to construction, and their flow response can be complex, exhibiting shear thinning (decreasing viscosity) and shear thickening (increasing viscosity) in different shear-rate regimes. Attractive forces between particles can cause a yield stress, requiring a minimum applied stress to induce flow. Concentrations approaching maximum packing can exhibit re-entrant behavior and discontinuous shear thickening, in which the shear viscosity suddenly increases by orders of magnitude. We have simulated the flow of concentrated suspensions of both Brownian and non-Brownian spheres in a Newtonian suspending fluid at the particle level by taking into account the apparent breakdown of classical fluid lubrication and the onset of particle-particle friction when applied stresses force flowing particles together. A simple friction model is sufficient to capture the observed physical behavior and provides an avenue to the formulation of macroscopic models for processing flows, but the breakdown of the continuum assumption implied by particle contact points to a need for exploration of the molecular nature of the tribological interactions.

.

621318 Multiscale Models for the Kinetics of Polymer Upcycling

Abstract:

Polymer upcycling is a potential solution to the abundance and persistence of plastic wastes in the natural environment. Starting from elementary steps and rate constants, we use pseudo-steady-state approximations and kinetic local density approximations to systematically transform mechanistic proposals into population balance models for time evolution of the molecular weight distribution. We examine homogeneous and heterogeneous catalysts, including catalysts that cut randomly, catalysts that cut processively, and catalysts that cut cooperatively. In the case of heterogeneous catalysis, we will discuss the critical importance of polymer adsorption isotherms. Finally, using experimental data on polyethylene upcycling, we show how these models can help to interpret experimental data, to test mechanistic proposals, and to extract kinetic parameters.

626040 Leveraging Nature's Mechanisms to Design Solutions to Engineering Challenges

Abstract:

Nature presents us with superior, if not optimal, evolved solutions to crucial properties, such as efficiency, scalability and resilience. Decoding these solutions can help us address similar problems in chemical engineering and interdisciplinary areas, where transformative technology is required to address pressing challenges in resource efficiency, energy, water, environment or health. However, because the context of nature and technology often differ, direct biomimicry based on superficial similarities does not guarantee an effective solution. A nature-inspired solution methodology will be presented that first draws from fundamental mechanisms underpinning desired properties, grouped thematically based on universal characteristics. It then applies these abstracted, nature-inspired concepts to design solutions to technological challenges with suitable adaptations to account for the different, concrete environment. We group our solutions according to four themes, based on ubiquitous fundamental mechanisms in nature: (1) hierarchical transport networks, (2) force balancing and nano-confinement, (3) dynamic self-organization, and (4) ecosystems, networks and modularity. This fundamentally grounded design approach allows us to systematize the discovery of new solutions to a range of chemical engineering problems related to process intensification (an approach we call NICE or nature-inspired chemical engineering)[1], but it is also apt to address problems in developmental bioengineering[2] and biomedical engineering[3]. An example is cancer immunotherapy[4]. As enabling properties in nature often result from emergent, complex systems behavior, we advocate a holistic, systems-based approach, departing from the usual reductionistic bottom-up or top-down approaches.

1. M-O. Coppens, 2021, Nature-inspired chemical engineering for process intensification. *Ann. Rev. Chem. Biomolec. Eng.* **12**. DOI:10.1146/annurev-chembioeng-060718-030249.
2. E.C. Goldfield and M.-O. Coppens, 2020, Developmental bioengineering: Recapitulating development for repair. *Molec. Sys. Design Eng.* **5**, 1168-1180.
3. A.S. Perera and M.-O. Coppens, 2018, Re-designing materials for biomedical applications: From biomimicry to nature-inspired chemical engineering. *Phil. Trans. Roy. Soc. A.* **377**(2138):20180268.
4. M.H.W. Chin, E. Gentleman, M.-O. Coppens and R. Day, 2020, Rethinking cancer immunotherapy by embracing and engineering complexity. *Trends in Biotechnology*. **38**(10), 1054-1065.

630378 Developing Multiscale Models of the Blood-Brain Barrier Interface for the Treatment of Alzheimer's Disease

Abstract:

The blood-brain barrier (BBB) is a selective permeability barrier that impedes the influx of potentially harmful blood-borne chemicals from entering the brain. Specialized physical barriers called tight junctions, formed by the endothelial cells lining the brain, act as intercellular gatekeepers in regulating the passive diffusion of molecules and ions into the brain. Although the BBB is vital for neurological functions, its selective permeability is a deterrent in getting drug molecules into the brain. The tight junction selectivity obstructs effective treatment of neurodegenerative diseases such as Alzheimer's and Parkinson's, which disproportionately affect older individuals. Since the segment of the US population older than 65 is expected to increase by 50% by 2030, the cost of care of these diseases is on the order of billions of dollars per year, finding ways to enable drug molecules to cross the BBB is urgently needed.

The claudin family of membrane proteins plays a crucial role in forming the tight junctions. Experimental imaging methods have successfully achieved high-resolution images of the self-assembled claudin strands in vitro. However, these length scales do not provide molecular-level resolution between interacting protein partners. Several computational studies have shown detailed molecular-level interactions, but high computational costs limited the analysis to small lengths and timescales than those observed experimentally. To bridge this knowledge gap, we developed two sets of computational algorithms: Protein Association Energy Landscape (PANEL) to obtain protein dimer interaction stabilities; and Tight Junction Strand Prediction Protocol (TJ-straPP) to predict the TJ strand architecture. Based on the claudin-claudin dimer stability and probabilistic data generated from PANEL, the TJ-StraPP algorithm predicts TJ strands' structure made up of thousands of claudin proteins reaching up to micron-scale output while preserving the atomistic-level details of the assembled claudin strands.

This work is enabling the discovery of sustainable, non-evasive, and thermodynamically favorable pathways to transport small drug molecules into the brain. This talk will provide an overview of the challenge blood-brain barrier and how multiscale simulations are contributing to the development of new treatment strategies for Alzheimer's disease.

631238 Exploring Non-Biological Foldamer Secondary Structure Using Tuneable Coarse-Grained Models

Abstract:

Non-biological foldamers are a promising new class of macromolecules which share similarities to classical biopolymers, such as proteins and nucleic acids. With designed monomers, foldamers enable researchers to incorporate novel chemistries into folding macromolecules. Designability of chemical moieties in foldamer monomers allows for the development of brand new highly specific macromolecules which can be designed with the final application in mind. Currently, designing novel foldamers is a non-trivial process, often involving many iterations of trial synthesis and characterization until a folded structure is observed. In this work, we aim to address these foldamer design challenges using computational modeling techniques. We develop CG PyRosetta, a coarse-grained folding software with tuneable foldamer models to prototype various foldamer monomers. Using CG PyRosetta, we can define a coarse-grain foldamer and search its configuration space to find its minimum energy structures. Through systematic variation of coarse-grained parameters we can investigate various folding hypotheses at the coarse-grained scale to help guide experimental design of foldamers. To demonstrate CG PyRosetta's ability to identify minimum energy structures we propose two preliminary folding hypotheses to test. Firstly, we examine how sidechain size of a 1 backbone and 1 sidechain model affects foldamer secondary structure. We find for simple models, foldamers with larger sidechains tend to adopt tighter helical structures, often governed by packing effects of the sidechains. Secondly, we explore the effects of monomer geometry on foldamer secondary structure through variation of the 1 backbone and 1 sidechain model's backbone angle. Using CG PyRosetta's framework we plan to explore more complex folding hypotheses, such as effects of backbone rigidity, solvation effects and sequence heterogeneity on foldamer secondary structure.

621463 [Invited Talk] Improving Accuracy of Systematic Coarse-Grained Simulations

Abstract:

Billion atom simulations are just now becoming possible in molecular simulation for nanoseconds. We've also crossed the millisecond barrier for simulating biomacromolecules. What's left? Unfortunately, a typical cell contains 100 trillion atoms. Even simulating something like a polymer nanoparticle (~100 million atoms) has timescales of interest far beyond nanoseconds. One way around the length-scale limitation is coarse-grained simulation. Coarse-graining requires two ingredients: (i) a mapping that determines how to group atoms into coarse beads and (ii) a force field that describes these interactions. In this talk, I will describe our recent progress on determining mapping operators, which has previously been an arcane topic guided by intuitiuon. We've developed novel theory, shown what role symmetry plays, and developed ML models that find mappings for arbitrary molecular systems. Finding the force field of a coarse-grained model is a rich field with a long history. Typically, it is broken into two types: top-down, where we choose the force field to reproduce an observed phonemenon in experiment; and bottom-up, where we draw upon the observed forces in a molecular simulation. I will describe our recent work on combining these approaches to create hybrid top-down/bottom-up models via the principle of maximum entropy.

629698 A Multiscale Approach to Analyzing the Van Hove Correlations of Water

Abstract:

The physical properties of water are of fundamental importance in innumerable fields. Despite the many modes of analysis used to describe water including viscosity, self-diffusion coefficients, and pair distribution functions (PDFs), there is still much to be understood regarding water and the molecular origins of its unique properties. Recently, several studies have calculated the time-dependent pair distribution function, also known as the Van Hove function (VHF), for the first time through a combination of inelastic x-ray scattering (IXS) and classical molecular dynamics (CMD) simulations. Through these analyses, water molecules were found to be correlated in both space and time.

Although CMD simulations produce qualitative agreement with the IXS response, the chosen models lack the details required to quantitatively reproduce the interactions of water. To do so, a multiscale approach with the inclusion of quantum mechanical and reactive approaches is required. *Ab initio* MD (AIMD) is a quantum mechanical approach that computes electronic structure calculations on the fly to generate forces for dynamics. Though computationally expensive, these calculations allow these methods to be unbiased to a given state or set of reference data. Density-functional tight-binding (DFTB) is an approximation to density functional theory that allows for routine larger scale simulations on nanosecond time scales. Reactive MD includes additional interaction terms to account for chemical reactions and hydrogen bonding. Here we present VHF results aimed at better understanding the robustness of molecular simulation techniques describing water at a multitude of time and length scales. Encompassed in this study are a range of classical, polarizable, and reactive force fields, as well as AIMD, and DFTB.

Iwashita, T., Wu, B., Chen, W.-R., Tsutsui, S., Baron, A. Q. R., & Egami, T. (2017). Seeing real-space dynamics of liquid water through inelastic x-ray scattering. *Science Advances*, *3*(12), e1603079. https://doi.org/10.1126/sciadv.1603079

Van Hove, L. (1954). Correlations in space and time and born approximation scattering in systems of interacting particles. *Physical Review*, *95*(1), 249–262. https://doi.org/10.1103/PhysRev.95.249

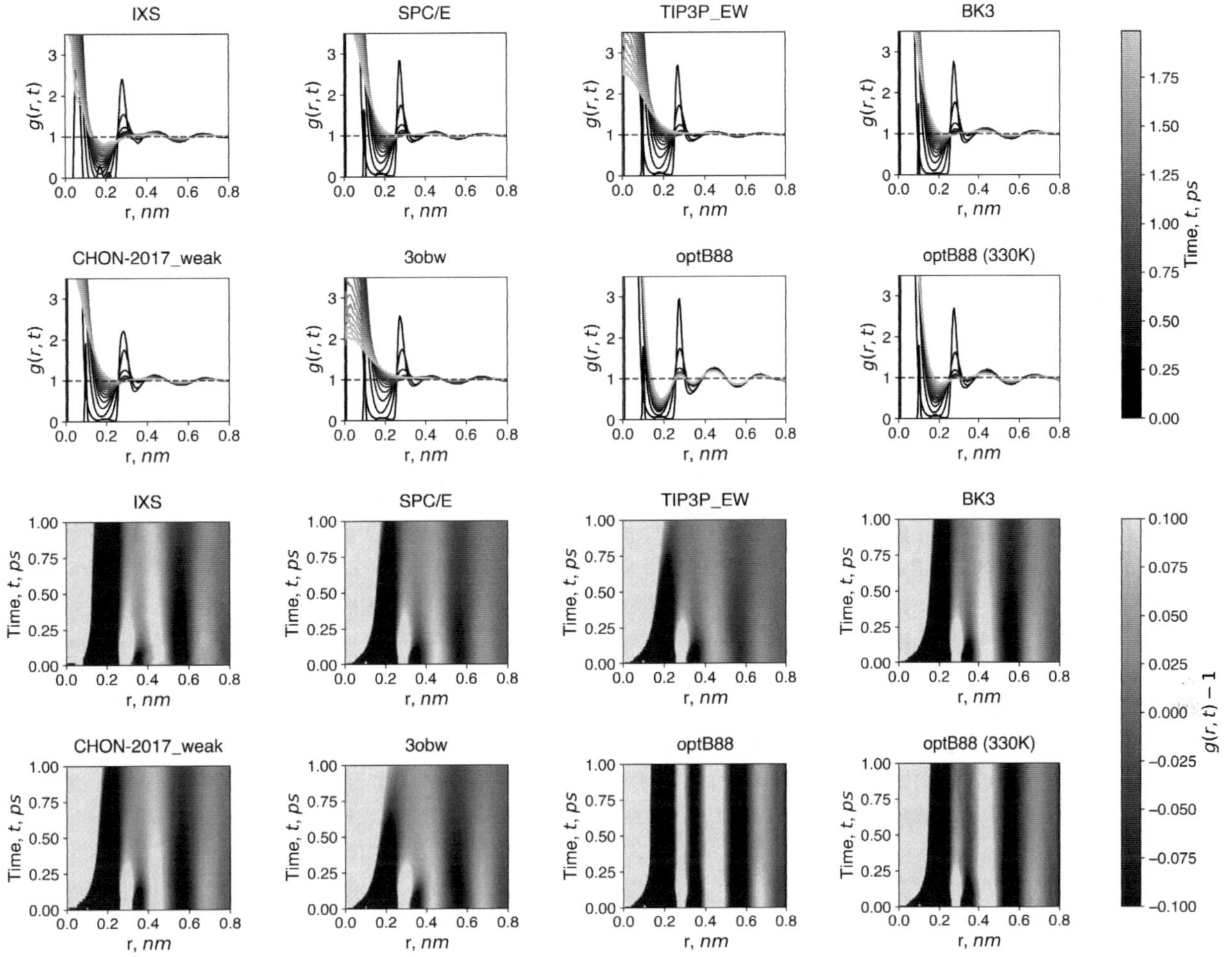
IXS
SPC/E
TIP3P_EW
BK3
CHON-2017_weak
3obw
optB88
optB88 (330K)
g(r, t)
r, nm
Time, t, ps
g(r, t) − 1

626584 Revealing Mechanisms of Pinholin Activation through Thermodynamic Integration and Mutational Analysis

Abstract:

Pinholin is a protein employed by bacteriophages, consisting of two transmembrane domains (TMDs) that drive cell membrane rupture. This is achieved by inducing the formation of small membrane pores to disrupt proton gradients, thus activating larger holin proteins. The mechanism of pinholin activation is well characterized and involves a conformational rearrangement in which TMD1 leaves the membrane to become exposed to the solvent. Then, TMD2 can oligomerize with neighboring pinholin proteins and form the membrane pore.

While the mechanism has been explored thoroughly, the thermodynamic contributions of each amino acid to the activation process are however unknown. Experimentally, many mutations have been characterized in terms of conformational shifts and the tendency of TMD1 to translocate between the membrane and the solvent. Spin label distance measurements from electron paramagnetic resonance have shown that mutations in A17 and S16 amino acids can cause early activation. From these experiments, a general structure emerges, but the exact molecular-level coordinates remain unknown. To generate a more detailed atomistic structure, molecular dynamics (MD) simulations have been conducted to explore conformational ensembles over timescales up to hundreds of nanoseconds. Molecular models were validated by comparing structure and dynamics to experimental data, showing close agreement between the two. However, the timescales of conformational rearrangement involving the translocation of TMD1 are too long for standard MD simulations to probe. To compensate for slow kinetics, an alternative thermodynamic model has been employed to compare the relative free energies of translocation for various experimentally characterized mutants. Thermodynamic integration in the form of alchemical mutations provides an alternate route to characterize the thermodynamic cycle of pinholin activation. Thus, through complementary experimental and computational mutational analysis, the thermodynamic contributions of specific amino acids to pinholin rearrangement can be probed relatively quickly. Within this thermodynamic integration framework, we have tested the extent to which (a) hydrophobic interactions between TMD1, TMD2, and the membrane and (b) hydrogen bond interactions between TMD1 and TMD2 affect the conformational stability of pinholin's active and inactive states. We foresee future applications of this methodology to other systems with inconveniently long time scales.

626949 Drug Resistance Predictions Based on Non-Equilibrium Alchemical Calculations

Abstract:

Resistance to small molecule drugs often emerges in cancer cells, viruses, and bacteria as a result of the evolutionary pressure exerted by the therapy. Protein mutations that directly impair drug binding are frequently involved in resistance, and the ability to anticipate these mutations would be beneficial in drug development and clinical practice. While atomistic simulation of drug binding events is computationally challenging and often intractable in practice, the use of *alchemical* (i.e., non-physical) thermodynamic cycles enables the efficient computation of thermodynamic properties, such as ligand binding affinity. In this talk, I will introduce *pmx*, a Python tool for the setup and analysis of alchemical calculations in *Gromacs,* and present the results of a retrospective study in which we tested the ability of a non-equilibrium protocol to predict protein mutations associated with inhibitor resistance in the cancer target Abl kinase. Results obtained with data-driven, structure-based approaches like *Rosetta* and machine learning will also be presented. Finally, I will show the results of a prospective evaluation of these approaches against novel Abl kinase mutations.

Reference: Aldeghi, M., Gapsys, V. & de Groot, B. L. Predicting Kinase Inhibitor Resistance: Physics-Based and Data-Driven Approaches. *ACS Cent. Sci.* 5, 1468–1474 (2019).

625176 Variational Autoencoders As a Unifying Framework for Molecular Coarse Graining, Back-Mapping, and on-the-Fly Learning of Efficient Monte Carlo Moves

Abstract:

Bottom-up coarse graining enables simulations of nanoscale processes, such as self-assembly, that occur on long time and length scales relative to intramolecular vibrations and reconfigurations. Typically, degrees of freedom are grouped together (e.g., the atoms of an amino acid side chain) or simply removed, leaving behind only effective interactions between remaining groups (e.g., implicit solvent). It is difficult, however, to implicitly model solvent-mediated interactions, such as hydrophobic attractions, that involve a complex, many-body response of solvent molecules to solutes. As a result, information loss is inevitable during the coarse graining process, with resulting structural and thermodynamic differences from the fine-grained model difficult to assess. We apply variational autoencoders (VAEs) to the problem of coarse graining and back-mapping, demonstrating that this framework enables Monte Carlo moves that pass through and sample within a learned coarse-grained space while exactly preserving the original fine-grained ensemble of interest. In fact, we show that the acceptance rates of these moves approach unity for a perfect VAE model. While this is never observed in practice, VAE-based Monte Carlo moves still enhance sampling of new configurations. However, assumptions concerning the form of the encoding and decoding distributions, in particular the extent to which the decoder reflects the underlying physics, greatly impact the performance of the trained VAE. In conjunction, we also demonstrate that the learned coarse-grained model itself depends on the forms of both the encoder and decoder, with far-reaching implications for coarse graining as well as utilization of autoencoders within molecular simulation.

625656 Temperature-Transferable Coarse-Grained Modeling with Relative Entropy

Abstract:

Systematic methods for bottom-up coarse-graining are essential tools for developing accurate multiscale models of complex systems. Relative entropy optimization can produce coarse-grained models that accurately capture interactions at a given set of thermodynamic conditions. However, such approaches can suffer issues related to the transferability and representability problems in coarse-graining. For example, coarse interactions appropriate for one temperature might lead to inaccurate behavior at another, while thermodynamic properties including energy might be incorrectly predicted even when structure is reproduced properly. These shortcomings are fundamental to systematic coarse-graining approaches that treat coarse force fields in terms of fixed potentials rather than temperature-dependent interaction free energies. As strong temperature-dependent behavior is a hallmark of many fluid and polymer systems of interest for multiscale modeling, inaccurate prediction of such dependence remains a critical problem for coarse-graining in general.

Here, we present a novel approach to resolving this temperature transferability problem with an extension of the relative entropy coarse-graining method based on joint probability distributions of configurations and energies. In this approach, relative entropy minimization at a single temperature allows rigorous decomposition of a coarse-grained potential of mean force into temperature-dependent energetic and entropic components. The resulting coarse-grained model can then be used to calculate an effective force field at any temperature. Furthermore, it can predict the distribution of energies in the underlying detailed system, and its temperature dependence, for the entire ensemble of possible coarse-grained configurations. We demonstrate our method here with a variety of systems including model fluids and molecular systems such as alkanes and water. Using atomistic simulations of these systems at only single temperatures, our method is able to predict configurational and energetic probability distributions over a range of temperatures. In the end, this systematic approach to temperature transferability and related representability issues stands in contrast to other *ad hoc* solutions to these problems, and has the potential to expand the range of systems that can be studied readily with coarse-grained models.

628281 Replica Exchange and Backbone Sampling Methods Improve Protein-Protein Docking By Mimicking Induced-Fit Pathways

Abstract:

Protein-protein interactions (PPIs) are involved in almost all biological processes in human health and disease. Predicting protein complex structures and the associated dynamics of protein interactions can reveal biological mechanisms and suggest intervention strategies. Since investigation of these structural complexes by experimental techniques can be often expensive, laborious and limited, computational modeling provides an alternative. Current computational methods are confounded by the many degrees of freedom of the binding-induced conformational changes in one or both protein partners. To overcome these limitations in capturing backbone motions, we developed a new, aggressive conformational sampling method that incorporates temperature and Hamiltonian replica exchange Monte Carlo (T-REMC and H-REMC) techniques within docking protocols in Rosetta. To mimic the induced-fit approach of protein binding, we sample backbone motions on residues comprising a putative interface on-the-fly, thereby enabling the algorithm to capture protein backbone flexibility while docking. Our updated method ReplicaDock 2.0 requires 300-500 CPU hours, depending on protein sizes, which is efficient compared to molecular dynamics-based docking approaches. In local docking of a benchmark set of 80 proteins with moderate to high flexibility (unbound to bound RMSD over 1.2 Å), ReplicaDock 2.0 achieves successful docking for 61% of moderately flexible complexes and 33% of highly flexible complexes. Additionally, highly flexible targets can be predicted to sub-angstrom accuracy when backbone sampling is biased towards residues comprising flexible loops or hinge domains, indicating that additional gains are possible when mobile protein segments are known.

625748 [Invited Talk] a Multiscale Simulation Approach for Formulation Design: Using Bottom-up Coarse Graining to Bridge All-Atom and Field Theory Simulations

Abstract:

Formulations are multicomponent mixtures featuring a range of species from small molecule additives and surfactants to large macromolecules. The design of formulations involves resolving the self-assembly and phase behaviors of formulations, which requires large scale simulations that can be prohibitively expensive to study by traditional particle-based molecular dynamics simulations. We address this challenge by taking advantage of statistical field theoretic simulations that have the ability to efficiently probe large length scale behaviors. There is no free lunch, however, as the predictive power of these field-theoretic simulations have traditionally been limited by a dependence on emergent (e.g. chi) parameters with non-obvious connections to the underlying chemistries. Using information theoretic and statistical analyses, we have developed new coarse graining techniques that take all-atom simulations as input and derive coarse-grained field theory parameters that faithfully preserve thermodynamic behavior. We are thus able to combine the high-resolution chemical detail of all-atom simulations and large-scale simulation capabilities of field-theoretic simulations in a systematic framework, enabling the de-novo in-silico exploration of formulation phase behavior. We demonstrate this approach on model formulations with surfactants like sodium dodecyl sulfate, exploring micellization and microphase behavior.

*Funded by BASF California Research Alliance

Polymer-guided interfacial assembly of nanoparticles into low-dimensional architectures

Gaurav Arya*, Yilong Zhou, Tsung-yeh Tang, and Brian Hyun-jong Lee

* Presenter, Department of Mechanical Engineering and Materials Science, Duke University, Durham, NC 27708, USA

ABSTRACT

In this talk, I will discuss how polymer-grafted nanoparticles at fluid-fluid interfaces can self-assemble into unique higher-order structure due to the three-way competition between interparticle interactions and the interfacial tensions of the nanoparticle with the two liquids. I will begin by demonstrating through molecular dynamics simulations the interfacial assembly of spherical nanoparticles into open clusters with tunable orientation and bond angles; semi-linear structures like serpentine strings and branched networks; and layered structures such as corrugated monolayers and square-ordered bilayers. Next, I will show how tailored Monte Carlo minimization methods along with analytical models of interparticle and interfacial interactions can be used to quickly map out the full repertoire of layered architectures achievable through this assembly approach. Lastly, I will show, also via molecular dynamics simulations, how interfaces can be used to orient shaped nanoparticles, especially those with faceted geometries whose subsequent assembly at the interface can yield an interesting class of porous architectures. Overall, this work demonstrates the power of polymer-guided, interfacial assembly for fabricating next-generation nanoparticle architectures with possible applications in plasmonics, optics, membranes, and catalysis where control over particle orientation and spatial arrangement is critical for function.

623511 Soft and Stretchable Energy Harvesting Using Metal/Gel Interfaces

Abstract:

Capacitors are widely used in electrical circuits due to their ability to store electrostatic energy and discharge current almost instantly when needed. Variable capacitors are those that can change capacitance – usually via geometric changes—to convert mechanical energy to electrical energy. Despite the headway, harvesting mechanical energy by varying capacitance is limited to sliding or oscillating the electrodes up and down in a systematic way due to their rigid nature. In addition, the relatively large gap between the electrodes in a traditional rigid capacitor results in low capacitance, limiting the charge storing capacity. To address these limitations, we utilize a so-called electric double layer capacitor. Such a capacitor forms at the interface between metal and electrolyte, due to the formation of oppositely charged layers form at the interface. Electrical double layers innately offer higher capacitance than conventional capacitors due to the small distance between the charged layers. We use liquid metal for the electrode, which allows the area of these interfaces to vary in response to mechanical forces. Thus, we were able to create variable area capacitors that can harvest mechanical energy. Since they are soft and deformable, they can convert all modes of mechanical input to electricity including pressing, stretching, bending, and twisting. These devices generate ~1 mW m^{-2} by harnessing energy from mechanical motion without the need for an external battery, unlike the conventional electrostatic energy harvesters. We have characterized the behavior of these devices as a function of several parameters including material properties, design architecture, and mechanical deformation. The devices perform as expected and the response of the devices to deformation matches a physics-based model. The liquid metal-based soft device generates an electrical signal when deformed and works underwater, which may be useful for oceanic energy harvesting as well as wind energy harvesting. Besides, these devices can harvest human motion thereby find applications in wearable electronics, healthcare systems like rehabilitation and prosthetics.

630815 Morphology and Charge Transport Predictions across Organic Photovoltaic Components Using Coarse-Grained Molecular Dynamics

Abstract:

Optimizing optical and electronic properties of organic photovoltaic active layers is important for realizing low-cost, high-efficiency solar power. Molecular simulations can guide selection of organic components and processing conditions if sufficiently large length scales and time scales can be accessed. We manage ensembles of multiscale molecular dynamics simulations of polymer electron-donor and both fullerene and non-fullerene electron-acceptor materials, making extensive use of MoSDeF tools signac, foyer, and mBuild. Coarse-grained representations and GPU acceleration are used to access long-time equilibration dynamics. Atomistic representations from the coarse models are used for calculating charge transfer dynamics between chromophores, in service of predicting charge mobility as a proxy for active layer quality. We find simplified coarse-grained representations to reliably predict experimentally-validated morphologies for broad classes of organic components and describe specifics for ITIC, PTB7, BDT-TPD, PCBM, and P3HT. Varying agreement between our predictions of charge mobility and measured experimental values reinforce charge assumptions of charge delocalization that are valid for small-monomer polymers, but which need improvement for large non-fullerene acceptors. The predictive capabilities of coarse-grained models suggest that as these simulations become more routine, adding non-equilibrium and rheological processing effects and improving the speed and accuracy of charge transport calculations will be the next frontiers in organic photovoltaic optimization.

631527 Exploring the Energy Landscape of Soft Glassy Systems

Abstract:

The dynamics and structure of systems beyond the glass and jamming transition have been of interest for a long time due to their importance in a wide range of fields. Molecular glasses are essential active layers in organic light-emitting diodes, while the phenomena of jamming is important in dense emulsions and granular flows. The slow dynamics and complexity of theses systems has restricted the ability to simulate them using conventional methods, particularly at low temperatures or high volume fractions. The properties of the low energy states are of particular importance to answering many fundamental questions surrounding the glass transition and in connecting simulation results to experiments. In this study, we use an energy landscape based approach to probe the low-energy inherent structures and provide an algorithm to drive the system to lower energy states. The algorithm is inspired by metadynamics, and proceeds as a high dimensional basin filing algorithm that allows us to explore the energy landscape of a model soft-sphere foam system. We demonstrate the ability of the algorithm to reach low energies while revealing interesting geometric characteristics about the nature of the landscape. We study how connected metabasins are organized relative to a starting energy minimum and their orientations relative to each other. Finally, we compare our method with conventionally used methods like simulated annealing and swap Monte Carlo and discuss the advantages and disadvantages of each.

632594 Optimized Generation of Initial Conformations for the Simulation of Amorphous Polymer Systems to Reduce Required Simulation Resources

Abstract:

Significant breakthroughs have been derived from the simulation of non-crystalline polymer phases including optimization of such polymers for mechanical, membrane, surface, biomedical and microelectronics applications. Unfortunately, these simulations require long simulation times when the molecular weight of interest is at or above the entanglement length of the polymer chain. Some reduction in the required simulation time is attainable through the use of simulation algorithms that efficiently sample conformation space, but an additional reduction can be obtained by producing an accurate initial conformation of the polymer system that is close to the equilibrated structure. This work investigates the effect of density on the initial conformation and the importance of long-range corrections in both cohesive energy and density. We utilize the Suter-Theodorou initial guess algorithm that employs a Rotational Isomeric States model for short range conformational choices coupled with long range self-avoiding corrections. However, we carry this out as a function of the density of the periodic cell. Depending on the nature of the molecular interactions (dispersion, polar, and hydrogen bonding), there appears to be an optimal density at which the initial conformation is generated. Generation of the initial guess at the experimental density can produce unrealistically high energy conformations. While generation at lower density can alleviate these high energy conformations, it also reduces the chain entanglement. This entanglement is significant because chain diffusion is slow, and insufficient entanglement requires extensive simulation time to recover. By manipulating density, we can optimize both energetics and entanglement to produce an initial conformation close to the equilibrium structure of the amorphous polymer. Additionally, we utilize an updated protocol for calculating the long-range energy and pressure corrections that extrapolates a series of reproduced periodic structures. This approach avoids the conditional convergence of the many of the commonly used methods for electrostatic energy correction.

632864 Adaptive Grid-Based Method for Mapping Cavity Connectivity in Thermal Crystals and Amorphous Materials

Abstract:

The cavities in ideal, zero Kelvin crystal structures are straightforwardly charactized by the largest sphere that can be placed in each void. For example, in an fcc crystal it is well-known that the largest sphere that can fit in an octahedral (O) site has a diameter 0.414 times the atomic diameter [1]. This method, however, does not account for the non-spherical shape of each void nor does it account for fluctuations in the size and shape of the cavity at finite temperatures. Additionally, it is not adaptable to amorphous materials in which the cavity centers are not easily identifiable.

We develop an adaptive grid-based method for computing the size and shape of cavities in thermal crystals and amorphous materials. We identify pathways connecting cavities by launching stochastic trajectories and compute the energy barrier for an absorbed atom to hop between connected cavities using the nudged elastic band (NEB) method. We calibrate our methodology by analyzing a zero Kelvin fcc crystal. The resultant cavity map has one O site and two tetrahedral (T) sites per atom, with each O site connected to eight nearest neighbor T sites. We demonstrate the sensitivity of our results to two adjustable parameters, the probe diameter and the number of trajectories. Finally, we present cavity maps for an ensemble of amorphous configurations for a binary Lennard-Jones mixture [2].

This work was performed under the auspices of the U.S. Department of Energy by Lawrence Livermore National Laboratory under Contract DE-AC52-07NA27344. Document release number LLNL-ABS-821430.

Figure. (a) An fcc unit cell showing O and T sites in blue and red, respectively. (b) The connectivity map for (a) showing that each O site is connected to 8 T sites. (c) An amorphous configuration showing six cavities. (d) The connectivity map for (c) showing how the cavities are connected.

References

1. Fukai, Yuh. The metal-hydrogen system: basic bulk properties. Vol. 21. Springer Science & Business Media, 2006.
2. Kob, Walter, and Hans C. Andersen. "Testing mode-coupling theory for a supercooled binary Lennard-Jones mixture I: The van Hove correlation function." *Physical Review E*5 (1995): 4626.

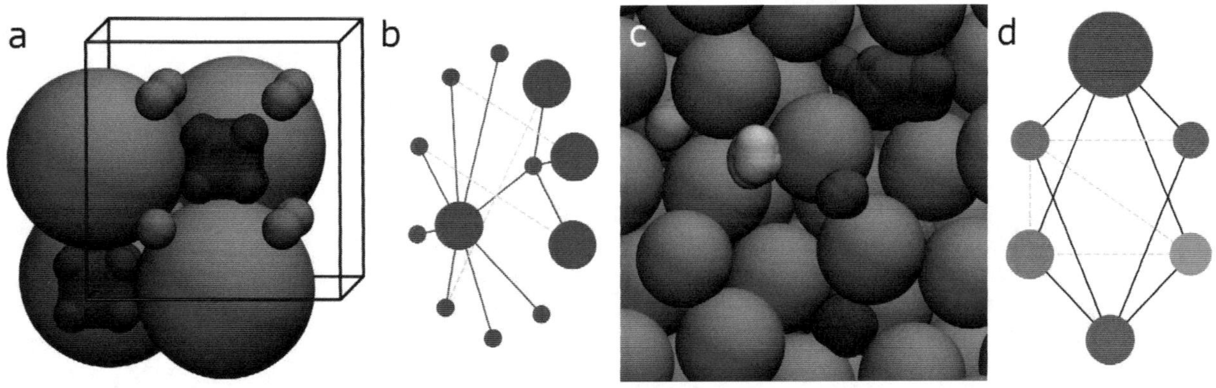

623338 Antigen Recognition at Immune-Cell Surfaces: Probing the Role of Mechanical Forces

Abstract:

One of the central problems in immunology involves molecular recognition at cell-cell interfaces. T cells and B cells directly engage other cells, and fascinating recent experiments have revealed that mechanical forces regulate processes by which they distinguish between self and foreign ligands. However, a unifying theoretical framework is missing. In this work, we develop hybrid computational models that account for key biophysical properties of immune-cell interfaces, including stochastic receptor-ligand binding kinetics, membrane mechanics, and actin-mediated forces on the membrane. We use these models to investigate how mechanical forces modulate the interactions of T cells and B cells with surface-presented antigens. We discuss our results in the context of antigen discrimination and highlight open questions in mechanoimmunology that theoretical approaches can help to address.

625687 Bridging Statistical Inference and Stochastic Modeling to Shed Light on Gene Regulation

Abstract:

Cell biological data is increasingly available at single-cell, single-nucleotide, and single-molecule resolution. Such experiments reveal often-unexpected levels of heterogeneity at these scales. In the Read lab, we use stochastic biochemical network modeling to study the origins of this heterogeneity and discover signatures in the noise that offer clues to molecular mechanisms.

I will present recent work in genomics and epigenomics where we are bridging "top-down" statistical inference with "bottom-up" stochastic mechanistic modeling to maximize the utility of big data. For example, stochastic models of enzymatic processes that confer epigenetic marks on DNA inform our interpretation of regional correlations derived from epigenome sequencing, with implications for epigenome stability.

621842 On the Molecular Diversification and Maturation of Neurons in the Visual System

Abstract:

A central question in neurobiology is how diverse neuronal types in the mammalian brain arise through an interplay of genetic regulation and experience-dependent mechanisms. I will present our recent collaborative efforts to address this question through a combination of high-throughput gene expression measurements in mouse models, machine learning and statistical inference of large scale datasets, as well as histological assays. Our focus will be two areas of the mammalian visual system - **the retina**, which is the part of the eye that initiates vision, and the **primary visual cortex**, where the first stage of higher-order visual processing occurs.

In the **retina**, we have studied how retinal ganglion cells (RGCs), the primary output neurons of the retina, differentiate into 45 distinct types during embryonic and postnatal life. We use a statistical inference approach rooted in the mathematical theory of optimal transport to identify fate associations from "snapshot" measurements of RGCs at different stages of development in mice. We find that RGCs diversify through a gradual process, and predict that precursors are likely to be multipotential. Importantly, by analyzing three distinct models where mice are deprived of vision, we also show that RGC diversification appears to be genetically hardwired.

In contrast, we find a much more significant role for vision in the molecular maturation of neurons in the **primary visual cortex**. It has been long known that the cortex subdivided into five anatomically "layers" (L2, 3, .. 6)containing distinct populations of cells. By analyzing the postnatal maturation of these cells based on single-nucleus gene expression measurements in mice, we find that neurons in the superficial layers (2/3/4) require visual input to acquire their molecular identities, while neurons in the deeper layer (5/6) do not. Unexpectedly, we also discovered that vision plays a role in "patterning" these neurons into distinct sublayers which correlate with projection specificities to higher visual regions. Through rigorous perturbation assays we show that visual deprivation during early postnatal life completely abrogates neuronal maturation in these superficial layers, and their associated anatomical patterns. Thus, we identify role for experience mediated gene expression programs in the anatomical patterning of this brain region.

These two interconnected but contrasting studies suggest a tempting hypothesis that neural diversity sensory systems is likely shaped largely by hardwired genetic mechanisms, while those in higher-order processing centers are likely to be shaped significantly through experience dependent plasticity.

628443 Multiscale Affinity Maturation Simulations to Elicit Broadly Neutralizing Antibodies Against HIV

Abstract:

Vaccines play a critical role in both preventing and eliminating global pandemics. This is especially true in the case of highly mutable pathogens like HIV and influenza, which present unique challenges to vaccine design. One promising strategy for designing vaccines against highly mutable pathogens is to target regions on the surfaces of the pathogenic proteins that are functionally important (e.g., receptor binding sites for adhering to/entering host cells), and thus cannot be so easily mutated by the pathogen. Antibodies that can bind to these conserved regions on pathogenic proteins are called broadly neutralizing antibodies, or bnAbs. BnAbs have now been isolated from people naturally infected with many different highly mutable pathogens. However, in most cases it remains unclear how to elicit these same antibodies by vaccination. In this presentation, we present an agent-based model of affinity maturation – the Darwinian process by which antibodies evolve against a pathogen – that, for the first time, enables the *in silico* investigation of real germline nucleotide sequences of antibodies known to evolve into potent bnAbs, evolving against real amino acid sequences of HIV-based vaccine-candidate proteins (Ags). These high-resolution simulations are enabled by the use of a fast and reasonably accurate free energy function to calculate the binding affinity between the evolved B cell receptor (BCR) sequences and Ags. We find that sequentially administering Ags that are increasingly diverse in the variable residues surrounding conserved residues will maximize bnAb production. Additionally, by analyzing and comparing crystal structures of bnAbs and germline antibodies bound to HIV-based Ags, we identified a novel mechanism through which bnAbs may evolve against the CD4 binding site of HIV. Lastly, we find that in response to specific vaccination protocols, the BCR population evolving *in silico* employs this same mechanism, recapitulating a key aspect of biology. In the future, our model may be used to iteratively design protein sequences that will maximize bnAb formation, thus forming the basis of vaccines against highly mutable pathogens.

627561 Inverse Design of Molecular Probes to Bind with Water Contaminants

Abstract:

Traditional water treatment systems are not designed to address many of the emerging classes of contaminants appearing in waterways and wastewater streams. Each water source has a unique mixture of solutes with varying degrees of environmental health and safety concern, and treating to remove them all is unlikely to be the most efficient strategy. Rather, it is desirable to design treatment methods capable of targeted, selective removal. To address the challenge of recalcitrant water pollutants, we take assistance from machine learning (ML) for navigating the large sequence and structure space of molecular probes by determining the key features that can enhance the selectivity and binding of targeted species. We describe our development of a platform combining deep representational learning, surrogate model construction, and Bayesian optimization to rationally traverse molecular design space and identify probes with high binding affinity and selectivity. We couple high throughput virtual screening using enhanced sampling all-atom molecular simulations with targeted experimental synthesis and testing within an integrated computational/experimental screening protocol. The optimal probes discovered by this approach will ultimately be used for detection and removal of organic pollutants from water.

632006 Excluded Volume Monte Carlo Simulations for Anisotropic Fluid Structure Predictions

Abstract:

High-flux neutron scattering is recently being used to determine the multidimensional structures of polyatomic and anisotropic molecules, providing novel insights into the structural patterns of local fluid structure. The experimental outcomes indicate complex molecular organization and behavior change with thermodynamic conditions. We have recently investigated two hydrocarbon fluids (benzene and 1-methylnaphthalene) adopting wide-angle neutron scattering of isotopically substituted molecules to experimentally determine local fluid structure. We observed that face-to-face structural motifs tend to show increased order (1-methylnaphthalene) or remain unchanged (benzene) when the temperature is increased. We attributed this apparently contradictory behavior as entropically-driven induced by the shape of the molecule, widely observed in hard-particles and colloidal systems. We developed a stochastic Monte-Carlo-based simulation tool to portray this entropically-driven ordering by simply estimating the free-volume, eliminating the molecular modeling high-computational costs and complexity associated with many-electron systems and consequent limitations to represent the emergent many-body entropic properties. The tool enabled predicting the structural motifs susceptible to increased (or unchanged) order with the temperature representing an emergent entropic effect in the system. The motifs with the highest free volume were observed to show enhanced ordering from neutron scattering results. The developed simple technique can be an ideal tool to predict entropically-driven ordering for other complex molecules (e.g., liquid crystals, organic photovoltaics, petroleum derivatives), specifically for systems difficult to test experimentally.

627748 Machine Learning of Retrosynthetic Disconnections and Reaction Outcomes: Influence of Reaction Template Characteristics

Abstract:

Template-based and template-free machine learning models have been shown to accurately predict retrosynthetic disconnections and forward reaction outcomes for organic synthesis. They usually take a reactant or product as input and propose a ranked list of reaction templates, i.e. chemical transformations, via multi-class classification. However, the performance of template-based models is heavily influenced by the size and canonicalization of the set of employed templates as well as the choice of evaluation metrics. First, extracting larger, more specific templates usually leads to a large number of unique templates for a given dataset, with some of the templates only being applicable to very few reactions. The performance of multi-class classification models usually decreases with an increasing number of classes and performs poorly on underrepresented classes. Thus, smaller, more general templates tend to increase model performance but may lead to poor suggestions from a chemical point of view. Finding the optimal level of template specificity is therefore an important objective. Second, non-exclusive templates add noise to the training data, unnecessarily increase the number of classes, and complicate model evaluation. Non-exclusivity can be caused by incomplete canonicalization and inconsistencies in the template extraction algorithm. Third, an inconsistent use of evaluation metrics in literature hampers the comparability of different models. For example, success may be defined as correctly recovering either the template extracted for a test reaction, or the reaction outcome after template application (reaction outcome being reactants in case of retrosynthesis, products in case of forward prediction). If non-exclusive templates exist in the employed set of templates, i.e. different templates lead to the same outcome, a discrepancy arises between the two measures of success. In addition, other metrics such as the fraction of recommended templates that are applicable, i.e. describe a chemical transformation that is actually possible for the given molecule, are often neglected.

Therefore, we systematically study the influence of template size, exclusivity and canonicalization on the performance of template-ranking algorithms as measured by different metrics on datasets of different sizes. Based on our findings, we composed a set of recommendations on the optimal template size and specificity for retrosynthesis and forward prediction models. We furthermore developed a fast, hierarchical correction scheme to filter out non-exclusive templates, which increased model performance considerably. Beyond our study of single-step reactions, we expect these findings to be useful for multi-step retrosynthesis pathway planning.

627818 Mechanistic Insights into HIV Common Escape Pathways from Broadly Neutralizing Antibodies Via Molecular Dynamics Simulations

Abstract:

The delivery of broadly neutralizing antibodies (bnAbs) via adeno-associated virus gene transfer has shown potential as a therapeutic for treating HIV patients [1]. However, despite the ability of bnAbs to neutralize diverse strains of a pathogen by binding to conserved regions on the surface of pathogenic proteins, studies have shown that HIV can still evolve mutations permitting escape from these bnAbs, thereby reducing the immune system's ability to control viral load [2]. Motivated by recent *in vivo* experiments characterizing common HIV mutational escape pathways from the VRC07 bnAb, we performed molecular dynamics (MD) simulations of these same components/mutations to determine the molecular basis of escape in each case. The homology modelling program Modeller and Visual Molecular Dynamics (VMD) were employed to generate glycosylated structures of strains REJOc and JRCSF of the HIV-1 envelope (Env) trimer, incorporated with the observed escape mutations and in complex with VRC07. Simulations were carried out using GROMACS 2018.3 for 40 ns, after which the average change in binding free energy was calculated using the molecular mechanics Poisson-Boltzmann surface area (MM-PBSA) method and a bootstrapping analysis. We found trends in the binding free energies calculated from MD to be in strong qualitative agreement with trends in the experimentally determined EC_{50} values. We also found that weakened electrostatic binding was the major driving force for escape in multiple instances. Furthermore, visual analysis of the simulation trajectories in VMD revealed many commonalities in the escape mechanisms across the observed mutations, including consistently disrupted interactions between VRC07 and the D-loop and V5-loop of each HIV strain. Lastly, we observed a unique role of glycans in assisting in escape from the antibody. Our results may assist in the future development of bnAb therapies that are more resistant to mutational escape, enabling complete suppression of various HIV strains.

1. Casazza, J. P. Durable HIV-1 antibody production in humans after AAV8-mediated gene transfer. In *Conference on Retroviruses and Opportunistic Infections (CROI) 2020, Oral abstract*(Vol. 41).
2. Scheid, J., Horwitz, J., Bar-On, Y. *et al.*HIV-1 antibody 3BNC117 suppresses viral rebound in humans during treatment interruption. *Nature* **535,** 556–560 (2016). https://doi.org/10.1038/nature18929

627876 Workflow Development for First-Principles Molecular Simulations in CP2K

Abstract:

Reproducibility in molecular simulations has become a prominent issue in recent years [1,2]. With the growing complexity of systems being studied, it is becoming more challenging for researchers to disseminate their computational details and methods in traditional forms. Moreover, modern simulation codes have a large number of control parameters beyond the underlying force field, and these parameters can affect the outcome of the simulations in a qualitative manner near a phase transition or in a quantitative manner when high data precision is desired. To tackle these problems, the *Mo*lecular *S*imulation *De*sign *F*ramework (MoSDeF) is being developed to increase the reproducibility and transparency of the simulations [2,3]. With MoSDeF, simulation workflows for complicated molecular simulations can be developed, which streamlines the process of setting up and performing the simulations. The workflows can also be used to automatically perform multiple simulations at different state points, decreasing the possibility of human error. Once the workflow is made available (e.g., as supporting information of a research article), it can be used by other researchers in the field to set up their simulations. Therefore, MoSDeF simulation workflows help create TRUE (Transparent, Reproducible, Usable by others, and Extensible) simulations.

We developed workflows for first-principles molecular dynamics (FPMD) and Monte Carlo (FPMC) simulations in CP2K, an open-source quantum chemistry package that can be used to perform simulations with the interactions described by Kohn-Sham density functional theory (KS-DFT). Our workflows make the process for setting-up, running, and analyzing certain types of first principles simulations convenient and straightforward.

In this poster, we present MoSDeF simulation workflows for two scientific applications.

Molecular dynamics simulations to probe the liquid-phase structure of dihalogens

Using the workflow developed for FPMD simulations in CP2K, we conducted MD simulations in the canonical ensemble using two different generalized gradient approximation exchange-correlation functionals to investigate the structure of liquid Cl_2, F_2, and ClF. The structure factor calculated from simulations for Cl_2 and F_2 shows good agreement with the experimental results. We developed a consistent and transferable set of criteria to detect halogen bonds that is exclusively based on the nuclear positions, making its application to force field-based simulations straightforward.

Gibbs ensemble Monte Carlo simulations for vapor–liquid equilibria and gas adsorption in BCR-704 zeolite

Argon and nitrogen adsorption isotherms are commonly used to characterize the pore size distribution in nanoporous materials. However, for certain materials, the argon and nitrogen adsorption isotherms do not yield interchangeable results. BCR-704 is a calcium aluminosilicate zeolite that is used as a reference material for zeolites with strong adsorption sites. From the argon, and nitrogen adsorption isotherm experiments for BCR-704, experimental measurements show that the relative pressure at which the onset of the steep increase in loading occurs for nitrogen is much lower than that for argon [4]. Reproducing the argon and nitrogen adsorption isotherms in BCR-704 with transferable force fields proved challenging [5]. FPMC simulations in the Gibbs ensemble were conducted to obtain an estimate for the saturated vapor pressures of argon and nitrogen and to investigate the adsorption of these gases in BCR-704. In addition, the spatial distributions of adsorbate molecules are compared to force-field-based simulations.

References

1. Baker M. 1500 scientists lift the lid on reproducibility. *Nature*. 2016;533(7604):452.

2. Thompson MW, Gilmer JB, Matsumoto RA, Quach CD, Shamaprasad P, Yang AH, Iacovella CR, McCabe C, Cummings PT. Towards molecular simulations that are transparent, reproducible, usable by others, and extensible (TRUE). *Mol. Phy*. 2020;118(9-10):e1742938.

3. Cummings PT, McCabe C, Iacovella CR, Ledeczi A, Jankowski E, Jayaraman A, PalmerJC, Maginn EJ, Glotzer SC, Anderson JA, Ilja Siepmann J, Potoff J, Matsumoto RA,Gilmer JB, DeFever RS, Singh R, Crawford B. Open-source molecular modeling software in chemical engineering focusing on the Molecular Simulation Design Framework. *AIChE* Journal. 2021;67(3):e17206.

4. Ross RB, Brennan JK, Frankel KA, Moore JD, Moore JD, Mountain RD, Ahmad R, Thommes M, Shen VK, Schultz NE, Siderius DW, Smith KD. Perfluorohexane adsorption in BCR-704 Faujasite zeolite benchmark studies for the seventh industrial fluid properties simulation challenge. *Fluid Phase Equilib*. 2014;366:141–145.

5. Bai P, Ghosh P, Sung JC, Kohen D, Siepmann JI, Snurr RQ. A computational study of the adsorption of n-perfluorohexane in zeolite BCR-704. *Fluid Phase Equilib*. 2014;366:146–151.

628050 DFT Study on Ag Loaded 2H-MoS_2 for Mechanism of Improved Photocatalytic Reduction of CO_2

Abstract:

Transition metals modified molybdenum disulfide to improve the performance of photocatalytic reduction of carbon dioxide has been receiving much attention. Herein, a novel high-efficiency photocatalytic composite Ag/2H-MoS_2 has been constructed and simulated using density functional theory (DFT) for unveiling the mechanism of improved photocatalytic reduction of CO_2 in our experimental research. Our calcualtions about band structure, electronic and optical properties indicate that the loading of Ag atoms enhances the photocatalytic performance of 2H-MoS_2 nanosheets by transforming the photogenerated electrons from the valence band of 2H-MoS_2 to the loaded Ag atoms. Furthermore, 20wt% Ag loaded on 2H-MoS_2 containing is the most suitable for the thermodynamic requirement of reducing CO_2 to CH_3OH/CH_4 among the catalysts with different Ag loadings, and the formation of *CHO in 20wt% Ag/2H-MoS_2 is the potential-determining step, whose Gibbs free energy reduces from 2.830 eV of 2H-MoS_2 to 0.925 eV. Meanwhile the thermochemical results prodict the best path for reducing CO_2 on such a photocatalyst being CO_2→*COOH→*CO→*CHO→*CH_2O→*OCH_3→*CH_3OH→CH_4. The photocatalytic performance in reducing CO_2 of pristine 2H-MoS_2 is therefore significantly improved by loading silver. This research provides a theoretical reference for transition metal modification 2H-MoS_2 nanosheets.

628451 Solidification Model for Silicon Crystal-Melt Interface in the Horizontal Ribbon Growth Process: A Molecular Dynamics Study

Abstract:

We implement a molecular dynamic (MD) based forced velocity solidification (FVS) model for silicon crystal growth from the melt. The continuous solidification at the crystal-melt interface between a heat source and sink was employed to provide a molecular understanding of single-silicon crystallization from horizontal ribbon growth—a cost-effective but not fully understood technique of manufacturing silicon wafers for solar cells. We track the evolution of crystals on the Si(111) surface at varying pulling speeds (0.005 – 0.05 Å/ps), studied the solidification front's thermal profiles, and their combined effects on the stability of the crystal-melt interface. We found that the thermal gradient across the solidification front influences the interface's position and latent heat dissipation speed across the interface. Finally, we discuss a molecular perspective on the role of heat fluxes in crystal and melt sides of the interface in stabilizing the crystal-melt interface at high pulling rates.

629470 First-Principles Study of the Role of Carbon Host in the Sodiation of Selenium

Abstract:

Sodium–selenium (Na–Se) batteries are considered as an emerging alternative for the traditional Li-ion batteries owing to the high theoretical volumetric capacity (~ 3250 mAh/cm^3) with high electronic conductivity of Se (~ 10^{-3} S/m). However, Na-Se batteries suffer from massive volume changes associated with the sodiation/desodiation and the formation of soluble polyselenides. Carbon composites could be employed as host materials to buffer the sodiation-induced expansion, provide conductive channels, and physically trap the high-order polyselenides. In this talk, we evaluate the role of carbon host in the sodiation of Se as the fundamental understanding of the sodiation of the Se-C composites is still limited. Based on first-principles simulations, we present comprehensive analyses on the sodiation of Se-graphene (C), in terms of the interfacial structure, charge transfer, diffusion characteristics of Na, and electronic properties. Our calculations demonstrate a distinct Na enrichment at the Se/C interface, donating electrons to both Se and C. The interfacial strength is enhanced with sodiation, which could lead to compressive stresses at the interface during Na incorporation and mechanical failures. Longer Se chains are more populated with the presence of graphene, as less reduced Se ions (due to the electron flows to graphene) are more likely to cluster. Our calculations demonstrate that the interfacial Na ions shows a higher mobility along the graphene surface, compared to the bulk diffusion in Se, which suggests the enhanced Na diffusion could promote the sodiation/desodiation rate. The density of states (DOS) analysis clearly illustrates that the Na–Se/C system shows n-type metallic behavior, which could provide facile electronic pathways for electrochemical reaction process of Na–Se batteries. This work highlights the role of carbon host in establishing the sodiation kinetics, which may provide a framework for the development of the Na–Se batteries.

629643 Optimization of High-Throughput Photocatalyzed Bimetallic Nanoparticles Using Online Active Learning

Abstract:

Development of cost-effective industrial solutions for renewable hydrogen generation would would have other significant uses: hydrogen has applications in ammonia synthesis, iron and steel production, and the production of many specialty chemicals, including polymers and synthetic fuels. The key barrier is in catalysis technology: although the platinum-catalyzed hydrogen evolution reaction is one of the most active catalytic processes ever discovered, platinum is extremely rare, and no cheap, stable catalyst has been developed with similar activity. To address this need, we developed a high-throughput experimental workflow capable of assessing the activity of bimetallic nanoparticles for photocatalytic hydrogen evolution. Metal ions are injected into an aqueous solution containing DMSO, triethanolamine, and an Iridium-based photocatalyst, and a blue LED induces a reducing environment. This precipitates nanoparticles that start to evolve hydrogen gas, which is tracked in real time by a chemosensory tape. Approximately 1 wellplate can be run per day, with each well containing metal ions at varying concentrations, and hardware restrictions limit each day to 4 unique metals. A online active learning framework was then created to optimize this experimental space for bimetallic wells not containing platinum or palladium, aided by a database of hydrogen adsorption energies. Several promising combinations were discovered, and although further characterization of the nanoparticles is forthcoming, active site adsorption energies give some explanation of trends in activity through the Sabatier principle. This framework demonstrates the ability of active learning techniques combined with high-throughput experimentation to quickly explore complex spaces. Future work is expected to extend this analysis to similar chemical systems, including the oxidation of simple alcohols.

629779 Structure-Property Relationships of Thiolate-Protected Metal Nanoclusters

Abstract:

Thiolate-protected metal nanoclusters (TPNCs) have attracted tremendous interest owing to their unique physicochemical properties, which enable applications from imaging cancer cells and bacteria to many forms of catalysis. TPNCs make up a unique class of atomically-precise nanomaterials that possess high stability at specific “magic size” compositions (i.e. $M_n(SR)_m$). Since TPNCs can be modulated by varying ligand and metal type, they are ideal systems to formulate structure-property relationships (SPRs) through systematic studies on the growing catalogue of structures. Herein, we employ Density Functional Theory and machine learning to develop SPRs that can accelerate TPNC design for targeted applications. First, we developed a simple, generalized framework that captures TPNC solubility behavior[1]. The work leverages the molecular nature of TPNCs to reveal that ligand type and structural symmetry dictate solubility. Of note, we then leveraged these molecular properties to rationalize crystallization behavior in a series of trimetallic TPNCs[2]. Next, we introduce structure-electronic-property relationships that capture electron affinity and ionization potential across the TPNC space[3]. Capturing size and ligand effects, the models rely only on simple structural characteristics, and thus avoid the need for ab initio calculations. Importantly, we apply these SPRs to an atom-by-atom evolution series of TPNCs, rationalizing the dopant-based stability that is observed under experiment[4]. Finally, we investigate composition and chemical ordering effects on the stability of heterometal-doped TPNCs. The work extends our previously developed genetic algorithm that predicts bimetallic nanoparticle stability[5] and sets the ground work for complete structural prediction of TPNCs[6]. Our new model captures explicit metal-ligand interactions to determine stable dopant concentrations and positions in TPNCs. Overall, our computational studies rationalize a large series of experimental observations.

1. Cowan, M. J.; Higaki, T.; Jin, R.; Mpourmpakis, G., Understanding the Solubility Behavior of Atomically Precise Gold Nanoclusters. *J. Phys. Chem. C* **2019,** *123* (32), 20006-20012.

2. Li, Y.; Cowan, M. J.; Zhou, M.; Taylor, M. G.; Wang, H.; Song, Y.; Mpourmpakis, G.; Jin, R., Heterometal-Doped M_{23} (M = Au/Ag/Cd) Nanoclusters with Large Dipole Moments. *ACS Nano* **2020,** *14* (6), 6599-6606.

3. Cowan, M. J.; Mpourmpakis, G., Structure–Property Relationships on Thiolate-Protected Gold Nanoclusters. *Nanoscale Adv.* **2019,** *1* (1), 184-188.

4. Li, Y.; Cowan, M. J.; Zhou, M.; Luo, T.-Y.; Song, Y.; Wang, H.; Rosi, N. L.; Mpourmpakis, G.; Jin, R., Atom-by-Atom Evolution of the Same Ligand-Protected Au_{21}, Au_{22}, $Au_{22}Cd_1$, and Au_{24} Nanocluster Series. *J. Am. Chem. Soc.* **2020,** *142* (48), 20426–20433.

5. Dean, J.; Cowan, M. J.; Estes, J.; Ramadan, M.; Mpourmpakis, G., Rapid Prediction of Bimetallic Mixing Behavior at the Nanoscale. *ACS Nano* **2020,** *14* (7), 8171–8180.

6. Cowan, M. J.; Mpourmpakis, G., Towards elucidating structure of ligand-protected nanoclusters. *Dalton Trans.* **2020,** *49* (27), 9191-9202.

629785 Gaining Mechanistic Insights into the Influence of O-Linked Glycosylation on Insulin Properties with Molecular Dynamics

Abstract:

Insulin has been widely used as a peptide drug to treat diabetes by promoting the absorption of glucose from the blood. In recent years, there has been an increased interest in the development of the insulin drug for oral administration to address the low patients compliance caused by frequent subcutaneous injections. However, the research in this field has been challenging due to insulin's low permeability across the intestinal epithelium upon oligomerization and the susceptibility to proteases in the digestive system of the human body, which lead to overall low absorption efficiency. O-linked glycosylation, as applied in a few recent experimental studies, has been considered one of the most promising approaches to address the challenge. Attachment of proper saccharides to the glycosylation sites in the tail region can enhance insulin's proteolytic stability and decrease its oligomerization propensity without sacrificing biological activity. However, with a lack of mechanistic insights into how the insulin properties are improved by O-glycosylation, the strategies for glyco-insulin design remain elusive. To gain a better understanding at the molecular level, we performed long (2 microseconds) molecular dynamics (MD) simulations of the insulin glycoforms investigated in previous experimental studies. These insulin glycoforms were based on five different crystal structures of insulin wildtype. We present an analysis of experimental observables, including cleavage site solvent accessible surface area, secondary structure propensity, and peptide flexibility to predict the proteolytic stability and dimerization propensity of glycoforms and use these observables to predict behaviors of proposed glycoforms.

630463 Si and C Models as the Basis for Regression-Driven Development of the Interface Force Field

Abstract:

The Interface Force Field (IFF) project centers around the construction of an all-atom force field and surface model database for simulation of nanostructures of 1 to 1000 nm size, including minerals, metals, oxides, and (bio)polymers.[1] To achieve this aim, the IFF project develops model systems that mimic experimentally observed behavior, which is validated through benchmarks such as density, surface energy, and melting point.

Via an ancillary electron method, we constructed a silicon (Si) model which reproduces a density of 2.312 g/cm^3 at 298 K, compared to 2.329 g/cm^3 as experimentally measured.[2] Our model is representative of experimentally observed Si melting temperature near 1687 K, largely owing to accurate description of surface energy which we demonstrate through the calculation of the Si(111) surface energy within 5% error of experimental estimates.[3] A similar model was constructed for carbon. Si and C were assembled to simulate the SiC layer that results from heterogeneous nucleation between C and molten Si during melt infiltration experiments. In the solid-state diffusion-controlled regime of the reaction occurring at the interface, the diffusivity of C through the initial nucleation layer of SiC governs further layer growth. We calculated the diffusivity of C through SiC layers of various thickness in order to investigate the effect of the time-dependent permeability on layer growth.

Where diffusivity is critical to reaction rate, as in the SiC system, a model that captures nanoscale system dynamics is necessary for micron-scale material simulation. Through accurate description of diffusivity in SiC, our Si and C ancillary electron models advance the capabilities of Interface Force Field for ceramic materials.

References:

(1) Heinz, H., Lin, T.-J., Kishore Mishra, R., & Emami, F. S. (2013). Thermodynamically Consistent Force Fields for the Assembly of Inorganic, Organic, and Biological Nanostructures: The INTERFACE Force Field. Langmuir, 29(6), 1754–1765. https://doi.org/10.1021/la3038846

(2) Assael, M. J., Armyra, I. J., Brillo, J., Stankus, S. V., Wu, J., & Wakeham, W. A. (2012). Reference Data for the Density and Viscosity of Liquid Cadmium, Cobalt, Gallium, Indium, Mercury, Silicon, Thallium, and Zinc. Journal of Physical and Chemical Reference Data, 41(3), 033101. https://doi.org/10.1063/1.4729873

(3) Tyson, W. R., & Miller, W. A. (1977). Surface free energies of solid metals: Estimation from liquid surface tension measurements. Surface Science, 62(1), 267–276. https://doi.org/10.1016/0039-6028(77)90442-3

630615 Altered Protein Dynamics Delineates the Oncogenic Potential of Various Kinase Mutations

Abstract:

Kinases are a class of proteins that play essential roles in cell signaling, differentiation, and proliferation. They are frequently mutated in cancer and are one of the largest therapy targets of specific inhibitors in clinical research. The activation status of mutated kinases in cancer can profoundly impact phenotypic outcomes not limited to tumor progression and drug sensitivity. To quantify these phenotypic outcomes through mutated kinase activities, optical approaches implemented to control cell signaling pathways by photoswitchable kinases have proved to be transformational by relying on specific gain-of-function mutations. To better understand this at the molecular level, the role of mutations in intrinsic kinase activity needs to be quantified. Free energy calculations obtained through enhanced sampling techniques of statistical mechanics have facilitated the understanding of structural stabilization of kinases and their mutated systems and provided a detailed view of the underlying structure-activity relationship. However, a quantification of the degree of alterations caused by the mutated systems to this structure-activity relationship and the resulting proximity between wild type and mutated kinase systems is not well studied. We implement a computational suite combining enhanced sampling techniques of Metadynamics and INDUS to investigate the role of mutations in altering protein dynamics.

Additionally, our Boltzmann weighted correlation approach efficiently quantifies these alterations in protein dynamics obtained through the free energy landscapes by sampling the conformational transition in kinase systems. Moreover, the suite also investigates the long timescale role of solvent water molecules in protein dynamics. Finally, we analyze our results with log P profiles obtained from Hydrogen Deuterium Exchange (HDX), an experimental technique to study protein dynamics to validate our model. Our method is a significant improvement over unbiased molecular dynamics techniques to study protein dynamics over experimental timescales and provides a molecular-level understanding of the role of mutations in conferring an oncogenic potential.

We acknowledge financial support from the NIH and computational support from the Flatiron Institute.

631895 Toward Data-Driven Approaches to Inverse Design: Fast and Accurate Evaluation of Gas Adsorption and Diffusivity in Nanoporous Materials

Abstract:

The recent advance in data-driven approaches to inverse design has emphasized the importance of accurate and efficient evaluation of the physical properties of materials underpinning their performance. In this poster, we demonstrate recent theoretical developments and high-performance computation rendering new capabilities to generate large-scale high-fidelity data for gas adsorption and diffusivity in nanoporous materials with atomistic models. The efficient computational tools form an important cornerstone for the inverse design of nanoporous materials either as adsorbents or membranes for gas separation.

632116 The Molecular Simulation Design Framework (MoSDeF): New Features to Improve User Experience and Support Reproducible Molecular Simulations

Abstract:

The Molecular Simulation Design Framework (MoSDeF) is a suite of Python libraries containing routines necessary for the programmatic and reproducible initialization of chemical systems for molecular simulation, with a major focus on molecular dynamics (MD) and Monte Carlo (MC) simulations[1]. This project is driven by our efforts to encourage and support TRUE (Transparent, Reproducible, Usable by others, Extensible) simulations as discussed by Thompson et al. [2]. Of the three major libraries in the MoSDeF suite, two of them, mBuild [3, 4] and Foyer [5, 6], have been used in numerous scientific projects to date [7-9]. As part of our continuing development of the MoSDeF tools, these two libraries have undergone major improvements over the past year to assist with the construction of chemical systems. For example, mBuild now incorporates RDKit to generate molecules from SMILES strings, providing better control and improved reproducibility of the system initialization process. The Foyer library has undergone major refactoring, supporting atom-typing and parameterization of generic graph structures. These changes make Foyer a more versatile tool, now supporting a wider range of input formats. A new backend data structure has been integrated called the General Molecular Simulation Object (GMSO)[10], which stores topological information for chemical systems and their interaction parameters. The GMSO library was designed with minimal assumptions about the type of chemical systems to be supported, as well as symbolic mathematics support for the many potential energy functional forms found in common force fields. This enables GMSO to support a wide range of systems, from small molecule organic liquids, to coarse-grained systems, crystalline structures, and others. All packages in the MoSDeF software suite follow software development best practices, which includes detailed documentation of implemented functions, unit testing and Continuous Integration (CI). The MoSDeF tools listed above are all available on conda-forge and installed using the Conda scientific software package manager.

References

1. MoSDeF: https://github.com/mosdef-hub.
2. M. W. Thompson, Justin B. Gilmer, Ray A. Matsumoto, Co D. Quach, Parashara Shamaprasad, Alexander H. Yang, Christopher R. Iacovella, Clare McCabe & Peter T. Cummings (2020) Towards molecular simulations that are transparent, reproducible, usable by others, and extensible (TRUE), Molecular Physics, 118:9-10.
3. C. Klein, J. Sallai, T. J. Jones, C. R. Iacovella, C. McCabe, and P. T. Cummings, "A Hierarchical, Component Based Approach to Screening Properties of Soft Matter", Foundations of Molecular Modeling and Simulation, **2016**, 79-92.
4. mBuild: https://github.com/mosdef-hub/mbuild.
5. C. Klein, A. Z. Summers, M. W. Thompson, J. B. Gilmer, C. McCabe, P. T. Cummings, Janos Sallai, Christopher R. Iacovella. "Formalizing atom-typing and the dissemination of force fields with foyer", Computational Materials Science, Volume 167, **2019**. https://doi.org/10.1016/j.commatsci.2019.05.026.
6. https://github.com/mosdef-hub/foyer.
7. A. Z. Summers, C. R. Iacovella, P. T. Cummings, and C. McCabe. "Investigating Alkylsilane Monolayer Tribology at a Single-Asperity Contact with Molecular Dynamics Simulation".

Langmuir 2017 *33* (42), 11270-11280. DOI: 10.1021/acs.langmuir.7b02479
8. A. Z. Summers, J. B. Gilmer, C. R. Iacovella, P. T. Cummings, and C. McCabe. “MoSDeF, a Python Framework Enabling Large-Scale Computational Screening of Soft Matter: Application to Chemistry-Property Relationships in Lubricating Monolayer Films”. Journal of Chemical Theory and Computation 2020 *16* (3), 1779-1793. DOI: 10.1021/acs.jctc.9b01183.
9. N. C. Craven, J. B. Gilmer, C. J. Spindel, A. Z. Summers, C. R. Iacovella, and C. McCabe , "Examining the self-assembly of patchy alkane-grafted silica nanoparticles using molecular simulation" , The Journal of Chemical Physics 154, 034903 (2021) https://doi.org/10.1063/5.0032658
10. GMSO: https://github.com/mosdef-hub/gmso.

632167 Managing a High-Throughput Screening Workflow with Open-Source Software: A Study of Tribological Properties of Thin Films

Abstract:

High-throughput screening studies heavily utilize automated workflow-management tools to scan over large parameter spaces. Here, we present a case study of a high-throughput screening study using molecular dynamics (MD) simulations that scan over monolayer film designs in search of those that reduce friction and wear for surfaces in contact. We will discuss the workflow used in the study to build and manage nearly 40,000 systems using the Molecular Simulation Design Framework (MoSDeF), as well as the processing of the data generated. MoSDeF includes tools to enable the automatic initialization of chemical systems [2] via two Python libraries, mBuild [3, 4] and Foyer [5, 6]. These tools automate the system initialization process and enable the batch creation of ten of thousands of systems. These systems can then be simulated with simulation codes under various ensembles before their final trajectories are analyzed and desired properties calculated using data analysis libraries such as MDTraj and MDAnalysis [7, 8]. Tools from the Signac framework were utilized to generate and manage the workspace and control the workflow, including system initialization, simulations, data analysis, and orchestrating the remote submission of the jobs [1]. The random forest regressor as implemented in the scikit-learn library was also used to determine the structure-property relationships (QSPR) for the data set [9, 10]. The project workflow is hosted on GitHub and with the necessary scripts to generate the systems, following the guidelines suggested by the TRUE standard [11].

References

1. Signac Framework: https://signac.io.
2. MoSDeF: https://mosdef.org.
3. mBuild: https://github.com/mosdef-hub/mbuild.
4. C. Klein, J, Sallai, T. J. Jones, C.R. Iacovella, C. McCabe, P. T. Cummings. (2016) A Hierarchical, Component Based Approach to Screening Properties of Soft Matter. In: Snurr R., Adjiman C., Kofke D. (eds) Foundations of Molecular Modeling and Simulation. Molecular Modeling and Simulation (Applications and Perspectives). Springer, Singapore. https://doi.org/10.1007/978-981-10-1128-3_5
5. Foyer: https://gitub.com/mosdef-hub/foyer.
6. C. Klein, A. Z. Summers, M. W. Thompson, J. B. Gilmer, C. McCabe, P. T. Cummings, J. Sallai, C. R. Iacovella. "Formalizing atom-typing and the dissemination of force fields with foyer". Computational Materials Science, Volume 167, 2019, Pages 215-227, https://doi.org/10.1016/j.commatsci.2019.05.026.
7. R. T. McGibbon, K. A. Beauchamp, M. P. Harrigan, C. Klein, J. M. Swails, C. X. Hernández, C. R. Schwantes, L. Wang, T. J. Lane, V. S. Pande. "MDTraj: A Modern Open Library for the Analysis of Molecular Dynamics Trajectories". Biophysical Journal, Volume 109, Issue 8, 2015, Pages 1528-1532, https://doi.org/10.1016/j.bpj.2015.08.015.
8. R. J. Gowers, M. Linke, J. Barnoud, T. J. E. Reddy, M. N. Melo, S. L. Seyler, D. L. Dotson, J. Domanski, S. Buchoux, I. M. Kenney, and O. Beckstein. MDAnalysis: A Python package for the rapid analysis of molecular dynamics simulations. In S. Benthall and S. Rostrup, editors, *Proceedings of the 15th Python in Science Conference*, pages 98-105, Austin, TX, 2016. SciPy,

doi:10.25080/majora-629e541a-00e.
9. Y. L. Pavlov. (2019) 'Random forests', Random Forests, pp. 1–122. doi: 10.1201/9780429469275-8.
10. scikit-learn: https://github.com/scikit-learn/scikit-learn
11. Matthew W. Thompson, Justin B. Gilmer, Ray A. Matsumoto, Co D. Quach, Parashara Shamaprasad, Alexander H. Yang, Christopher R. Iacovella, Clare McCabe & Peter T. Cummings (2020) Towards molecular simulations that are transparent, reproducible, usable by others, and extensible (TRUE), Molecular Physics, 118:9-10, DOI: 10.1080/00268976.2020.1742938

633016 Mathematical Modeling of the Observed Neutralization Threshold of Antibodies Against Sars-Cov-2

Abstract:

Neutralizing antibodies for SARS-CoV-2 commonly bind to the receptor binding domain (RBD) on the surface of the trimeric viral fusion Spike (S) protein, to prevent the virus from attaching to its cellular target angiotensin-converting enzyme 2 (ACE2) to gain entry into human cells. Neutralization assays show there is a strict neutralization threshold for SARS-CoV-2 virus particles at 0.001 μm/mL of antibody. To examine this phenomenon, a mathematical model incorporating the association and dissociation rates of RBD-targeting antibodies was developed. The number of bound antibodies was then translated into a corresponding degree of viral infectivity. Our model shows that at a concentration of 0.001 μm/mL of antibody, despite a sufficient number of antibodies available to bind to the virus, the antibodies are unable to fully neutralize the virus. Therefore, the dynamics of binding, governed by the association and dissociation of the antibodies from the S protein, results in the strict neutralization cutoff observed in the experiments. An overarching finding of this work is that neutralization thresholds for multiple viruses can be predicted using such a mathematical model that describes viral infectivity as it relates to the number of bound antibodies. For example, this model can also be employed to understand the origins of the antibody neutralization threshold against human immunodeficiency virus (HIV), where it is also observed. A more general understanding of the specific antibody concentrations at which neutralization thresholds occur for different diseases could guide the design of future vaccines against these and other diseases.

624810 Computational Studies of the Phase Transitions and Self-Assembly of Thermoresponsive Peptide-Based Biomaterials

Abstract:

Peptide based biomaterials have garnered significant attention recently due to their ability to respond to external stimuli such as heat, light, and pH, thus allowing for controllable release of therapeutics for drug delivery applications and tunable mechanical and chemical properties for tissue engineering and in vitro cell culture applications. While experimental synthesis and characterization of biomaterials can be time consuming and limited in terms of resolution, simulations allow for efficient screening of broad design spaces while also giving insight into the molecular mechanisms and driving forces governing the complex phase behavior and assembly of responsive biomaterials. For simulations to produce experimentally relevant predictive design rules, there is a need to develop molecular models that can capture the thermodynamics and self-assembly of biopolymers at large enough length scales and time scales which are applicable to experiments. In this poster I will present my doctoral thesis work using computational tools, such as atomistic and coarse-grained molecular dynamics (MD) simulations, and machine learning to study responsive, peptide-based materials, such as elastin-like peptides (ELP), collagen-like peptides (CLP), and ELP-CLP bioconjugates. I will highlight our recent studies on the lower critical solution temperature (LCST)-like phase behavior of ELPs and ELP-CLP conjugates, the melting transitions of CLPs with natural and non-natural amino acids, and the self-assembly of CLPs into triple helices, fibrils, and supramolecular networks. Furthermore, I will discuss our recent efforts in using feedforward neural networks to predict supramolecular assembly and aggregate sizes of CLPs with the goal of designing peptides which display multiscale assembly. Overall, this work highlights the benefits of using a suite of computational techniques, including MD simulations and artificial neural networks, to streamline the discovery of self-assembling biopolymers with tunable physicochemical properties.

624852 Accelerating on-the-Fly Active Learning of Catalyst Simulations Using Large Scale Pretrained Models

Abstract:

Given the current energy and climate crises, new technologies must be developed to enable sustainable harvesting and usage of energy. Many proposed solutions, such as hydrogen fuel cells or the reduction of carbon dioxide to fuels, require the discovery of new heterogeneous electrocatalysts to be viable. Since physical experiments cost significant time and materials, high throughput computational screening methods are required to accelerate discovery. The Open Catalyst 2020 (OC20) dataset and community challenges has set forth tasks for models to improve upon if they are to serve as effective screening tools. Recently sophisticated models have been developed to perform initial structure to relaxed structure (IS2RS) tasks, at higher throughput than is possible than with traditional density functional theory (DFT) methods. However, while these models are fast, they are not quite accurate enough to be used reliably for IS2RS tasks. We take an alternative approach to this task, by transferring the knowledge gained by these pretrained models to an active learning framework for accelerating DFT relaxations. The active learning framework has been shown to be an effective method of accelerating molecular relaxations, and by combining it with the knowledge learned by a pretrained model we show that its performance may be further improved, with greater accuracy than the underlying model.

624899 Computational Analysis of Confined Deep Eutectic Solvents and Ionic Liquids for Separations of Carbon Dioxide from Methane

Abstract:

Classical molecular dynamics simulations and Monte Carlo simulations were performed to investigate gas separations of carbon dioxide from methane by deep eutectic solvents (DES) confined in graphite pores of 2 nm and 5 nm in width. The DES studied was choline chloride + ethylene glycol (ethaline) with 1:2, 1:4, and 1:8 molar ratios. The gas bulk mixture to be separated consisted of carbon dioxide and methane with a 5: 95 molar ratio at a temperature of 318 K and pressures on the order of 100 bar. The separation performance of these systems in terms of the solubility selectivity, diffusion selectivity, and permselectivity of the confined DES will be presented, discussed and compared against the performance of the bulk DES. Our results will also be rationalized by examining local density profiles, radial distribution functions and interaction energies.

625199 Molecular Dynamics Simulations of the Inhibition of HIV and Host Cell Machinery By Antiretroviral Drugs

Abstract:

It is estimated that around 1.2 million people in the U.S. are living with HIV today, for which there is no known cure[1]. A large fraction of the ageing HIV population is currently experiencing some loss of neurocognitive function, a hallmark symptom of HIV-Associated Neurocognitive Disorder, or HAND, due to the virus' attack on the central nervous system[2]. Antiretroviral (ARV) drugs are commonly used to treat the non-neurological symptoms of HIV infection, but an accumulating body of evidence points to the ability of many types of ARVs to promote or even mediate neurocognitive decline in the patients that take them. For example, protease inhibitors (PIs) are one of the most widely prescribed types of ARVs, which act to inhibit the protease of HIV, thus preventing the virus from fully maturing to be able to multiply and infect other cells[3]. However, numerous experimental studies have now shown that PIs also inadvertently inhibit the proteasome in multiple human cell types, including neurons, causing activation of a stress response pathway called the unfolded protein response (UPR) and ultimately leading to cell death[4]. Yet, the precise molecular details of how PIs inhibit the proteasome and how this mechanism may differ from that employed by PIs to inhibit HIV protease remain unclear.

To this end, we performed molecular dynamics (MD) simulations to investigate the inhibitory interactions between a wide range of PIs and both the protease of HIV and the human 20S proteasome. Molecular docking simulations were first employed to identify initial binding locations and conformations of each PI on the two proteins, followed by MD and subsequent binding free energy calculations using the molecular mechanics Poisson-Boltzmann surface area (MM-PBSA) method. We observe consistent differences in the calculated binding free energies and molecular-level interactions that govern PI-protease versus PI-proteasome binding, across the many PIs studied. Our results suggest that exploiting these differences during the immunotherapeutic design process may be a promising pathway towards the development of new PIs that still robustly inhibit HIV machinery while leaving host machinery intact.

1. Centers for Disease Control and Prevention. HIV Surveillance Report, 2018 (Updated); vol. 31 (2020).
2. Clifford, D. B.; Ances, B. M. HIV-Associated Neurocognitive Disorder. *Lancet Infect. Dis.*2013, *13* (11), 976–986.
3. Gulnik, S.; Erickson, J. W.; Xie, D. HIV Protease: Enzyme Function and Drug Resistance. In *Vitamins & Hormones*; Elsevier: San Diego, CA, 2000; Vol. 58, pp 213–256.
4. Pajonk, F.; Himmelsbach, J.; Riess, K.; Sommer, A.; McBride, W. H. The Human Immunodeficiency Virus (HIV)-1 Protease Inhibitor Saquinavir Inhibits Proteasome Function and Causes Apoptosis and Radiosensitization in Non-HIV-Associated Human Cancer Cells. *Cancer Res.*2002, *62* (18), 5230–5235.

625602 Predicting Rate Constants for Catalyzed Reactions with Machine Learning

Abstract:

Catalyzed reactions lie at the core of our fight against climate change, from their use to provide new renewable energy sources to the removal of CO2 from the atmosphere. However, computing catalytic reactivity atomistically is unfeasible for many reactions of interest as computational cost scales exponentially with system size [1]. Supervised machine learning has been proposed as a method to increase the speed with which predictions of reaction rates can be made [2]. However, one of the critical roadblocks to developing useful machine learning models to predict reactivity is developing a dataset to train the model with [3]. We are currently generating a dataset of reaction rate constants for ground state organic chemistry reactions using density functional theory CI-NEB calculations of minimum energy paths and aim to extend it to include reactions in the presence of catalysts. In this context, we are working to design input features which come at a low computational cost yet include information on the dynamics of the reaction and lead to a low loss in prediction. Here we will show our recent work on identifying and designing these features using our current dataset and previously published datasets [2, 4].

References

[1] Friesner, R. A. Ab initio quantum chemistry: Methodology and applications. *PNAS* **10,** 6648–6653 (2005)

[2] Komp, E. & Valleau, S. Machine Learning Quantum Reaction Rate Constants. *J. Phys. Chem. A* **124**, 8607–8613 (2020).

[3] Goldsmith, B. R., Esterhuizen, J., Liu, J. X., Bartel, C. J. & Sutton, C. Machine learning for heterogeneous catalyst design and discovery. *AIChE J.* **64**, 2311–2323 (2018).

[4] Winther, K. T., Hoffmann, M. J., Boes, J. R., Mamun, O., Bajdich, M., & Bligaard, T. (2019). Catalysis-Hub.org, an open electronic structure database for surface reactions. *Scientific Data*, **6**, 1–10 (2019).

625603 Predicting Uncertainty in Supervised Machine Learning Predictions of Chemical Kinetics

Abstract:

Computing reaction rate constants atomistically becomes prohibitively expensive as the size of the system grows.[1,2] For this reason, some recent work has focused on training supervised machine learning (ML) algorithms to predict rate constants and activation energies without requiring large ab initio calculations.[3–6] These approaches have shown to be successful in predicting target kinetics properties, however they do not provide a measure of error with each new prediction; only training and validation losses are available. In our previous work[5], a DNN was trained to predict the product of reaction rate constants and partition functions with a test mean absolute error of 1.1%. While this was encouraging, individual predictions had errors as high as 33.5%. It was thus clear that anticipating individual prediction errors was necessary to inform design choices. Here we present our recent work which employs modified generative adversarial networks (GANs)[7] to estimate the uncertainty of each predicted reaction rate constant product. The model was trained on the database created for the original work.[5] We show that the consideration of individual prediction error can drastically impact rate constant predictions on a target system.

1. Ju, L. P., Han, K. L. & Zhang, J. Z. H. Global dynamics and transition state theories: Comparative study of reaction rate constants for gas-phase chemical reactions. *Journal of Computational Chemistry* vol. 30 305–316 (2009).

2. Sumathi, R. & Green, W. H. A priori rate constants for kinetic modeling. *Theoretical Chemistry Accounts* vol. 108 187–213 (2002).

3. Grambow, C. A., Pattanaik, L. & Green, W. H. Deep Learning of Activation Energies. *J. Phys. Chem. Lett.* **11**, 2992–2997 (2020).

4. Nandi, A., M. Bowman, J. & Houston, P. A Machine Learning Approach for Rate Constants. II. Clustering, Training, and Predictions for the O(3P) + HCl → OH + Cl Reaction. *J. Phys. Chem. A* **124**, 5746–5755 (2020).

5. Komp, E. & Valleau, S. Machine Learning Quantum Reaction Rate Constants. *J. Phys. Chem. A* **124**, 8607–8613 (2020).

6. Ulissi, Z. W., Medford, A. J., Bligaard, T. & Nørskov, J. K. To address surface reaction network complexity using scaling relations machine learning and DFT calculations. *Nat. Commun.* **8**, 14621 (2017).

7. Lee, M. & Seok, J. Estimation with Uncertainty via Conditional Generative Adversarial Networks. arXiv:2007.00334v1 (2020).

625879 Transforming Automated Quantum Chemistry Calculation Workflows with Machine Learning: Towards Faster and More Accurate Chemical Discovery

Abstract:

Machine learning (ML) has begun to accelerate chemical discovery by providing advances in efficiency needed to overcome the combinatorial challenge of computational materials design. In ML-assisted chemical discovery, an automated quantum chemistry (QC) calculation workflow is used to generate datasets to train surrogate models, which serve as alternatives to rapidly explore a large chemical space to identify candidate materials. However, current automated QC calculation workflows with density functional theory (DFT) as workhorse leads to many attempted calculations that are doomed to fail and brings biases/inaccuracy to the training data that may be out of the domain of applicability of DFT. This includes many compelling functional materials and catalytic processes that are difficult because of their complex electronic structure, such as systems involving strained chemical bonds, open-shell radicals and diradicals, or metal–organic bonds to open-shell transition-metal centers. We address these challenges of computation efficiency and accuracy by integrating ML approaches into conventional DFT-based QC workflows. We build two types of classifiers to predict the likelihood of calculation success: 1. prior to calculations and 2. on-the-fly during calculations. The prior to calculations classifier is a near zero-cost model that rapidly filters out candidate calculations most likely to fail, while the on-the-fly model monitors and terminates an already running calculation if it is predicted to fail with high confidence. Together, these classifiers save half of the computation resources. We also develop multiple types of classifiers to predict the presence of strong correlation, which is usually a sign of a system being out of the domain of applicability of DFT. Our models only require calculations at DFT cost and can classify which systems in a dataset will require more expensive but accurate wave function theory calculations, leading to overall high fidelity of the entire dataset. Since electronic structure information is encoded as the inputs, our models are readily transferable to larger systems and systems with unseen elements. All these classifier models represent the first efforts toward autonomous workflows that move past the need for expert determination of the robustness of DFT-based materials discoveries.

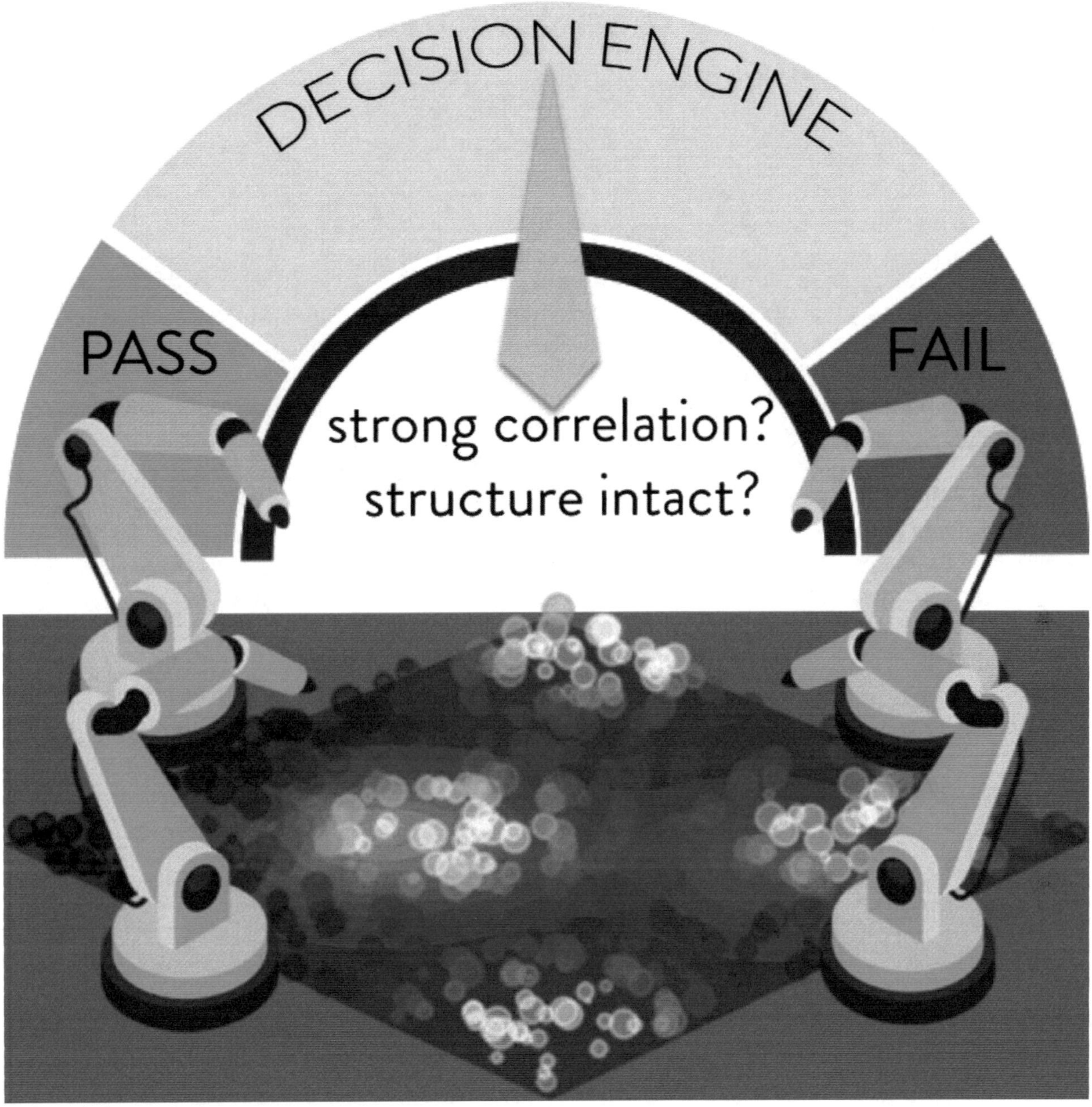
DECISION ENGINE
PASS
FAIL
strong correlation?
structure intact?

626333 Improving Feature Selection Methods for Heterogeneous Catalysis

Abstract:

We developed a novel feature selection (FS) method, called iterative Bayesian additive regression trees (iBART), for deriving robust physical descriptors in heterogeneous catalysis. Traditional FS approaches first collect the chemical properties of each constituent in the system as the primary features, and then apply a feature engineering (FE) step to construct complex, non-linear secondary features. These processes can be repeated indefinitely to generate features of arbitrary complexity. FS is then used to identify the features in the resulting candidate pool that best predict the property of interest. This approach suffers from the fact that the FS step must operate on an enormous feature space that contains millions of candidate features that are highly correlated by construction. This leads to a high rate of false positives, while also requiring significant computational resources. We developed the iBART method to address these issues, and we show that it generates more accurate models at reduced computational cost compared with other state-of-the-art methods for FS. This is accomplished by iteratively applying the FS step (i.e., Bayesian additive regression trees, BART) to first select the best-performing features before proceeding to the FE steps. This iterative FS/FE process can be repeated indefinitely without causing the candidate feature pool to grow beyond the limits imposed by available computational resources. Thus, high-quality descriptors can be constructed in a reduced feature space containing hundreds of candidates, as opposed to millions or billions. The enhanced performance of iBART is demonstrated here by applying it to both a simulated data set (i.e., statistical simulation) and to our previously published data sets describing metal atom adsorption on oxide surfaces. We find that iBART produces more accurate descriptors than those found in our previous work using traditional FS methods. This provides a powerful FS tool that can be implemented on laptop computers instead of requiring high-performance computing resources.

623550 Electronic Structure Modeling of Electric Field Driven Catalysis and Electrocatalytic Kinetics

Abstract:

Presence of charge separation in catalytic microenvironments can induce strong electric fields at microscopic length scales. Reaction chemistries involving polar transition states can be strongly influenced by such molecular scale electric fields. In addition, the involvement of polar solvent molecules can further enhance such effects and alter the mechanisms/selectivity of a complex reaction system. Electronic structure methods, including density functional theory (DFT), allow us to examine elementary step reaction trajectories and the redistribution of electrons along a reaction path. Given the value of the electric field***(F)*** and the reaction trajectory, the effect of fields on barriers can be estimated by the differential dipole-field stabilization ***($\Delta\mu F$)*** without requiring a charged surface simulation. 2nd order effects can be included by additionally estimating the polarizability***($\partial\mu/\partial F$)*** through applied field calculations. However, tuning/understanding these electric fields on a microscopic scale to benefit catalysis require fundamental understanding of correlation between externally tunable variables (e.g., applied potential, type of solvent, type of surface, etc.) and the microscopic fields and models to explore the possibilities in the reaction environments.

Electrochemical systems are notoriously challenging for modelling these influences due to presence of a micro-scale charge separation, referred to as the interfacial double layer, with the dual impact of solvation and electric field effects. Approximations are needed to efficiently address these influences computationally. With a Helmholtz model, the magnitude of these electric fields can be related directly to applied bias and approximate width of double layer and the potential of zero charge. However, these estimates are clearly dependent on the choice of model parameters (width of double layer, potential of zero charge and parameters in an explicit/implicit solvation model). Therefore, testing the sensitivity of activation barrier estimates to these parameters and their correlation is essential to making mechanistic conclusions. In this poster, we will share results on the field sensitivity of barrier estimates with a series of electrochemical interface models, and apply these to examine elementary reactions relevant to the oxygen reduction reaction, ammonia electrosynthesis, and carbon dioxide electro-reduction.

621384 Estimation of the Binary Interaction Parameters of the Anrtl Model Using Molecular Simulations

Abstract:

Process simulations play an important role in the design and optimization of an adsorption process for a given application. The mathematical expressions relating the various thermodynamic variables form the backbone of a process simulator. One such thermodynamic model for mixed-gas adsorption equilibria is the aNRTL model [1]. The model takes into account non-idealities in the adsorbed phase due to interactions of the adsorbate molecules of a binary gas mixture with an adsorbent. The arguments leading to the derivation of the molar excess Gibbs free energy and the subsequent calculations of the component activity coefficients are analogous to the NRTL model [2]. Thus, the physical significance of the aNRTL model is justified because it is based on the local composition concept [3]. The model uses one adjustable parameter called the binary interaction parameter which gives the difference in the Gibbs free energy of interaction of the adsorbate molecules with an adsorption site relative to thermal energy. The binary interaction parameter can be regressed from experimental adsorption data of binary gas mixtures. It can also be predicted from the experimental pure component adsorption data using framework of the Thermodynamic Langmuir Model [4].

Unavailability or unreliability of experimental datasets in terms of missing uncertainties and fewer data points affect the regression process, thus impacting model predictions. Therefore, it is extremely important to develop a methodology to estimate such parameters that does not make use of experimental adsorption data. Such an approach has been developed for the estimation of binary interaction parameters of the NRTL model using molecular dynamics simulations [5]. In this study, we follow a similar approach to estimate the binary interaction parameters for a pair of adsorbate molecules in MOFs such as Cu-BTC and UiO-66. We have determined the amounts adsorbed in the MOFs using Monte Carlo simulations in the Grand canonical ensemble at various conditions of pressure and temperature. This allowed us to validate the various force fields used by comparing the predicted equilibrium amounts adsorbed with experimental data. Distinct adsorption sites for a given binary pair were identified from the computed trajectories of the molecules using molecular dynamics simulations in the isothermal, isobaric ensemble. Following arguments similar to the ones used by Ravichandran et al. (2018) [5] and Brandini and Prausnitz (1982) [6], we have derived equations relating the binary interaction parameters to the molecular parameters of the adsorbate-adsorbent interactions in the first coordination shell of each distinct site. The effective molecular diameters and radius of the first coordination shells were computed from the radial distribution functions of the adsorbate-site pair. The effective interaction strength parameters for adsorbate-site pairs were determined from the potential of mean force calculations. The binary interaction parameters computed from molecular simulation gave an accurate prediction of the binary adsorption equilibria for the systems studied.

References:

1. Kaur, Harnoor, Hla Tun, Michael Sees, and Chau-Chyun Chen. "*Local composition activity coefficient model for mixed-gas adsorption equilibria.*" *Adsorption*25, no. 5 (2019): 951-964.
2. Renon, Henri, and John M. Prausnitz. "*Local compositions in thermodynamic excess functions for liquid mixtures.*" *AIChE journal*14, no. 1 (1968): 135-144.
3. Wilson, Grant M. "*Vapor-liquid equilibrium. XI. A new expression for the excess free energy of mixing.*" *Journal of the American Chemical Society*86, no. 2 (1964): 127-130..
4. Chang, Chun-Kai, Hla Tun, and Chau-Chyun Chen. "*An activity-based formulation for*

Langmuir adsorption isotherm." *Adsorption*(2019): 1-12.
5. Ravichandran, Ashwin, Rajesh Khare, and Chau-Chyun Chen. "*Predicting NRTL binary interaction parameters from molecular simulations.*" *AIChE Journal*64, no. 7 (2018): 2758-2769.
6. Brandani, V., and J. M. Prausnitz. "*Two-fluid theory and thermodynamic properties of liquid mixtures: General theory.*" *Proceedings of the National Academy of Sciences*79, no. 14 (1982): 4506-4509.

622077 DFT-Based Solvent Screening for Temperature Swing Solvent Extraction of High Salinity Brines

Abstract:

Recently, temperature swing solvent extraction (TSSE) has been proposed by Boo et al.,[1-2] which is a membrane-less and non-evaporative desalination technology.[3] TSSE utilizes solvents with temperature-dependent water solubility for the selective extraction of water over salt from saline feeds with low-grade heat.[1-2, 4] As reported in previous studies, solvent polarity is a key parameter used to quantify solvent extraction performance. For example, a low-polarity solvent with temperature-dependent water solubility can extract large amounts of water from a hypersaline feed to induce salt precipitation.[2] However, others found a higher surface polarity of the solvent facilitates more water absorption and contributes to solvent contamination from the salt ions.[5]

The key for the solvents used in TSSE for brine desalination is large water solvation variability within a relatively small temperature window (e.g., ~300 K +/- 25 K). We have performed molecular dynamics (MD) simulations, as well as experimental evaluation of different amine- and imidazole-based solvents in order to determine their TSSE performance. However, due to the large number of potential candidates and time requirements for MD simulations or experimental evaluation, first-principles density functional theory (DFT) calculations can provide initial solvent screening and provide rational design rules. Previously, we use similar approaches to predict CO_2 solubility in ionic liquids, based solely on the ionic polarity characteristics of the individual molecules.[6] Herein, we are using electrostatic potential properties obtained from the solvent van der Waals surfaces to quantify the polarity of these solvents, along with solvation free energy calculations, in order to understand the TSSE structure-property-performance relationships and provide guidelines for solvent design for brine desalination.

References:

1. Boo, C.; Winton, R. K.; Conway, K. M.; Yip, N. Y., Membrane-Less and Non-Evaporative Desalination of Hypersaline Brines by Temperature Swing Solvent Extraction. *Environ. Sci. Technol. Lett.* **2019,** *6*, 359-364.
2. Boo, C.; Billinge, I. H.; Chen, X.; Shah, K. M.; Yip, N. Y., Zero Liquid Discharge of Ultrahigh-Salinity Brines with Temperature Swing Solvent Extraction. *Environ. Sci. Technol.* **2020,** *54*, 9124-9131.
3. Davidson, R.; Smith Jr, W.; Hood, D. W., Structure and Amine-Water Solubility in Desalination by Solvent Extraction. *J. Chem. Eng. Data* **1960,** *5*, 420-423.
4. Luo, T.; Bajpayee, A.; Chen, G., Directional Solvent for Membrane-Free Water Desalination—A Molecular Level Study. *J. Appl. Phys.* **2011,** *110*, 054905.
5. Kim, M.; Choi, O. K.; Cho, Y.; Lee, J. W.; Cho, A. E., Elucidation of the Desalination Mechanism of Solvent Extraction Method through Molecular Modeling Studies. *Desalination* **2020,** *496*, 114704.
6. Liu, X.; O'Harra, K. E.; Bara, J. E.; Turner, C. H., Solubility Behavior of CO_2 in Ionic Liquids Based on Ionic Polarity Index Analyses. *J. Phys. Chem. B* **2021,** 10.1021/acs.jpcb.1c01508.

622988 Theory and Simulation Studies of Structure and Thermodynamics in Polymer Blends and Polymer Nanocomposites with Directional Interactions

Abstract:

Macroscopic properties of polymer blends and polymer nanocomposites (PNCs) depend on their morphology which is governed by the miscibility between constituent polymers in polymer blends or by the dispersed/aggregated state of particles in PNCs. One way to tune morphology is by introducing favorable interactions between components of the blend/nanocomposite; these interactions between the components of the blend/nanocomposite could be either isotropic (i.e., van der Waals) or directional (e.g., hydrogen bonds, pi-pi stacking). Although, directional interactions like hydrogen bonds offer a promising potential route to precisely tune structure within polymeric materials, computational studies on polymers with directional interactions have been limited to atomistic simulations that explicitly represent hydrogen bonding atoms and their interactions making them expensive for simulation of large length and time scale assembly in polymers. Conversely, simulations with coarse-grained (CG) models are able to capture the large length and time scales associated with polymer chains but cannot capture the effects arising due to the specificity and directionality of interactions without using complex anisotropic interaction potentials to model directional interactions. To overcome these challenges in modeling of directional interactions, I will present our development of a CG model that captures the short length and time scales of directional and specific interactions as well as the large length and time scales associated with polymer chains in polymer blends and PNCs. In this poster, I will describe the details of the CG model that allow us to produce effectively anisotropic attractive interactions while still using isotropic interaction potentials. I will then share how we have used the CG model in molecular dynamics (MD) simulations to contrast the local structure and morphology in PNCs with directional interactions to that of PNCs with isotropic interactions. I will also share how using this model in MD simulations, Polymer Reference Interaction Site Model (PRISM) theory, and machine learning (ML) we have quantified the complex phase behavior in immiscible homopolymer blends with dominant hydrogen bonding interactions.

625143 Molecular Dynamic Investigations of Thermodynamic and Transport Properties of Trinidad and Tobago Asphaltenic Heavy Oils

Abstract:

Introduction

Trinidad and Tobago heavy oils contain asphaltenes which can precipitate unpredictably and cause process disruptions such as coking during pyrolysis, pipe blockage and fouling of equipment and unwanted stabilization of oil-water emulsion[1]. The mechanisms underlying the above are not well understood and therefore cannot be taken into account at the design stage and can impede plant utilization. Asphaltenes comprise many species of molecules with high carbon to hydrogen ratios than normal crude oil: they contain aromatic cores, hetero-atoms and aliphatic chains. Their characterization present challenges and yet such information process response to process conditions is crucial to process assurance.

Crude oil in which these asphaltene are contained are complex mixtures making it important to know how the asphaltene molecules interact with the solvent-mixture.

This research, aims to use Molecular Dynamics (MD) to shed light on asphaltene-asphalene, asphaltene-solvent and asphaltene-modifier interactions. Such interactions can be expected to influence asphaltene properties in Trinidadian-heavy oils under transportation, and processing conditions. This technique can also be used to optimize the addition of modifiers and to guide the manipulation of operational conditions which will mitigate the nuisance and risk they pose.

Figure 1: Types of Asphaltene Molecules

The objectives are to do the following:

1. Validate the use of GROMACS, a Molecular Dynamics software to calculate the trajectories of molecules and to use them to predict important transport properties of asphaltenes (Diffusion Coefficient & Viscosity), and predict thermodynamic properties of asphalt solution such as density and asphaltene molecular association.
2. Track the interactions which lead to the formation of clusters of asphaltenes.
3. Study the influence of the composition of solvent on clustering
4. Complete the design of an asphaltene molecule whose physical properties match those of asphaltenes encountered in heavy oils in Trinidad and Tobago.

METHOD:

The atomistic MD is used to represent bonded atoms in molecules by balls held together by springs. Newtonian mechanics is used to calculate molecular trajectories under Potential Fields using the GROMACS software. The motion of the molecules is determined by the application of Newton's laws in potential field described by the **O**ptimized **P**otential for **L**iquid **S**imulation **A**ll **A**toms (**OPLS_AA**).

The **OPLS_AA** (eqn. (1)), with appropriately chosen parameters has been used to successfully model hydrocarbon liquids like Trinidad's heavy oils[2]. It comprises potentials due to: covalent bonds, Lennard-Jones forces, electrostatic forces due to partial charges on polar groups.

The Particle-mesh Ewald (PME) algorithm is used to calculate electrostatic interactions. The periodic boundary condition (PBC) and the Maxwell-Boltzmann distribution of the initial velocities are adopted. We use the Nose–Hoover thermostat and the Parrinello–Rahman barostat to control temperature and pressure in constant volume, constant temperature (NVT) simulations.

The system

The simulation box of size **7nm x 7nm x 7nm** is filled with a chosen number of molecules of model asphaltene (see structure C in Figure 1) and remainder being the appropriate solvent or solvent mixture (e.g. heptane, toluene etc.) and additives such that the solute is 7wt/wt%.

Simulation Steps

The steps in the simulation are shown in the flowchart above. Asphaltene molecules are inserted in simulation box and solvent added to create a configured system. A check is carried out to see if there is net a charge and ions added to neutralize the mixture. The energy of the initial configuration in the system is minimized using steepest descent method. Next, equilibrate to the desired temperature and pressure by running the NVT ensemble (constant number of molecules, volume, and temperature) for 100ps. Then the simulation is run at constant number of molecules, temperature and pressure (NPT ensemble) for 100ps. The next stage is called Production MD.

Production Molecular Dynamics

A simulation time step of 2.0 fs was used

MD simulation lasted 40ns

Cut off for non-bonded interaction was fixed at 1.0nm

Berendsen thermostat and the Parrilello-Rahman barostat setting were used with time step of 2fs.

Radial Distribution Function (**RDF**) was used to analyse cluster formation.

The Results and Discussion

Diffusion Coefficient

Figure 2: Einstein's Plot: Mean square Displacement between Molecules vs Time

The diffusion coefficient of asphaltene in heptane were estimated from the trajectories of the molecules using Einstein's law[3] in plotting the Mean Square Displacement (MSD) between molecules as a function of time as shown in the above Figure. The predicted values are comparable to measured values.

Viscosity

The TCAF routine in GROMACS[4] was used to estimate time constants in the decay of the transverse-current autocorrelation functions of the fluctuations in velocity gradients in the liquid and the viscosity over timeframes inferred from them. In Fig. 3 the viscosity is plotted against k, a function of the decay constant. It shows viscosity converges to values which agree with experimental data.

Figure 3: Plot of Viscosity vs Parameter k

Cluster Analysis

Molecules separated by 3.5 Å or less are considered associated and thus are part of a cluster. Fig. 4 below, shows the distance apart of two molecules in events, as a function of time. In vacuum, all the events recorded are below the red line and indicate all molecules are clustered, as expected. However, in the adjacent figure, the molecules tracked have not clustered in the presence of solvents molecules.

Figure 4: The distance between asphaltene molecules one and two as a function of time.

The picture frames of asphaltene molecules in various configurations in vacuum annealing experiment were recorded: to begin with all molecules are apart, nearly all the molecules have clustered during a simulation run for 5 ns of the molecules in vacuum. The molecules are stack laterally indicating interaction of π electrons above and below the plane of the molecules.

The picture frame in Fig. 5A show the molecules in heptane after 10 ns of production NVT simulation: some molecules are unclustered and others are clustered in pairs and in trimers. Some molecules are attached horizontally but displaced indicating π bonding between the aromatic cores; others are attached edgewise to the aromatic core. Edgewise attachments are not observed in MD runs in vacuo but may have been made possible due to solvation by heptane. The arrangements of the molecules in the clusters are expected to have implications for cluster growth, agglomeration, flocculation and other phase transitions.

Asphaltene in Heptane Asphaltene in "Crude" Dispersed Asphaltene in toluene

Figure 5A Figure 5B Figure 5C

Figure 5C shows the simulation box containing toluene and asphaltene molecules. Again some toluene molecules have been edited out for clarity. Unlike the simulation of asphaltene in heptane the formation of clusters of asphaltene molecules does not occur. We propose that the mechanism for reducing the probability of cluster formation (stacking) is steric in nature. The toluene molecules orient themselves transversely to the asphaltene molecules thereby making it unlikely that the asphaltene molecules are able to stack. This seems to explain why toluene is an effective solvent for asphaltenes.

A histograms of cluster sizes are recorded. The largest cluster contains 5 molecules however the relative occurrence is lowest compared to the monomer.

Figure 6 shows the probability of an asphaltene molecule belonging to clusters of varying sizes including the monomer after the simulation time of 100ns. The maximum cluster size contained 5 molecules. The proportion of molecules belonging to a cluster reduced from 36% for the monomer to 12.5% for the cluster with 5 molecules.

Figure 6

The Visual Molecular Dynamics software can be used to view the trajectories of the MD simulation over 40ns of 12 molecules of asphaltene in heptane (7wt/wt%). Clustering events are visible.

SUMMARY AND CONCLUSION

The use of GROMACS for atomistic MD simulation of asphaltene in vacuo and in heptane, toluene and "crude" are validated. The transport and thermodynamic properties can be evaluated from the trajectories to yield values in agreement with published experimental data. The phenomena of molecular aggregation and clustering in heptane is observed. Two types of stacking are observed – flat stacking and edgewise association with a cluster. In heptane only a minority of molecules remain unassociated. These results confirm that MD simulation can be used as a tool for investigating risks to flow assurance posed by asphaltenes and their risk mitigation.

FUTURE WORK

1. Complete the design an Asphaltic molecule to mimic better the properties of the cocktail of Trinidad and Tobago Asphaltenes.
2. Simulate more accurately the composition and processing conditions of Trinidad heavy oils to predict the inclusion of designer additives which can modify the properties of these oils favourably.
3. Parallize the computing facilities in the laboratory to improve computing power at minimal cost.

REFERENCES

1. M. Spiecker, KL Gawrya, PK Kilpatrick, J. Colloid & Interface Sci. 267, (2003) 178-193

2. Adrian Lutchman, F. Addo-Yobo. UTT Research Symposium September 2018
3. Kubo (1966), Report Progr. Phys 29, 255-259
4. Bruce J. ***Palmer***. ***Rev***. ***E***49, 359
5. Gromacs User Manual version 5.0.4, gromacs.org

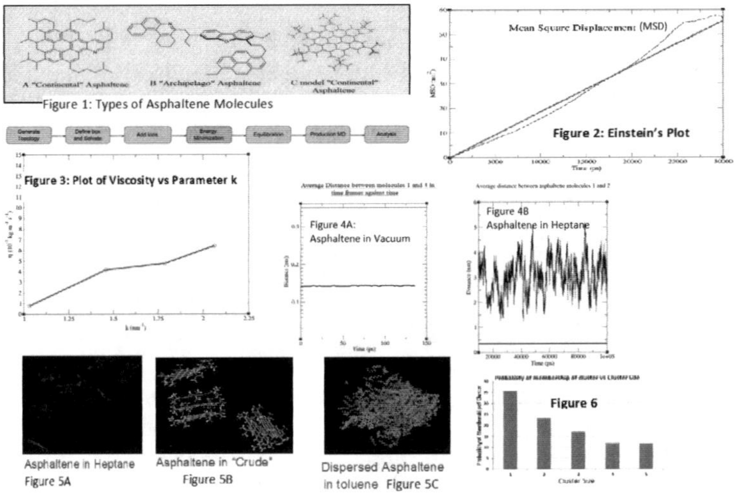

631683 Multimodal Integrated Modelling for COVID-19 Health Risk Management

Abstract:

Through the one-year experience of the pandemic, it has been evident that managing efficiently the public health risk of COVID-19 requires not only the ability to assess the extent of the contagion spread in the community, but to be able to predict the spread and the consequent pressure on the health services, and at the same time respect the social and financial viability kernel of the community and the local economy. Although strict and long-term lockdown is the most efficient practice for minimizing the contagion spread, the greatest challenge for the successful management of such a long-term sanitary crisis is the effective application of targeted interventions and closing of specific activities that are considered as the ones that contribute mostly to widespread contagion. On the other hand, it is important to be able to quantify the effect of the available artillery for combating the pandemic, including both vaccination, as well as non-pharmacological interventions, such as increased number of testing/rapid tests, indoor air purifiers, targeted containment measures and closing of specific activities, as well as the effect of seasonality on climate and transmission dynamics. In addition, the transmission dynamics of new mutants and the way they affect the contagion spread dynamics, or eventually the immunity have to be accounted as well. Accounting for all of the above, and aiming at the efficient management of the COVID-19 pandemic, we have developed a multi-modal computational tool for the evaluation of the public health risk from the COVID-19 epidemic in Greece, Italy and USA and we have evaluated the effectiveness of different non-pharmacological intervention scenarios for public health risk management. The computational tool for public health risk management from COVID-19 is called CORE: COVID Risk Evaluation model. It includes an advanced model of the spread of the epidemic, which is an evolution of the most advanced SEIR models available (Blackwood and Childs, 2018), also taking into account the implementation dynamics of non-pharmacological interventions such as virus detection testing geared towards the general population or targeted sub-population groups, circulation restriction and population containment measures (Maier and Brockmann, 2020), in the evolution of the dispersion (described as compartment X) and the final health risk assessment of the affected population. The SEIR-X dispersion model, which is the dispersion computation engine, has been extended to a multi-state population model to describe in detail the different possible states of the population based on the typology and severity of the symptoms of the disease (SEIR-X multi-state model, or SEIR-Xms). These estimates are then used as input to an already validated multi-stage SEIR-X model (CORE), which supports the management of the COVID-19 health risk by capturing reliably its dynamics, including the impact of risk reduction measures. The integrated model is already part of the OECD COVID-19 computational toolbox.

A key component for the identification of the effect of the targeted containment measures is the ability to properly account for the effect various activities may have on the effective contact among various population groups, accounting for their sociodemographic progfiles (i.e. age, occupation etc). Towards this aim, contact rates for the different population groups within the CORE model, are estimated using Agent-Based Modelling (ABM). ABM is a simulation technique that allows to explore and understand phenomena, where independent entities interact together, forming emerging behavior. While the direct representation of individuals' actions is organisationally difficult, captures this process by managing information at the level of the autonomous decision-makers, called "agents" (Sarigiannis et al., 2018). The agents are software objects that have internal states and are programmed to react and act in their environment while following a set of behavioural rules. By simulating actions and interactions at the individual level, the diversity that exists among agents can be detected, as rise is given to the behaviour of the system as a whole (Macal and North, 2005, De Marchi and Page, 2014). The use of ABM modelling, allowed us to deliver estimates of the effective contact rate of the general population, as well as of the various educational grades, as well as several daily activities such as supermarkets, bars and restaurants, public transportation, health services, retails and large malls. Thus, by describing dynamically the interactions of the key components of the contact matrix, we were able to capture the non-linearities associated with changes in closing or opening of specific activities, e.g. the effect of opening the retail vs opening secondary education. All the above, allowed us to be able to evaluate various scenarios of containment and closing or opening of activities, supporting the efficient planning for combating COVID-19 spread dynamics. It is important to highlight, that CORE predictions accounted for all the dynamically changing external conditions, including the ones that go beyond human behaviour, such as climate and the presence of mutants with increased transmissibility in the community, as well as measures that effectively contribute to contagion control such as the opportunities provided by rapid tests and self-testing, the continuously increased rate of vaccination and the use of disinfection devices in public places and transportation. Another very interesting scenario that has been analysed by the CORE model for ensuring the financial sustainability and at the same time controlling the pandemic, was the management of vaccination prioritization, to occupants that highly contribute to contacts within the contact matrix. At the same time, beyond the successful analysis of containment scenarios, we were able to predict in long term the spread dynamics and the respective burden in the health services at specific milestones, such as the impact of touristic flows in the summer and the development of the second wave in early August 2020, the effect of the new mutations (of higher transmissibility) in the third wave in 2021, and the projections for the forthcoming summer.

Our most recent scenario analysis results demonstrate that the contribution of self-tests made frequently (once a week for the general population and twice a week for targeted sub-populations at higher risk of contagion) to reducing the spread of the pandemic is crucial. However, although the simulations have shown that that a large number of self-tests frequently made is capable of helping to control the pandemic, opening specific activities has to be allowed progressively. It has also been shown that at the current situation, where only 10% of the population has acquired artificial immunity through vaccination, self-tests are the most precious ally in combating this pandemic, while after 3 months, the combination of artificial immunity, with the summer climate conditions that do not favour transmission will help in pandemic control. The long-term prediction of the COVID-19 dynamics in Greece till the end of June, accounting for the currently proposed containment and openings scenario, vaccination and self-testing rate is illustrated below.

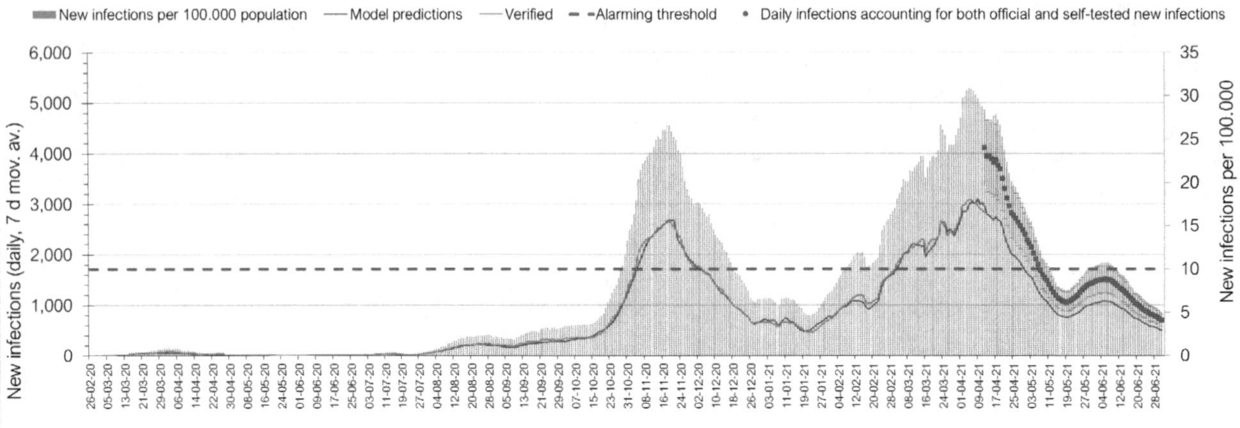
New infections per 100.000 population
Model predictions
Verified
Alarming threshold
Daily infections accounting for both official and self-tested new infections
New infections (daily, 7 d mov. av.)
New infections per 100.000
6,000
5,000
4,000
3,000
2,000
1,000
0
35
30
25
20
15
10
5
0

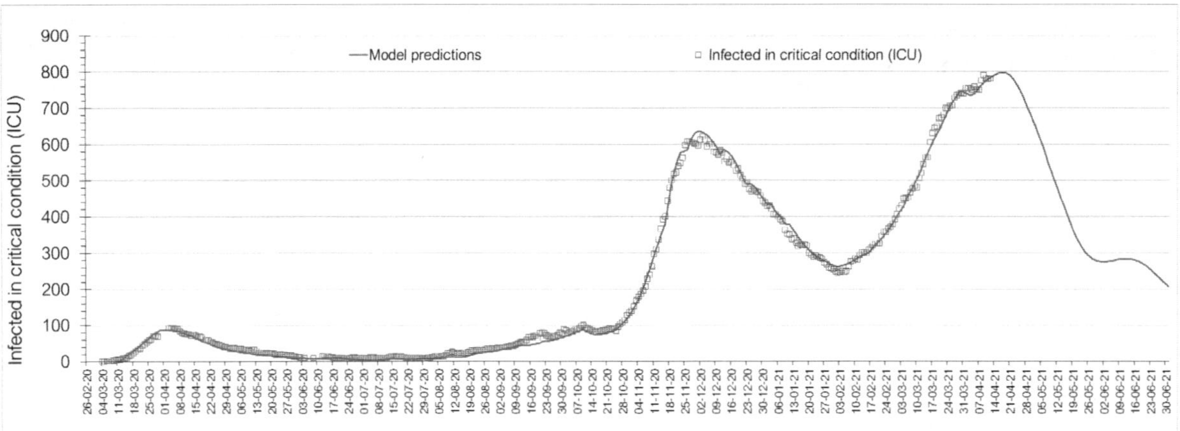
Model predictions
Infected in critical condition (ICU)
Infected in critical condition (ICU)
900
800
700
600
500
400
300
200
100
0

626537 Epidemic Models with 'Time Since Infection': Efficient and Flexible Modeling Tools

Abstract:

Epidemic models are useful tools in the fight against infectious diseases, but their usefulness is limited (in part) by their ability to accurately describe the underlying disease dynamics. The most accurate epidemic models are typically the most computationally expensive, and "compartment models" offer the most popular compromise between speed and accuracy. In a compartment model, an infected person progresses through a series of artifical 'stages' or 'compartments' (e.g. exposed, infectious, recovered), and the resultant equations are easily solved by standard tools for ordinary differential equations. A more realistic model would describe disease dynamics as a continuous function of time since infection (TSI), but the computational cost of a TSI model is typically assumed to be large by comparison.

In this talk, we share our recent work on TSI models. First, we improve upon existing TSI models by using a 'filter' to partition the infection population into discrete compartments, as and when such measurements are necessary for informing policy decisions (e.g. predicting hospitalizations, deaths, etc.). Second, we provide a more efficient numerical method for solving the equations of a TSI model with spectral accuracy. Given this numerical approach, we find that TSI models are now cost-competitive with the standard 'compartment' strategy for many applications.

632805 Is Curvature Better? Examining the Structure of Nanoparticles Containing Curved Carbons

Abstract:

Carbon-based molecules can become curved when one or more non-hexagonal rings are embedded within their hexagonal lattice. These polar fullerene-like molecules possess steric and electronic properties different from their planar counterparts that make them good candidates for many applications including microporous materials for gas storage, organic electronic devices such as imaging probes and batteries, sensors, and micelles for targeted nanomedicine. Development of these technologies requires an understanding of the dynamic nanostructure of particles containing curved carbons, which has not yet been well explored.

This work uses advanced molecular modelling to explore the properties of nanoparticles containing curved carbons. Intermolecular interactions were described using the recently developed curPAHIP potential able to capture the enhanced interactions of curved aromatics. Large timescales and temperature ranges were sampled to provide insight into the dynamic behaviour of curved aromatics in homogeneous systems as well as those containing molecules of different sizes and ratios, cations, and planar molecules. These results provide new information on structures and properties of nanoparticles containing cPAHs, including density, molecular structure, and surface composition. This provides valuable understanding of the energetic and structural properties of nanoparticles with potential in many applications.

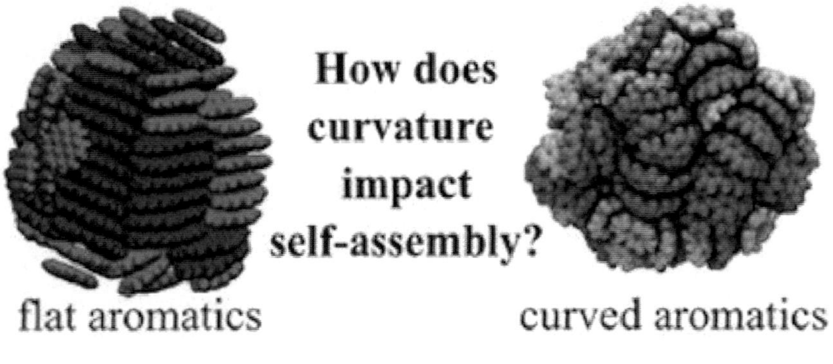

632858 Chain Topology Structure Simulation of Linear Low-Density Polyethylene Based on Graph Theory and Monte Carlo Algorithm

Abstract:

Chain topology structure simulation of linear low-density polyethylene based on graph theory and Monte Carlo algorithm

Wang Jie, Yao Yang*, Jingyuan Sun, Xiaoqiang Fan, Zhengliang Huang, Jingdai Wang, and Yongrong Yang

State Key Laboratory of Chemical Engineering, College of Chemical and Biological Engineering, Zhejiang University, Hangzhou, 310027, China

**Correspondence author, e-mail: yao_yang@zju.edu.cn*

The highly branched topology structure of linear low-density polyethylene (LDPE) is directly relevant to its product properties, such as melt index, rheological, etc. In order to obtain a better description of the radical polymerization process, a topology structure simulation of linear low-density polyethylene chain based on graph theory and Monte Carlo algorithm is proposed in our study.

By converting reaction speed to the random sampling probability problem, Kinetic Monte Carlo algorithm is capable to predict accurate molecular weight distribution (MWD), copolymer composition distribution (CCD), short-chain branching distribution (SCBD), and long-chain branching distribution (LCBD). As simulation of chain full topology structure of linear low-density polyethylene could be the basement of many other applications or research, it can set our imagination about high pressure polyethylene polymerization free. The first problem researchers encountered is how to symbolize the chain topology structure, as the chain structure is not a digit number but a kind of connection graph. Some specially designed methods are developed for this problem, but they are dedicated to certain problem or based on some temporary trick, rather than elegant theory method via mathematics and graph theory. To solve this problem, we studied chain topology structure simulation of LDPE using graph theory and Monte Carlo algorithm, which could provide better accuracy and speed.

The algorithm we proposed could obtain more accurate simulation result comparing with traditional ones. For demonstrating the difference, we take an industry LDPE tubular reactor as example, and establish new Monte Carlo model and old method for the same simulation input data. Both methods could provide the molecular weight distribution, which is consistent with experiment data. But Monte model with graph theory show better accuracy of molecular chain length distribution, short-chain branching distribution, and long-chain branching distribution. Comparing with traditional method, our algorithm obtains bigger main carbon chain length, the difference is more significant with molecular weight growing. Because of the difference of main carbon chain length distribution, the SCBD and LCBD also show great difference in two method, long chain branch is shorted in our simulation method.

The combination of computer theory and chemical technology knowledge inspire us a lot and give new life to old simulation method.

Key words: Polymer chain structure, Graph theory, Polymers, Algorithms

Reference

[1] Meimaroglou, Dimitrios, and Costas Kiparissides. "A novel stochastic approach for the prediction of the exact topological characteristics and rheological properties of highly-branched polymer chains."

Macromolecules 43.13 (2010): 5820-5832.

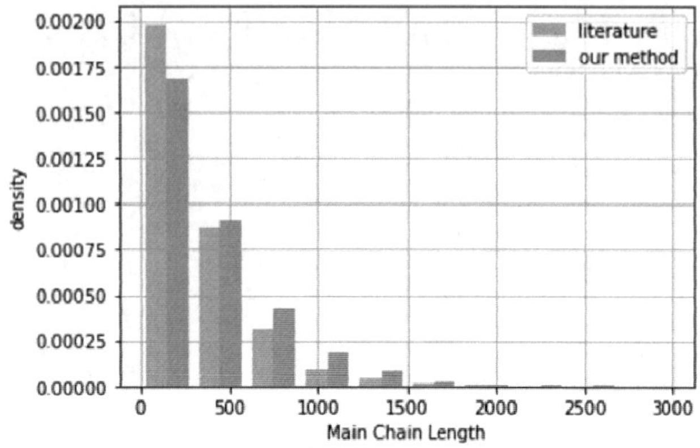

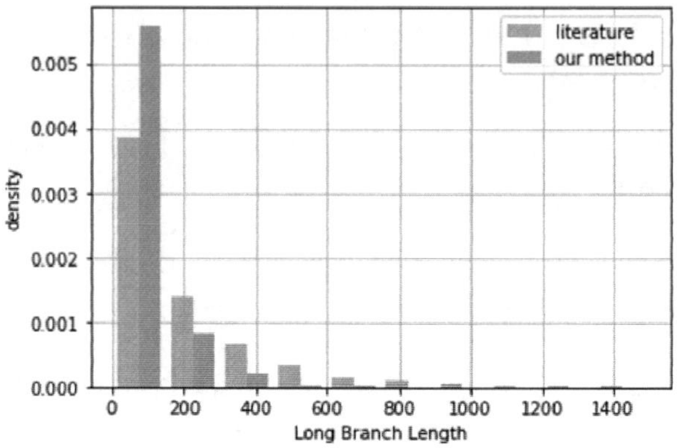

622808 Molecular Dynamic Investigations of *E. coli* and *V. Cholerae* Fadl Homologs

Abstract:

Vibrio cholerae is the enteric bacteria responsible for the disease cholera. While *V. cholerae* is naturally found in warm brackish waters where freshwater meets seawater, additonally it has been found in freshwater lakes and rivers far removed from any seawater source[1]. One of the interesting abilities of *V. cholerae* is its ability to uptake a larger range of long-chain fatty acids (LCFAs) than some other enteric bacteria – such as *E. coli.* LCFAs are believed to play an important role in bacterial ecology, with demonstrated effects on stress response, pathogenicity, and virulence[2]. We hypothesize that *V. cholerae* possesses multiple structurally unique fatty acid transporters to recognize a wide range of LCFAs, and that this capability is critical to its ability to spread disease. In order to investigate this hypothesis, we analyzed the predicted structures and LCFA docking profiles of the three most prevalent fatty acid transporter homologs in *V. cholerae.*

Using the singularly expressed *E. coli* FadL protein as template, we performed a Basic Local Alignment Search Tool (BLAST) search to find homologs in *V. cholerae.* We found 17 unique proteins across the three virulent serotypes and the three most prevalent FadL homologs (NP_233248, NP_230687, and NP_230688) were chosen for further study. We then used I-TASSER5.1 standalone to fold the protein sequences and generate 3D structural models of each of the selected *V. cholerae* FadL homologs. The generated structures were compared to the known X-ray crystal structure of *E. coli* FadL (RCSB ID: 1T16) and several key structures outlined by van den Berg[3] were identified (Figure 1). The L3 loop, L4 loop, S3 kink, and N terminal hatch domain were found to be aligned, but much of the *E. coli's* characteristic loop structures showed deviations between the homologs. To better simulate cell conditions, each structure was inserted into a model *V. cholerae* outer membrane [outer leaflet of *V. cholerae* LPS (with 15 O-antigen units) and an inner leaflet of 67% phosphatidylethanolamine (PE) and 33% phosphatidylglycerol (PG)] using CHARMM-GUI Membrane Builder[4]. Each trial of the four FadL/membrane systems were equilibrated for a minimum of 250 ns. A large array of FAs with various unsaturation ranging from 16:1 to 22:6 as well as the original detergent molecules found in the 1T16 X-ray crystal structure were chosen as ligands to be docked; these ligands are: Lauryldimethylamine oxide (LDAO), Palmitoleic acid (16:1), α-Linolenic acid (18:3α), γ-Linolenic acid (18:3γ), Dihomo-γ-linolenic acid (20:3), Arachidonic acid (20:4), Eicosapentaenoic acid (20:5), Tetraethylene glycol monooctyl ether (C8E4), and Docosahexaenoic acid (22:6).

The docking simulations predicted 25,000 possible ligand conformations spread over 250 unique protein conformations. These conformations were compiled and examined for recurrences in certain areas of each protein. Several sites were identified as common docking locations and were analyzed. In the case of the crystal structure 1T16, docking appeared to follow the convention of placing the ligands in the low affinity binding site, the high affinity binding site, and the S3 kink as outlined by van den Berg[3]. Specifically, the majority of docking of the high affinity binding site was restricted to the lower chain FAs, which *E. coli* is known to uptake, while the higher chain FAs were found in the low affinity binding site and the S3 kink. In the case of the membrane equilibrated FadL, only the smallest ligand, LDAO, was found to bind to the high affinity site. This implies that the high affinity binding site was not open, and is instead activated/opened through some selective uptake mechanism. In the case of NP_230687 and NP_230688 (similar *V. cholerae* homologs), the uptake channels appear to be non-selective until the N terminal hatch domain. These channels seem to be nonselective in regards to FA length and unsaturation, supporting the fact that *V. cholerae* serotypes have been shown to uptake longer chained and polyunsaturated FAs[2]. The channels end at the N terminus hatch domain right before the S3 kink. This is similar to the proposed N terminal hatch for *E. coli* proposed by van den Berg[3]. It is thought that *E. coli* has a N terminus selection mechanism near the S3 kink, and this highly conserved structure found in each homolog appears perform that same function for both *V. cholerae* homologs NP_230687 and NP_230688. Strangely, *for V. cholerae* homolog NP_233248, there was no discernible channel and binding did not occur within the FadL's β barrel. This could be due to a conformational structure mechanism that was not initiated by general equilibration, but for the other homologs tested, this did not seem to be the case.

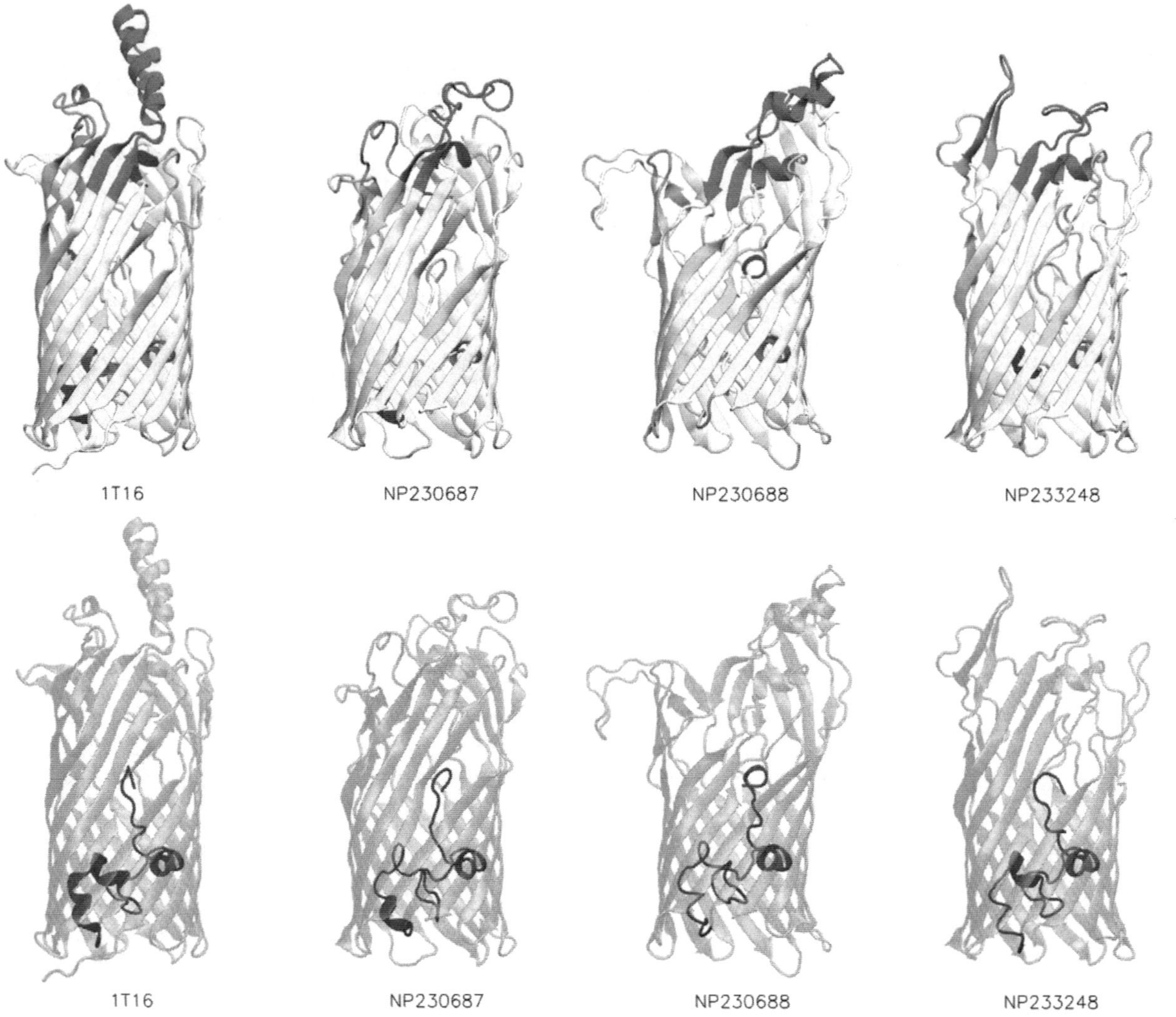

Figure 1. Folding results of FadL from I-TASSER. Top - the coloration of FadL is secondary structure, with the exception of the L3 loop (red), L4 loop (orange), and S3 kink (green) for reference. The β barrel strands are lined up well, and the S3 kinks are apparent in all the homologs generated. The L3 and L4 loops are somewhat formed, but the V. cholerae homologs' extra cellular loops do not mirror the original 1T16 crystal structure. Bottom – the FadL homologs internal N-terminus structure (blue) all appear to follow a similar folding scheme. The secondary structure detection algorithms tend to detect α and 3_{10} helices in some homologs but then detects turns and coils in others - even with very similar backbone shape.

621281 Clean, High Quality Low Emission Transportation Fuels with Fischer-Tropsch Synthesis: A Mesoscale Study of Transport Processes in Confined Systems

Abstract:

The Fischer-Tropsch synthesis (FTS) is a polymerization reaction used extensively in the Gas-to-Liquids (GTL) process to transform synthesis gas (H_2 and CO)[1] into clean, high quality low emission transportation fuels. The main FT reaction products, namely water, wax and small amounts of oxygenates (e.g. alcohols < 10 wt %), form a mixture through which the dissolved reactants diffuse, reach the catalytic nanoparticles and react. The transport properties of these mixtures, particularly inside catalyst pores, is a topic of significant interest for the petrochemical industry. Key factors in ensuring FT reactor's activity and stability is the selection of the catalyst (e.g. Co) and the support material. Among the typical catalyst carriers (eg. TiO_2, Al_2O_3), graphene and its derivatives have emerged as alternative supports in recent years.[2] Unfortunately, FTS catalysts deactivate over time. Among the factors contributing to deactivation, excess water reaction conditions are of particular interest, since – under this regime – sintering of catalytic nanoparticles is observed, which is responsible for FT reactor's reduced lifetime and increased operational cost. The degree of catalyst deactivation increases with increasing water partial pressure caused by high water loads.[3] Oxygenates in low concentrations were found to be responsible for increased active metal aggregation, especially under the conditions of high-water concentration with high CO conversion.[4] Studies remain inconclusive on the role of oxygenates and excess water in the loss of commercial catalyst activity and these issues remain as open questions.[5]

In order to gain a better understanding of the phase behavior of confined wax – water – alcohol mixtures in hydrophilic or hydrophobic environments at reaction conditions, our efforts concentrated on the *n*-C_{28} – H_2O mixture at 473.15 K inside pristine graphene (G) and graphene oxide (GO) pores by means of Coarse Grained Molecular Dynamics (CGMD) simulations with MARTINI.[6] Dodecan-1-ol was used to study the effect of long-chain alcohols on the wax – water mixture inside mesopores. Our simulations show that CG approaches capture the mixture's phase separation, the component's diffusivity as a function of the distance from the pore center and that dodecan-1-ol's preference towards the wax being mostly located at the *n*-C_{28} – H_2O interface. CGMD simulations with DFT derived parameters for Co, show that the Co NP does not affect phase separation of the mixture inside the pores. However, extensive coverage by water on the Co NP is observed. Given the experimental difficulties in probing the relevant mechanisms at this scale, our results showcase that CGMD using the MARTINI force field can be employed to study FTS related processes at this level[7,8] and are expected to open new pathways in the investigation of the effect of NPs on catalyst support interfaces in the presence of FTS relevant mixtures.

References

1. Wood, D.A., Nwaoha, C., Towler, B.F., *J. Nat. Gas Sci. Eng.*, 9, 196-208, 2012.

2. Karimi, S., Tavasoli, A., Mortazavi, Y., Karimi, A., *Appl. Catal. A*, 499, 188-196, 2015.

3. Bezemer, G.L., Remans, T.J., van Bavel, A.P., Dugulan, A.I., *J. Am. Chem. Soc.*, 132, 8540-8541, 2010.

4. Oh, J.-H., Bae, J.W., Park, S.-J., Khanna, P.K., and Jun, K.-W., *Catal. Lett.*, 130, 403-409, 2009.

5. Papavasileiou, K.D., Vasileiadis, M., Michalis, V. K., Peristeras, L. D., Economou, I. G., *Natural gas processing from midstream to downstream*, pp. 463-497, Wiley, 2019.

6. Marrink, S.J., Risselada, H.J., Yefimov, S., Tieleman, D.P., de Vries, A.H., *J. Phys. Chem. B*, 111, 7812-7824, 2007.

7. Papavasileiou, K.D., Peristeras, L. D., Bick, A., Economou, I. G., *J. Phys. Chem.* B, 123, 6229-6243, 2019.

8. Papavasileiou, K.D., Peristeras, L. D., Bick, A., Economou, I. G., *Energy Fuels*, , 35, 4313–4332, 2021.

632839 Ensemble Dependence of Mechanical Relaxations and Their Underlying Correlation Functions

Abstract:

Mechanical properties such as modulus and viscosity have their origin in the interactions among molecules and the extent that molecules relax, either under load conditions or in response to spontaneous stress fluctuations while at equilibrium. Improving the mechanical response of "real world" materials could be enabled by understanding how the presence of additives and adjuvants modifies their properties and relaxations. Classical molecular simulations enable developing physical insights about these mechanisms that underlie the mechanical response. The contributions of individual molecules can be tabulated explicitly, which can aid in interpreting experimental data for systems in which unique contributions cannot be distinguished. The presence of multiple components in a real-world material introduces realism yet complicates the analysis process by adding compositional variability. In a simulation, convergence of numerical results can be plagued by slow relaxations and sizable signal-to-noise barriers that are imposed by large relative nanoscale fluctuations. In this work, we consider the extent that stress time autocorrelation functions depend on the ensemble that is used to carry out the simulations. The storage and loss moduli that can be obtained directly from equilibrium stress fluctuations within molecular dynamics simulations of model bitumens are used as an example. Numerical methods that apply moving averages in the time and frequency domains improve the clarity of the complex modulus results. Use of a window function to screen the time-domain data provides further improvements. Comparisons of stress relaxation functions obtained using ensembles at constant temperature and pressure, temperature, and energy reveal if ensemble-dependent artifacts are present. Very large data sets are required to achieve a sufficient frequency range while simultaneously averaging over the few stress inputs that are available at each time step. Care during computation of the correlation functions is required to achieve both speed and accuracy while maintaining distinct time separations across long simulations. Examples that are enabled by sufficiently accurate results, such as time-temperature superposition across moderate frequency ranges, will be discussed.

631839 Development of Accurate Coarse-Grained Models of Polar Organic Solvents

Abstract:

Organic solvents are key components in a variety of hybrid material applications including polymer synthesis and self-assembly, nanoparticle film stabilization, colloid studies etc. Polar organic solvents like dimethyl acetamide (DMA) and dimethyl sulfoxide (DMSO) are majorly used for biochemical synthesis including polysaccharides and peptides. N-methyl-2-pyrrolidone (NMP) has been used in several fields of industrial applications including electronics, petroleum and pharmaceutical industries. Tetrahydrofuran (THF) is used as a solvent for surface coatings, adhesives as well as printing inks. It is also reported that solvent systems, consisting majorly of high boiling polar organic solvents like N,N-Diethylformamide (DEF) and 1,3-dimethyl-2-imidazolidinone (DMI), play an important role in MOF synthesis as well as in deciding the morphology of MOFs. Thus, in order to gain molecular insights into many of these applications using coarse-grained (CG) molecular dynamics (MD) simulations, it is necessary to have accurate solvent models, capable of replicating the process of MOF self-assembly. Here we have utilized an approach that integrates MD simulations with particle swarm optimization (PSO) to accelerate the development of polarizable and nonpolarizable CG models of six of the aforementioned polar organic solvents. Using a combination of beads previously parameterized in the group as well as a few newly defined CG beads with 2:1 or 3:1 mapping schemes, the bonded and non-bonded force-field (FF) parameters for all CG models were optimized to reproduce experimentally or atomistically obtained properties of density, enthalpy of vaporization, surface tension. The models were further tested at varying temperatures and system sizes to study their transferability. We also performed uncertainty quantification of parameters and properties using the Bayesian method.

624440 Molecular Simulation Study of Deep Eutectic Solvent-Water Liquid-Liquid Interfaces

Abstract:

We investigate the relationship between compositions of hydrophobic deep eutectic solvent (DES) and the molecular properties of DES-water interfaces using molecular dynamics simulations. Hydrophobic deep eutectic solvents composed of organic molecules have emerged as promising candidates for extraction applications in aqueous environments. The performance of a hydrophobic deep eutectic solvent depends on the properties of DES-water interfaces. However, it remains unclear how the DES-water interfacial properties relate to the DES compositions. To shed light on this question, we investigate the structural and dynamic properties of molecules at the DES-water interfaces formed between eight hydrophobic DES solutions and water. The simulation results show that the preferential orientation of DES compounds and water molecules form a thermodynamic barrier at the interfaces. The distinct hydrogen bond stability of DES and water solutions also contribute to a kinetic barrier at the interfaces.

631208 Relative Resolution: Implementation in Lammps of the Coulomb Potential

Abstract:

Recently, we introduced a novel algorithm for molecular simulations termed Relative Resolution (RelRes) [1]. Its basic idea is that near molecules, which are near to each other, are described by a detailed Fine-Grained (FG) model, and that molecules, which are far from each other, are described by a simplified Coarse-Grained (CG) model. Importantly, the FG and CG potentials are parameterized analytically by a multipole approximation, and the appropriate distance for switching between the two models is roughly in the middle between the primary and secondary coordination shells of the molecules. In turn, we already implemented this algorithm in LAMMPS for the Lennard-Jones potential: We retrieved the structural and thermal behavior of various organic systems, whilst achieving a fivefold improvement in computational efficiency as compared with conventional approaches in molecular simulation [2]. Currently, we are automating RelRes in LAMMPS for the Coulomb potential, and we are again achieving successful findings for charged systems. Most importantly, this LAMMPS algorithm is ready for use by other researchers.

[1] Aviel Chaimovich, Christine Peter, and Kurt Kremer (JCP, 2015).
[2] Mark Chaimovich, Aviel Chaimovich (JCTC, 2021).

629742 Deep Learning Molecular Force Field for Gaseous Adsorption in Metal-Organic Frameworks with Open-Metal Sites

Abstract:

Open-metal-site Metal−Organic Frameworks (MOFs) have exhibited promising performance in a variety of energy-related applications, particularly in gas separations. To facilitate the future design of new MOFs as well as to shed light on their atomic-level adsorption mechanism, molecular simulations play a critically important role. A prerequisite in molecular simulations is to properly describe the interactions between the adsorbates and the frameworks in order to ensure reliable predictions of relevant thermodynamic and transport properties. Although generic force fields have been shown to generate reasonable predictions, their failure to correctly capture the adsorption of CO_2 in open-metal-site MOFs necessitates the need for developing accurate potentials. Machine learning (ML) has proven to be a powerful approach to make accurate predictions given unseen data. While it has been recently applied in the field of MOFs to accelerate the discovery of materials, its application to describe adsorbate-adsorbent interactions has not yet been explored.

In this study, we have for the first time developed a deep neural network (DNN) potential to model adsorbate-adsorbent interactions in open-metal-site MOFs and demonstrated its potential in adsorption simulations. Our DNN model is based on physical interactions-guided features that utilize terms in the classic force fields and the pair-distance. The training and test sets were generated using state-of-the-art density functional theory (DFT) simulations. Both non-polar (e.g., CO_2) and polar (e.g., H_2O and CO) molecules are considered and studied in this work. Using Mg-MOF-74 as a case study, with as few as 1000 adsorbed configurations with their reference energies computed by DFT, sophisticated DNN models can be established and are capable of accurately describing the adsorption energy of gases. We have also demonstrated the promising computational efficiency of such DNN potential in the calculations of adsorption properties such as Henry's constants in porous materials. The approach presented herein is expected to be also applicable for MOFs without open-metal sites or porous materials in general. Overall, this study is anticipated to pave the way towards the future development of highly accurate molecular potential for use in molecular simulations to accelerate materials discoveries.

Group Contribution and Machine Learning Approaches to Predict Abraham Solute Parameters, Solvation Free Energy, and Solvation Enthalpy

Yunsie Chung,[†] Florence H. Vermeire,[†] Haoyang Wu,[†] Pierre J. Walker,[†,‡] Michael H. Abraham,[¶] and William H. Green[*,†]

†*Department of Chemical Engineering, Massachusetts Institute of Technology, Cambridge, MA, 02139, U.S.A*

‡*Department of Chemical Engineering, Imperial College London, SW7 2AZ, United Kingdom*

¶*Department of Chemistry, University College London, 20 Gordon Street, London WC1H OAJ, United Kingdom*

E-mail: whgreen@mit.edu

Abstract

We present a group contribution method (SoluteGC) and a machine learning model (SoluteML) to predict the Abraham solute parameters, as well as a machine learning model (DirectML) to predict solvation free energy and enthalpy at 298 K. The proposed group contribution method uses atom-centered functional groups with corrections for ring and polycyclic strain whilst the machine learning models adopt a directed message passing neural network. The solute parameters predicted from SoluteGC and SoluteML are used to calculate solvation energy and enthalpy via linear free energy relationships. Extensive data sets containing 8366 solute parameters, 20253 solvation free energies, and 6322 solvation enthalpies are compiled in this work to train the models. The three models are each evaluated on the same test sets using both random and substructure-based solute splits for solvation energy and enthalpy predictions. The results show that the DirectML model is superior to the SoluteML and SoluteGC models for both predictions and can provide accuracy comparable to that of advanced quantum chemistry methods. Yet, even though the DirectML model performs better in general, all three models are useful for various purposes. Uncertain predicted values can be identified by comparing the 3 models, and when the 3 models are combined together, they can provide even more accurate predictions than any one of them individually. Finally, we present our compiled solute parameter, solvation energy, and solvation enthalpy databases (SoluteDB, dGsolvDBx, dHsolvDB) and provide public access to our final prediction models through a simple web-based tool, software package, and source code.

1 Introduction

Information on solvation free energy aids in the selection of viable solvents in chemical processes such as the synthesis of organic molecules,[1,2] optimization of purification processes,[3] and pollutant level management.[4] The solvation Gibbs free energy (ΔG_{solv}) of a solute in a solvent is directly related to that solute's partition coefficient between the gas and solvent phase. This property is typically reported at room temperature and can be a valuable feature for the prediction of the solute's liquid-liquid partition coefficient and solid solubility in organic solvents. For process optimization, ΔG_{solv} is required at the specified process temperature and in a variety of solvents. Recently, we reported a strategy to calculate ΔG_{solv} of a dilute neutral solute in organic solvents at different temperatures.[5] Using only the solvation free energy and solvation enthalpy at 298 K and solvent's temperature-dependent density, ΔG_{solv} for temperatures between 298 K and the solvent's critical temperature can be calculated along the solvent's saturation curve in a fast and automated manner. For this work, we aim to provide improved predictions of $\Delta G_{\mathrm{solv}}(298\,\mathrm{K})$ and $\Delta H_{\mathrm{solv}}(298\,\mathrm{K})$, which can be used for the calculation of ΔG_{solv} at elevated temperatures, an easy-access tool for our predictive models, and new databases for $\Delta G_{\mathrm{solv}}(298\,\mathrm{K})$ and $\Delta H_{\mathrm{solv}}(298\,\mathrm{K})$.

The interest in solvation free energies dates back many years[6] and has led to the development of numerous predictive methods. These range from molecular dynamics and quantum chemistry methods to empirical or data-driven approaches. Quantum chemistry methods distinguish themselves between explicit and implicit solvent representations. Commonly used quantum chemistry methods are based on the implicit polarizable continuum model for solvent representation, such as the SMx methods developed by Cramer, Truhlar, and coworkers[7–9] and the COSMO(-RS) models proposed by Klamt and coworkers.[10–12] These first-principle methods are useful for the calculation of solvation properties of new solvent-solute combinations, but they are computationally expensive and require comprehensive and challenging searches for all relevant conformers of the considered solvent and solute molecules. Empirical or data-driven approaches, on the other hand, allow for the fast prediction of solvation properties. The main bottleneck for these methods is the scarcity and quality of available experimental data. As more data become available, more studies focus on the application of empirical and data-driven models to the prediction of solvation-related properties such as ΔG_{solv}. In this study, we focus on improving predictions for $\Delta G_{\text{solv}}(298\,\text{K})$ and $\Delta H_{\text{solv}}(298\,\text{K})$ with several empirical and data-driven models, more specifically using the commonly employed linear solvation energy relationship (LSER), a group contribution method, and state-of-the-art graph-convolutional message passing neural networks.

The LSER equation used in this work is the one developed by Abraham et al.[13,14] The equation is built based on earlier attempts to relate the free energy of solvation to molecular descriptors by quantitative structure-property relationships (QSPR), for example the solvatochromic comparison method by Kamlet and Taft.[15] The relationship developed by Abraham and coworkers is given in Eq. 1. It combines Abraham solute (E, S, A, B, L) and solvent (c, e, s, a, b, l) parameters through a linear equation for the determination of the gas-liquid partition coefficient K. The Gibbs free energy of solvation can be directly calculated from the partition coefficient through Eq. 2.

$$\log_{10} K(298\,\text{K}) = c + eE + sS + aA + bB + lL \tag{1}$$

$$\Delta G_{\text{solv}} = -RT \ln K \tag{2}$$

The LSER is further explored by Mintz et al.[16] for the prediction of $\Delta H_{\text{solv}}(298\,\text{K})$ as shown in Eq. 3. The solute parameters in Eq. 3 are the same Abraham solute parameters as those used in the Abraham LSER in Eq. 1, while new solvent parameters (c', e', s', a', b', l') are used.

$$\frac{\Delta H_{\text{solv}}(298\,\text{K})}{1\,\text{kJ}\,\text{mol}^{-1}} = c' + e'E + s'S + a'A + b'B + l'L \tag{3}$$

Each of the solute parameters used in the LSER is related to the physical property of a solute and can be either determined experimentally or regressed from experimental values of partition coefficients. E is the solute excess molar refractivity, S is the solute dipolarity/polarizability, A and B are the overall hydrogen bond acidicity and basicity, respectively, and L is the logarithm of the gas-hexadecane partition coefficient.[14,17] The solute parameters can be also used to estimate liquid-liquid partition coefficients, solid solubilities, heat capacities, enthalpies of sublimation, and liquid phase hydrogen abstraction reaction rates.[18–21] The solvent parameters, on the other hand, are largely treated as empirical parameters and obtained from fitting to experimental data.

Several methods have been proposed to estimate some or all of the solute parameters from molecular structure,[22–30] among which the group contribution (GC) approach is widely used. Platts et al.[24,25] were the first to devise the group contribution scheme that can predict all solute parameters. They developed 81 functional group fragments for predicting E, S, B, and L and 51 fragments for predicting A. The ACD/Absolv software[31] adopted the Platts-type fragments and further optimized the fragments for their module. ACD/Absolv has shown to give reasonably good estimates for many compounds, but it had large prediction errors for certain classes of molecular structure, such as highly halogenated compounds, triazoles, and bridged ring structures.[32] Brown et al. developed a different set of fragments or substructures using the iterative fragment selection approach, in which the fragments are selected by using k-fold cross validation.[33,34] They used solute parameter data of around 3700 compounds and built an open-access GC model that is available through the UFZ-LSER database.[29] Their model showed good predictive performance for the L parameter,[34] but the performance on the other solute parameters has not been reported.

As more comprehensive databases become available, researchers have started exploring the use of machine learning (ML) for the prediction of $\Delta G_{\mathrm{solv}}(298\,\mathrm{K})$. Initially, these efforts focused on hydration free energies using the FreeSolv database.[35] This database is often used as a benchmark to compare model architectures in chemical property prediction.[36,37] The Minnesota solvation database (MNSol)[38] contains solvation free energies for a wide range of solvents. Hutchinson and Kobayashi[39] were the first to use this database to account for different solvents in a neural network by using the DeepChem framework[40] with functional class fingerprints that have solvent-specific features. More recent works have been published based on a larger database Solv@TUM.[41] Pathak et al.[42] proposed a chemically interpretable graph interaction network model comprised of a message passing, an interaction, and a prediction phase using 6239 unique solvent-solute combinations extracted from Solv@TUM and FreeSolv data sets. Lim and Jung[43] used the same data sets and developed MLSolvA, a ML architecture that computes pairwise atomic interactions from the solvent and solute atomistic feature vectors and makes prediction by summing up the interactions. As part of our previous work, Vermeire and Green[44] presented a transfer learning approach in which the model was pre-trained on 1 million quantum calculations and further fine-tuned on 10145 solvation free energy experimental data. The main purpose of that work was to demonstrate how transfer learning from quantum chemical data improves performance on small dataset sizes and on out-of-range sample predictions. Overall, the model achieves a mean absolute error of 0.21 kcal/mol on a random test split.

In this work, we adopt three different approaches to predict $\Delta G_{\mathrm{solv}}(298\,\mathrm{K})$ and $\Delta H_{\mathrm{solv}}(298\,\mathrm{K})$, two of which predict Abraham solute parameters and then calculate solvation properties using the LSERs. The third approach obtains $\Delta G_{\mathrm{solv}}(298\,\mathrm{K})$ and $\Delta H_{\mathrm{solv}}(298\,\mathrm{K})$ directly for the specified solvent-solute pair. For the prediction of $\Delta G_{\mathrm{solv}}(298\,\mathrm{K})$ we chose not to make use of our previously published transfer learning model, since the purpose of this work is to extensively compare different approaches to experimental data-driven methods. Extensive databases of the solute parameters, $\Delta G_{\mathrm{solv}}(298\,\mathrm{K})$, and $\Delta H_{\mathrm{solv}}(298\,\mathrm{K})$ have been compiled for this work and all models are trained on experimental data. The performance of the three approaches is assessed for solvent-solute combinations that are considered out-of-sample with respect to the training data. We specifically use the random and substructure-based solute splits such that none of the solutes or selected solute substructures in the test sets appear in the training sets. One of the novelties of this work is that the performance of all the methods is evaluated on exactly the same test sets for different splits. For the final prediction of solvation properties, we advise to combine all three methods as they provide a different level of accuracy and interpretation of the contribution of several physical phenomenon to the calculated values. We also provide an easy-accessible tool for our predictive models that can be used to calculate the solute parameters, $\Delta G_{\text{solv}}(298\,\text{K})$, and $\Delta H_{\text{solv}}(298\,\text{K})$. Lastly, we provide new databases for the solute parameters, $\Delta G_{\text{solv}}(298\,\text{K})$, and $\Delta H_{\text{solv}}(298\,\text{K})$ compiled and curated from different sources.

2 Methods – Databases

Three different approaches are used to predict $\Delta G_{\text{solv}}(298\,\text{K})$ and $\Delta H_{\text{solv}}(298\,\text{K})$: (1) a GC method for the solute parameters in the Abraham equation **(SoluteGC)**, (2) ML for the prediction of the same solute parameters **(SoluteML)**, and (3) ML for the direct prediction of solvation properties **(DirectML)**. The advantage of using these three different methods for the prediction of solvation properties is that they each provide a different level of information on the physical contributions to the calculated values. Whereas the DirectML method is a black-box, the SoluteML model provides solute parameters with physical meaning, and the SoluteGC model relates those parameters to chemical substructures.

The three approaches start from two different experimental data sets: one with solute molecules and their solute parameters **(SoluteDB)**, and one with solvent-solute pairs and values for $\Delta G_{\text{solv}}(298\,\text{K})$ and $\Delta H_{\text{solv}}(298\,\text{K})$ **(dGsolvDB*x*, and dHsolvDB)**. The SoluteGC and SoluteML methods predict the solute parameters from SMILES[45] of the solute compounds. The solvation free energy and enthalpy are subsequently computed from the predicted solute parameters and empirically fitted solvent parameters through the LSERs (Eqs. 1 and 3). The DirectML method takes SMILES of the solvents and solutes as input and predicts the solvation free energy and enthalpy. Figure 1 depicts the prediction flowchart of the three models starting from the two data sets. The details on the individual models and information on the construction and splitting of the two data sets are given below.

2.1 Data Collection and Compilation

The experimental data for the solute parameters (E, S, A, B, L), $\Delta G_{\text{solv}}(298\,\text{K})$, and $\Delta H_{\text{solv}}(298\,\text{K})$ are collected from various sources. All data except those collected from the Minnesota solvation database are open-source and published as a part of this work. The data are limited to neutral solute compounds containing H, C, N, O, S, P, F, Cl, Br, or I atoms and nonionic liquid solvents in this work. The standard state of 1 $\text{mol}\,\text{L}^{-1}$ gas phase and 1 $\text{mol}\,\text{L}^{-1}$ liquid phase is used for ΔG_{solv} and ΔH_{solv}.

The solute and solvent information in the collected databases are given in various representations, such as CAS numbers, InChI,[46] SMILES, chemical names, and 3D coordinates. The data are standardized by converting the given identifiers to both SMILES and InChI strings

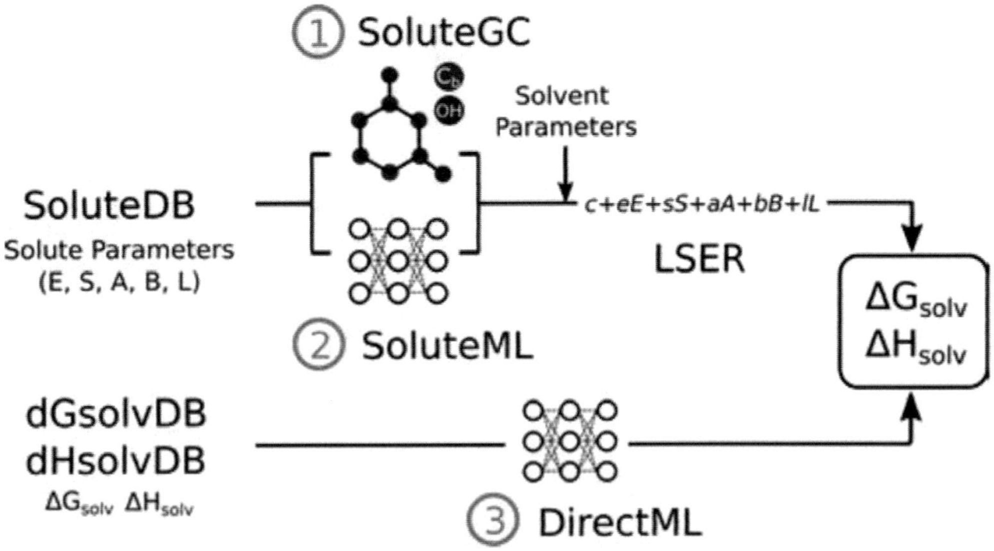

Figure 1: Schematic overview of $\Delta G_{\text{solv}}(298\,\text{K})$ and $\Delta H_{\text{solv}}(298\,\text{K})$ prediction with the three methods (SoluteGC, SoluteML, and DirectML) starting from the three data sets (SoluteDB, dGsolvDB, and dHsolvDB)

using PubChemPy,[47] CIRPy,[48] RDKit,[49] and ARC.[50] All available chemical identifiers are converted and the resulting InChI strings are compared to eliminate incorrect or ambiguous naming. The compounds with specified stereochemistry are converted to isomeric SMILES. The non-standard InChI containing a fixed-hydrogen layer is used instead of the standard InChI to distinguish tautomers. Those InChIs are used as unique identifiers to identify different solutes or solute-solvent pairs in the data sets. When multiple data points are found for the same solute or solute-solvent pair, the mean values of the Abraham solute parameters, $\Delta G_{\text{solv}}(298\,\text{K})$, or $\Delta H_{\text{solv}}(298\,\text{K})$ are used and the standard deviations are calculated. No data are removed based on the standard deviations as most of the data are found to have relatively low standard deviations.

Abraham solute parameters. The data statistics of the in-house Abraham solute parameter database (SoluteDB) are summarized in Table 1. The number of data points per parameter varies as some solutes have missing parameters. The number of Abraham and Mintz solvent parameters used in this study are provided in the table as well. These solvent parameters are empirically fitted based on the collected solute parameters, $\Delta G_{\text{solv}}(298\,\text{K})$, and $\Delta H_{\text{solv}}(298\,\text{K})$ data. The details on the fitting method, the fitted solvent parameters, and the molar weight distribution of the solutes in the data set can be found in the Supporting Information Sections 1 and 2.1.

Table 1: Data summary for the in-house Abraham solute parameter database (SoluteDB) and the number of fitted solvent parameters. The total number of data points (N total), mean values, standard deviations (std. dev.), and minimum and maximum (min, max) values are presented.

solute parameter	N total	mean	std. dev.	min	max
E	8163	0.97	0.81	-1.51	6.87
S	7654	1.14	0.78	-1.60	10.97
A	8159	0.21	0.35	0.00	6.80
B	7395	0.66	0.61	0.00	7.97
L	7038	6.77	3.79	-1.20	49.98
Total number of solutes				8366	
Number of solvents with fitted Abraham solvent parameters (e, s, a, b, l, c for ΔG_{solv})				195	
Number of solvents with fitted Mintz solvent parameters (e', s', a', b', l', c' for ΔH_{solv})				66	

Solvation free energy and enthalpy. A summary and analysis of the compiled solvation free energy and enthalpy data sets are given in Table 2. The solvation free energy data are acquired from the Minnesota solvation database (MNSol),[38] CompSol database,[51] FreeSolv database,[35] and published work by Abraham, Acree and coworkers[52–83] and compiled together as a dGsolvDB1 data set. FreeSolv database is obtained from MoleculeNet.[36] Additionally, we have nearly 4000 gas-water partition coefficient data ($\log K_{\mathrm{w}}$) from the in-house database. While adding these data significantly increases the size of the data set and the number of solute compounds considered, it also causes the solvation free energy data to be more heavily biased towards water as a solvent. Therefore, a separate data set, dGsolvDB2, is prepared that includes the dGsolvDB1 data and the aqueous solvation free energy data converted from the in-house $\log K_{\mathrm{w}}$ data set.

We also convert octanol-water partition coefficient data ($\log P_{\mathrm{ow}}$) from the in-house database and solvent-water partition coefficient ($\log P$) data from OCHEM,[84] DrugBank,[85] PHYSPROP,[86] and published work by Abraham, Acree, and coworkers[52,53,57–75,78–82,87] to solvation free energies. For the solute species with $\log K_{\mathrm{w}}$ and $\log P$ data available, the gas-solvent partition coefficient ($\log K$) can be calculated from Eq. 4.

$$\log(K) \approx \log(P) + \log(K_{\mathrm{w}}) \tag{4}$$

$\log K$ is subsequently converted to the solvation free energy through Eq. 2. Note that Eq. 4 assumes that $\log P$ is measured in dry solvents, *i.e.* the solvents do not dissolve into one another while in contact. This approximation would introduce an additional uncertainty to the data set, especially for polar solvents in contact with water. Moreover, including the in-house $\log P_{\mathrm{ow}}$ data causes the data set to be biased toward 1-octanol as a solvent. Thus, another separate data set, dGsolvDB3, is prepared that includes the dGsolvDB2 data and the solvation free energy data converted from the log P_{ow} and log P data. As Table 2 shows, the dGsolvDB3 data set has the most data but nearly a half of the solvents correspond to water and 1-octanol. All three solvation energy data sets (dGsolvDB1 - dGsolvDB3) are used separately to build DirectML models, and the test set errors are compared to determine which data set gives the best prediction.

Solvation enthalpy data are collected from Acree Enthalpy of Solvation data set[88] and CompSol database.[51] Self-solvation data (solvation of a compound in itself) are included in both solvation energy and enthalpy data sets. It can be seen from Table 2 that the majority of the solvents in the data sets only have a single entry that corresponds to the self-solvation datum in the final database. As a result, if self-solvation data are excluded, the number of solvents would reduce to 302 and 142 solvents in the solvation free energy and enthalpy data sets, respectively. The molar weight distribution of the solutes in each data set is provided in the Supporting Information Section 2.1.

Table 2: Summary of the data sets for solvation free energy (dGsolvDB) and enthalpy (dHsolvDB), including the total number of data points (N total), the number of solutes (N solutes), the number of solvents with and without solvents that only appear in self-solvation (N solvents), and a list of the most commonly found solvents. (CCl4: carbon tetrachloride, DMF: dimethylformamide.)

data set (included data)	N total	N solutes	N solvents (excl. self-solv)	top 5 solvents (% data)
dGsolvDB1 (ΔG_{solv}[35,38,51–83])	12202	2387	1459 (302)	water (11.6 %) 1-octanol (3.2 %) hexadecane (2.1 %) heptane (1.7 %) hexane (1.7 %)
dGsolvDB2 (ΔG_{solv},[35,38,51–83] in-house log K_w data)	16180	5991	1459 (302)	water (33.3 %) 1-octanol (2.4 %) hexadecane (1.6 %) heptane (1.3 %) hexane (1.3 %)
dGsolvDB3 (ΔG_{solv},[35,38,51–83] in-house log K_w data, in-house log P_{ow} data, log P[52,53,57–75,78–82,84–87])	20253	5991	1459 (303)	water (26.6 %) 1-octanol (21.1 %) hexadecane (1.3 %) ethanol (1.1 %) heptane (1.1 %)
dHsolvDB (ΔH_{solv}[51,88])	6322	1665	1432 (142)	cyclohexane (3.5 %) methanol (3.2 %) benzene (3.1 %) CCl4 (2.8 %) DMF (2.8 %)

Comparison of the LSER estimates and experimental data. Based on the collected data sets, the errors associated with the LSER are evaluated prior to the construction of the prediction models. Solvation free energies and enthalpies are calculated from the Abraham and Mintz LSERs using the experimental solute parameters (SoluteDB) and the fitted solvent parameters, and the calculated values are compared to the $\Delta G_{\text{solv}}(298\,\text{K})$ and $\Delta H_{\text{solv}}(298\,\text{K})$ data sets listed in Table 2. The parity plots and errors are shown in Figure 2. It can be seen that the LSER has small $\Delta G_{\text{solv}}(298\,\text{K})$ error and relatively higher $\Delta H_{\text{solv}}(298\,\text{K})$ error. These errors are due to both the aleatoric uncertainty of the experimental data and the fact that the linearity of the LSER cannot capture the non-linear relationship between the solute parameters and solvation energy/enthalpy. Since the LSER estimates of both solvation free energy and enthalpy in Figure 2 are calculated using the same experimental solute parameters, it is likely that the model error associated with the Mintz LSER (Eq. 3) is higher than that of the Abraham LSER (Eq. 1). It is also possible that the solvation enthalpy data (dHsolvDB) have higher experimental uncertainty than the solvation energy data (dGsolvDBx), but we are unable to assess the experimental uncertainties of the data sets used in this work as the majority of solvent-solute pairs only have a single data point. The SoluteGC and SoluteML models that predict $\Delta G_{\text{solv}}(298\,\text{K})$ and $\Delta H_{\text{solv}}(298\,\text{K})$ through the LSER will, at their best performance, have these underlying errors when compared to the dGsolvDBx and dHsolvDB data sets.

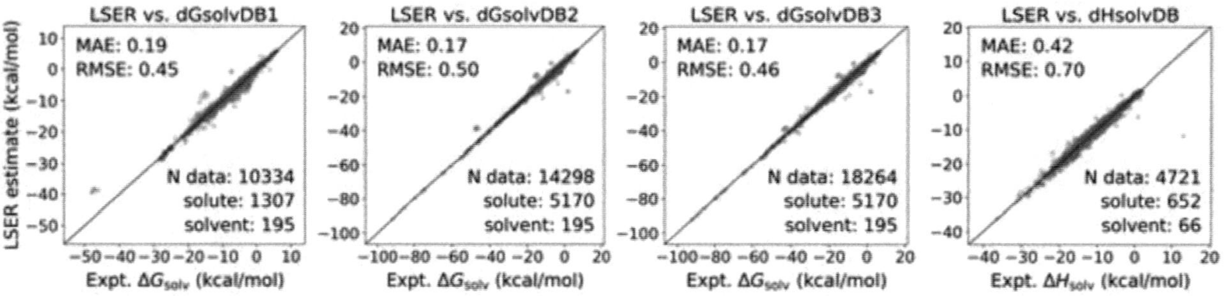

Figure 2: Comparison between the experimental $\Delta G_{\text{solv}}(298\,\text{K})$ and $\Delta H_{\text{solv}}(298\,\text{K})$ data and the estimations from the LSER. The LSER estimations are calculated using the Abraham solute parameter database and the fitted solvent parameters. The experimental data from dGsolvDB1, dGsolvDB2, dGsolvDB3, and dHsolvDB are compared to the LSER estimates. The mean absolute error (MAE) and root mean square error (RMSE) are reported in kcal/mol.

2.2 Data Split for Model Comparison

As summarized in Tables 1 and 2, a total of four data sets are prepared for the prediction of solvation free energy: (1) SoluteDB, (2) dGsolvDB1, (3) dGsolvDB2, and (4) dGsolvDB3. The final models that are available through the easy-access tool are trained on all of the data available. However, to evaluate and compare the performance of the models, each data set is split into a ~90 % training/validation and a ~10 % carefully selected test set. The test set is constructed such that the solute compounds from the test set do not overlap with the training and validation set. For example, if compound A is chosen as a test set solute, all solvent-solute pairs that have compound A as a solute are included in the test sets and excluded from the training and validation sets. The test set solutes are selected in a (1) random and (2) substructure-based manner while maintaining a ~90/10 % data split. The substructure-based splits are employed to test the out-of-range performance of the models on new classes of solute molecules. All solute compounds that contain any of the selected functional groups and scaffolds are included in the test set for our substructure splits. Those splits are comparable to the Murcko scaffold-based splits that are commonly used in molecular property predictive models for drug discovery.[36,37] The substructures are manually selected to maintain a ~90/10 % split for all data sets. The substructures are represented in SMARTS,[89] and substructure search on solute compounds is done using RDKit. Examples of chosen substructures are benzoic acid, adamantane and phenanthrene scaffolds, and a trifluoromethyl group. The list of all substructures used to split the data sets is presented in the Supporting Information Section 2.2. The data sets used for the prediction of solvation enthalpy (SoluteDB and dHsolvDB) are split in the same fashion.

The training and testing sets we have prepared so far are designed for evaluating the model performance on unseen solutes. To investigate the performance of the DirectML models on unseen solute and unseen solvent pairs, 10 and 8 solvents are removed from the training sets of dGsolvDB and dHsolvDB, respectively, and placed in the test sets. The solvents with Abraham and Mintz solvent parameters are selected to allow the solvent-wise error comparison with the SoluteGC and SoluteML models. Moreover, the solvents are selected such that the number of solutes in each training set remains unchanged even after the chosen solvent data are excluded from the training set. The list of excluded solvents is provided in the Supporting Information Section 2.3, and detailed data statistics of all training and test sets used in this study is presented in the Supporting Information Section 2.4.

3 Methods – Models

3.1 SoluteGC: Group Contribution Method for Abraham Solute Parameter Prediction

The SoluteGC model is built as part of Reaction Mechanism Generator (RMG),[90,91] open source chemical kinetic modeling software package. RMG uses the Benson-type[92] group contribution method for gas phase thermochemistry estimations and has over 2000 groups in the database. For the SoluteGC model, RMG's gas phase group contribution scheme is adopted to include the Abraham solute parameters, and missing groups that are important to the solvation data sets are added.

RMG's GC method estimates thermochemistry by dividing a molecule into atom-centered (AC) functional groups and summing the contribution from all groups. Additionally, RMG implements ring strain correction (RSC) and long distance interaction (LDI) groups to account for more advanced structural effects that cannot be captured by the atom-based approach. The SoluteGC model follows the same scheme to calculate the solute parameters as shown in Eq. 5

$$E, S, A, B, \text{ or } L = \sum_{i=1}^{N_{\text{atom}}} \text{AC}_i + \sum_{j=1} \text{RSC}_j + \sum_{k=1} \text{LDI}_k \tag{5}$$

where N_{atom} is the number of heavy atoms in a molecule and RSC and LDI corrections are applied for each ring cluster and long distance interaction group found in a molecule, respectively. For more details on the GC scheme, the reader is referred to the dedicated work by Gao et al.[90] and Han et al.[93] as well as the RMG documentation.[94]

While most molecules follow Eq. 5, halogenated molecules are treated differently in the SoluteGC model: all halogen atoms are first replaced by hydrogen atoms, then the GC estimate is made on the replaced structure, and halogen corrections are lastly added for each halogen atom to get the final GC prediction as shown in Eq. 6

$$\begin{aligned} E, S, A, B, \text{ or } L &= \sum_{i=1}^{N_{\text{atom}*}} \text{AC}_i + \sum_{j=1} \text{RSC}_j + \sum_{k=1} \text{LDI}_k + \sum_{l=1}^{N_{\text{halogen}}} \text{Halogen}_l \\ &= (E, S, A, B, \text{ or } L)_{\text{replaced compound}} + \sum_{l=1}^{N_{\text{halogen}}} \text{Halogen}_l \end{aligned} \tag{6}$$

where the subscript 'replaced compound' denotes the compound whose halogen atoms are replaced by hydrogen atoms, and $N_{\text{atom}*}$ and N_{halogen} represent the number of non-halogen heavy atoms and the number of halogen atoms in a molecule, respectively. The halogen groups are halogen-centered functional groups and defined by neighboring atoms including other halogens that are bonded to the same atom. We take this unique approach because it allows one to use the experimental solute parameter data of a replaced compound and simply apply halogen corrections to get more accurate estimates for a halogenated compound. While this approach is inspired by various works,[95,96] it is primarily based on the hydrogen bond increment (HBI) method devised by Lay et al.,[95] in which a radical correction is applied to a saturated compound datum to get a thermochemistry estimate for a radical compound.

Once the types of functional groups to be considered and the GC relationships are established (Eqs. 5 and 6), a set of functional groups relevant to our data set needs to be created before fitting the associated group values. The functional groups for the SoluteGC model are constructed largely based on the existing groups within RMG. The groups that do not appear in the Abraham solute parameter database (SoluteDB) are removed, and new AC, RSC, and halogen groups that frequently appear in the database but are missing from RMG, mainly halogen, phosphorus, and heterocyclic groups, are created. The number of groups used for the final SoluteGC model is listed in Table 3.

Table 3: Description and the number of groups used for the SoluteGC model.

Group Category	Number of Groups	Description
AC-regular	729	Atom-centered functional groups. Applied for each non-halogen heavy atom
AC-halogen	193	Halogen correction (halogen-centered functional groups)
RSC-ring	116	Monocyclic ring strain correction
RSC-polycyclic	179	Polycyclic ring strain correction
LDI-cyclic	29	Aromatic ortho, meta, para correction
LDI-noncyclic	16	Gauche interaction correction

Finally, the group values are fitted to the experimental data using the ridge regression method from Scikit-learn package.[97] Ridge regression is chosen as a fitting method because it gave overall the lowest error on our test sets compared to other regression methods such as ordinary least squares, lasso, and elastic net. Ridge regression[98] is a linear least squares regression with L2 regularization and can help prevent overfitting. The hyperparameter of the regression is tuned using 10-fold cross-validation on each training set prepared from SoluteDB in the earlier section. The final SoluteGC model implemented in RMG is fitted using the entire data set, omitting the very small molecules where GC approach is unsuitable. All molecules with more than two heavy atoms are used to fit the group values. The group definitions and the number of data used to fit each group can be found on GitHub as part of the RMG-database at `https://github.com/ReactionMechanismGenerator/RMG-database/tree/master/input/solvation/groups`. Detailed information on how the groups are defined and organized in the RMG-database is given in the Supporting Information Section 3.1.

3.2 SoluteML and DirectML: Machine Learning for Solute Parameters, Solvation Free Energy, and Enthalpy Prediction

Two machine learning models are developed for the prediction of solvation free energy and enthalpy. The first deep neural network ensemble (SoluteML) is trained on the database with the Abraham solute parameters (SoluteDB). The average prediction of this ensemble of neural networks is combined with the solvent parameters through the LSER (Eqs. 1 and 3) to calculate $\Delta G_{\mathrm{solv}}(298\,\mathrm{K})$ and $\Delta H_{\mathrm{solv}}(298\,\mathrm{K})$. The second ensemble of deep neural networks (DirectML) is trained on dGsolvDB and dHsolvDB and used to predict the solvation properties directly. For both machine learning models, the final models are trained on the complete data sets. For the purpose of comparing model performance in the remainder of this work, the models are trained on the different splits (see Section 2.2).

The model architecture for both models used in this work are based on the state-of-the-art chemical property prediction software Chemprop.[37] This software uses a directed message passing neural network (D-MPNN), a type of graph convolutional neural network, to convert atom and bond feature vectors to a molecular latent representation. The embedded molecule is subsequently sent through a second neural network for the property prediction task. For more details on the D-MPNN, the reader is referred to the dedicated work by Yang et al.[37] The software package can be found on GitHub (`https://github.com/fhvermei/chemprop_solvation`). Specifics related to this work are discussed below.

For the SoluteML model, the training set consists of solute SMILES as input and their corresponding solute parameters as output to the neural network. Minor adjustments are made compared to the original version of Chemprop to include atom, bond, and molecular features specific to solvation. A summary of the used atom and bond features is given in Table 4. Different sets of additional molecular descriptors are tested for training the neural network. These are concatenated with the molecular latent representation generated by the D-MPNN. We find that using all the available 2D-RDKit molecular descriptors (a vector of 200 features automatically generated by RDKit[37]) yields overall the best performance for the SoluteML model, and therefore they are chosen as the additional molecular features.

For the DirectML model, the original version of Chemprop is adapted to the application of multiple input molecules, a solvent-solute pair. The solvent and solute SMILES strings are each converted to a latent representation by a separate D-MPNN, after which the embeddings are concatenated prior to the second neural network for property prediction. The same atom and bond features are used for the D-MPNN as for the SoluteML model (see Table 4). Similar to SoluteML, different sets of additional molecular descriptors are tested to improve the performance of the model. In this case, the selected molecular features are the RDKit-calculated octanol-water partitioning coefficient and Van der Waals Surface Area (VSA) combined in RDKit's `SlogP_VSA_` descriptor. The molecular feature vectors for a solute and a solvent are concatenated with the latent representation after the D-MPNN and prior to the second neutral network for property prediction.

Different hyperparameters are used for the SoluteML and the DirectML models. For each model, the hyperparameters are selected based on a search algorithm that includes optimization of the neural network for different hyperparameter combinations. The search procedure combines the algorithms available in the software package Hyperopt[99,100] with manual intervention to balance the size of the neural networks with the gain in accuracy. The hyperparameters that are optimized include the depth and hidden size of the D-MPNN, the number of layers and hidden size of the neural network for property prediction, the learning rates, and the batch size. An overview of the selected hyperparameters for the different models is given in the Supporting Information Section 3.2.

The models are trained with 5-fold cross-validation. For each of the different splits, the test set is pre-defined as detailed in Section 2.2. During the 5-fold cross-validation, a different

Table 4: Atom and bond features used for the directed message passing neural network.

Atom features	Type	Bond features	Type
Atomic number	One-hot	Bond type	One-hot
Total neighbor number	One-hot	Conjugation	One-hot
Formal charge	One-hot	In ring	One-hot
Connected hydrogen atoms	One-hot	Stereo	One-hot
Hybridization	One-hot		
Lone pairs	One-hot		
H-bond donor	One-hot		
H-bond acceptor	One-hot		
Ring size	One-hot		
Aromaticity	One-hot		
Electronegativity	Binary		
Molar mass	Binary		

random validation set of 10 % is generated and distinguished from the training set to determine at which iteration (or epoch) to stop training the model. Within each fold, an ensemble of 5 different models is generated by Glorot initialization of the model parameters[101] with different seeds. All models are used to make predictions for the pre-defined test set. The predictions of each individual model are averaged to calculate the prediction of the model ensemble. Additionally, the variance on the predictions of the individual models is used as a measure for the model uncertainty (*i.e.* epistemic uncertainty) on the individual data points.

3.3 Prediction Using Existing Methods

The performance of our models is compared to the following quantum chemistry (QM), ML, and GC methods from literature: SMD,[9] COSMO-RS,[12] the solvation free energy ML model by Lim and Jung (MLSolvA),[43] the transfer learning model by Vermeire and Green[44] (transfer learning), the solvation enthalpy ML model by Jacquis et al.,[102] and the solute parameter GC method from the UFZ-LSER database (UFZ-LSER).[29] The COSMO-RS calculations are performed in-house at the BP86/TZVPD-FINE level of theory using the software COSMO*therm*.[103] These calculations are done for the solvent-solute pairs whose pre-calculated quantum chemical COSMO data are available in the COSMO*base* database.[104] Note that no additional quantum chemical calculations are done in this work. The solvation enthalpy at 298 K is computed by calculating solvation free energies at 297, 298, 299 K, estimating the temperature gradient at 298 K from the three data points, and using the relationship $\Delta H = \Delta G - T\frac{\mathrm{d}\Delta G}{\mathrm{d}T}$. The GC method from the UFZ-LSER database[29] is used to predict the solute parameters. These predicted parameters are combined with our in-house solvent parameters to compute the prediction errors of the UFZ-LSER GC method. Since the training data used for the regression of the solute parameters in the UFZ-LSER database are unknown, the errors are evaluated using all of our data. Only the solute molecules with more than two heavy atoms are used to evaluate the UFZ-LSER GC method as the GC method is not suitable for small molecules. For the remaining methods, SMD[9] and the ML models,[43,44,102] the reported errors from literature are used for comparison. Note that these errors are reported on a different test set than the one used in this work.

4 Results and Discussion

In the subsequent section, we evaluate the performance of the three models (SoluteGC, SoluteML, and DirectML) on 10 % test sets for both random and substructure-based solute splits. The comparison is done for the test splits with unseen solutes, and additionally for the DirectML model, on unseen solute and unseen solvent pairs. Because the size of the test set is different for each model, only overlapping test data of the compared models are considered in our comparison. For the comparison of the three models, the test solvents are limited to those with Abraham or Mintz solvent parameters since the SoluteGC and SoluteML models can be evaluated on only those solvents (see Table 1). The results on the entire test set data of each model can be found in the Supporting Information Section 4, which includes the results on the solvents without the solvent parameters for the test sets of the DirectML model. Furthermore, only the prediction of the solvation free energy and enthalpy is discussed in the main text. The prediction results on the individual solute parameters are presented in the Supporting Information Section 4. At last, the performance of our models is compared to that of the existing quantum chemistry, ML, and GC models.

The model performance is analyzed by comparing parity plots, the mean absolute error (MAE), and the root-mean-square error (RMSE), both in kcal/mol. Because the scale of the solvation free energies and their errors differ between some of the compared test sets, we also compare the relative error using the percent MAE (PMAE) defined as:

$$\text{PMAE} = \left| \frac{\text{MAE}}{\text{test mean}} \right| \cdot 100\ \% \tag{7}$$

The test mean represents the average experimental value in the test set. PMAE is used instead of the mean relative error, which is a mean of the individual absolute errors divided by their experimental values, because several solvation free energy and enthalpy have experimental values close to zero. This leads to inflation of the mean relative error, which would not be representative of the model performance.

4.1 Comparison of the DirectML Models Trained on Different Solvation Free Energy Data Sets

Before comparing the performance of the different models (SoluteGC, SoluteML, and DirectML), we start by comparing the performance of the DirectML models that are trained and validated on three different solvation free energy data sets (dGsolvDB1, dGsolvDB2, dGsolvDB3) to determine which model gives the best prediction. As explained in Section 2.1 and summarized in Table 2, the data sets differ in what kind of solvation data is included (ΔG_{solv}, in-house $\log K_{\text{w}}$ data, and $\log P$ data). The size of the data set increases from dGsolvDB1 to dGsolvDB3, including more solutes, but the solvents become more skewed towards water and 1-octanol as $\log K_{\text{w}}$ and $\log P$ data are added. To select the best performing model, in this section, only overlapping test data between the three DirectML models trained with the three data sets are compared. Because each data set is the subset of the next larger data set (dGsolvDB1 $\subset$ dGsolvDB2 $\subset$ dGsolvDB3), the overlapping test data only include the ΔG_{solv} data from dGsolvDB1 and do not include in-house $\log K_{\text{w}}$ and $\log P$ data. The results on the entire test set of each model are given in the Supporting Information Section 4.1.

The test errors of the three models on the overlapping test set data are shown on the parity plots in Figure 3. For a random split, the performance of the three models is very similar. The model trained on dGsolvDB3 gives the lowest MAE of 0.29 kcal/mol, which is only 0.02 kcal/mol lower than the model trained on dGsolvDB1. The difference between the models is a bit more pronounced in the substructure split. The model trained on dGsolvDB3 gives the lowest MAE of 0.81 kcal/mol, which is 0.06 kcal/mol lower than the model trained on dGsolvDB1. The result indicates that adding both $\log K_{\text{w}}$ and $\log P$ data to the training and validation data set slightly improves solvation energy predictions compared to training and validating the models on only solvation free energy data. Yet, considering that the total number of data and the number of solutes in dGsolvDB3 are nearly twice as many as those in dGsolvDB1, the improvement is not as significant as expected. In the Supporting Information Figure S3, a similar comparison is done using the complete (10 %) test set of each respective data set rather than only the overlapping data. In this case, the model trained, validated, and tested on dGsolvDB1 gives the lowest MAE of 0.31 kcal/mol compared to 0.40 kcal/mol for the model trained, validated, and tested on dGsolvDB3 for a random split. For the substructure split, the difference is less significant with the lowest MAE of 0.87 kcal/mol for dGsolvDB1 closely followed by 0.89 kcal/mol for dGsolvDB3. If the PMAE is compared instead, the dGsolvDB1 and dGsolvDB3 models both have the lowest PMAE of 4.9 % for the random split and the dGsolvDB3 model has the lowest PMAE of 9.2 % for the substructure split compared to the PMAE of 10.1 % for the dGsolvDB1 model.

These observations are in line with the conclusions made in earlier work by Vermeire and Green.[44] The model performance improves with an increasing amount of data in the training and validation set; however, the extent to which the model performance can be assessed is limited by the experimental (or aleatoric) uncertainty in the test set. For the same test set, the performance of the three models on a random split is similar since the aleatoric limit of assessing the performance on that test set is reached. For the substructure split, there is still a slight improvement in performance observed by the addition of more data to the training and validation set. For the comparison against the complete (10 %) test set of each data set in the Supporting Information, it is expected that the test splits of the data

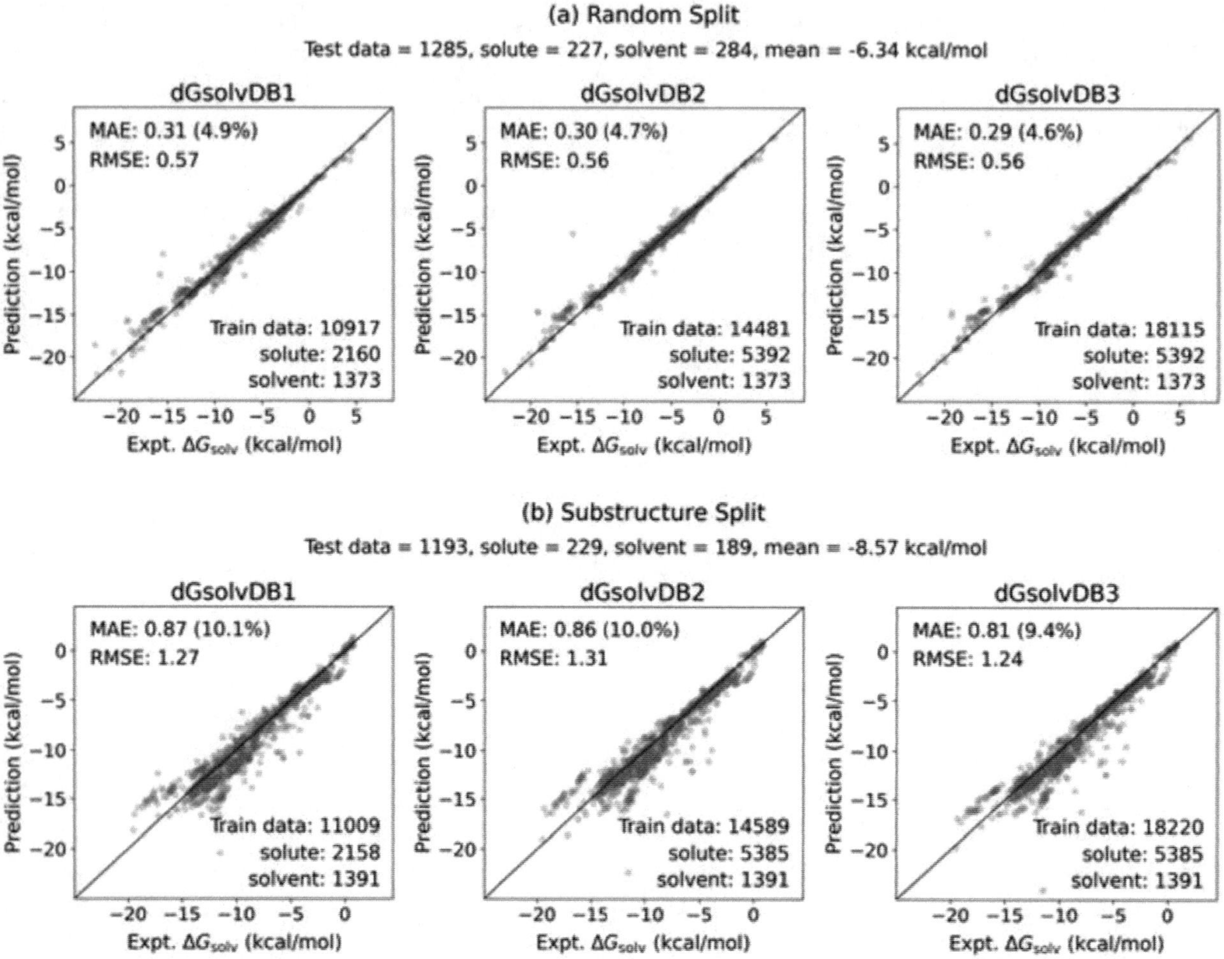

Figure 3: Parity plots of the DirectML models trained and validated using 3 different solvation free energy data sets (dGsolvDB1, dGsolvDB2, and dGsolvDB3) that differ in the type and amount of solvation data considered. The plots only show the overlapping test data in each split type. The MAE and RMSE are in kcal/mol, and the PMAE is given in parenthesis. Information on the training and test data are given on the figures together with the overall errors. Note that the test sets are prepared using random and substructure-based solute splits, and therefore none of the test set solutes appear in the training and validation sets.

sets containing $\log K_w$ and $\log P$ data have a higher associated experimental uncertainty. The experimental uncertainty in solvation free energies of neutral compounds is typically estimated as 0.2 kcal/mol,[105,106] but the solvation free energies measured in water ($\log K_w$) often have higher uncertainties,[107] which we believe are due to larger magnitudes of the values and disagreements between data reported by different sources. The conversion of $\log P$ data to solvation free energies also comes along with an additional uncertainty because of the assumption that solvents are not in contact with one another. As a result, the model trained and validated on dGsolvDB1 seems to have better performance if we compare the MAE evaluated on the complete test set. However, this conclusion cannot be directly drawn because of (1) the additional uncertainty in the other test sets and (2) the different range of the magnitudes of solvation free energies in the test set data. The latter is compensated for by comparing the PMAE of the different models instead of the MAE or RMSE.

Overall, including more data in the training and validation set is beneficial as it lowers the prediction error when assessed on the overlapping test set data, and thus dGsolvDB3 is chosen as the optimal solvation free energy data set for this work and comparison to other methods. We also compare the solvent-wise test errors of the three DirectML models in the Supporting Information Figure S4, and it is found that having data bias towards two opposing solvents like water and 1-octanol does not cause any particularly high errors in other solvents. However, since the accuracy of the three models is not significantly different, it is possible that different data splits or data sizes can yield contrasting results.

4.2 Comparison of the Three Prediction Approaches

4.2.1 Performance on Unseen Solutes

The test set errors for the solvation free energies $\Delta G_{\mathrm{solv}}(298\,\mathrm{K})$ predicted by the SoluteGC, SoluteML, and DirectML models are presented in Figure 4. The three different models are tested on unseen solutes, and only overlapping test data of the three models are compared in the figure. All solvents in the overlapping test data appear in the training set of the DirectML model, and the SoluteGC and SoluteML models use the empirical solvent parameters, and therefore, the results shown in Figure 4 can be considered as the predictive performance on pairs of unseen solutes and trained solvents. The DirectML model is trained and validated using the data set dGsolvDB3 (including ΔG_{solv} data, in-house $\log K_{\mathrm{w}}$ data, and $\log P$ data), which is selected based on earlier results from Section 4.1. Note that the overlapping test data in Figure 4 are different from those in Figure 3, and therefore the test errors of the DirectML model from the two figures are slightly different each other.

For both the random and substructure solute splits, the DirectML model achieves the best predictions and the SoluteGC model gives the highest error. While the DirectML model performs better than the SoluteML model, there is no significant difference; the MAE differs less than 0.1 kcal/mol for both splits. The SoluteML and DirectML models have likely reached close to the aleatoric limit of the experimental data in the random split, and we believe that the relative underperformance of the SoluteML model in comparison to the DirectML model is due to the inherent error caused by the linearity the LSER as discussed in Section 2.1. Even though the SoluteML model has about 2100 more solutes in the training set than the DirectML model has, the additional solute data are not able to compensate for the underlying error.

The three models are compared for the prediction of solvation enthalpies $\Delta H_{\mathrm{solv}}(298\,\mathrm{K})$ on

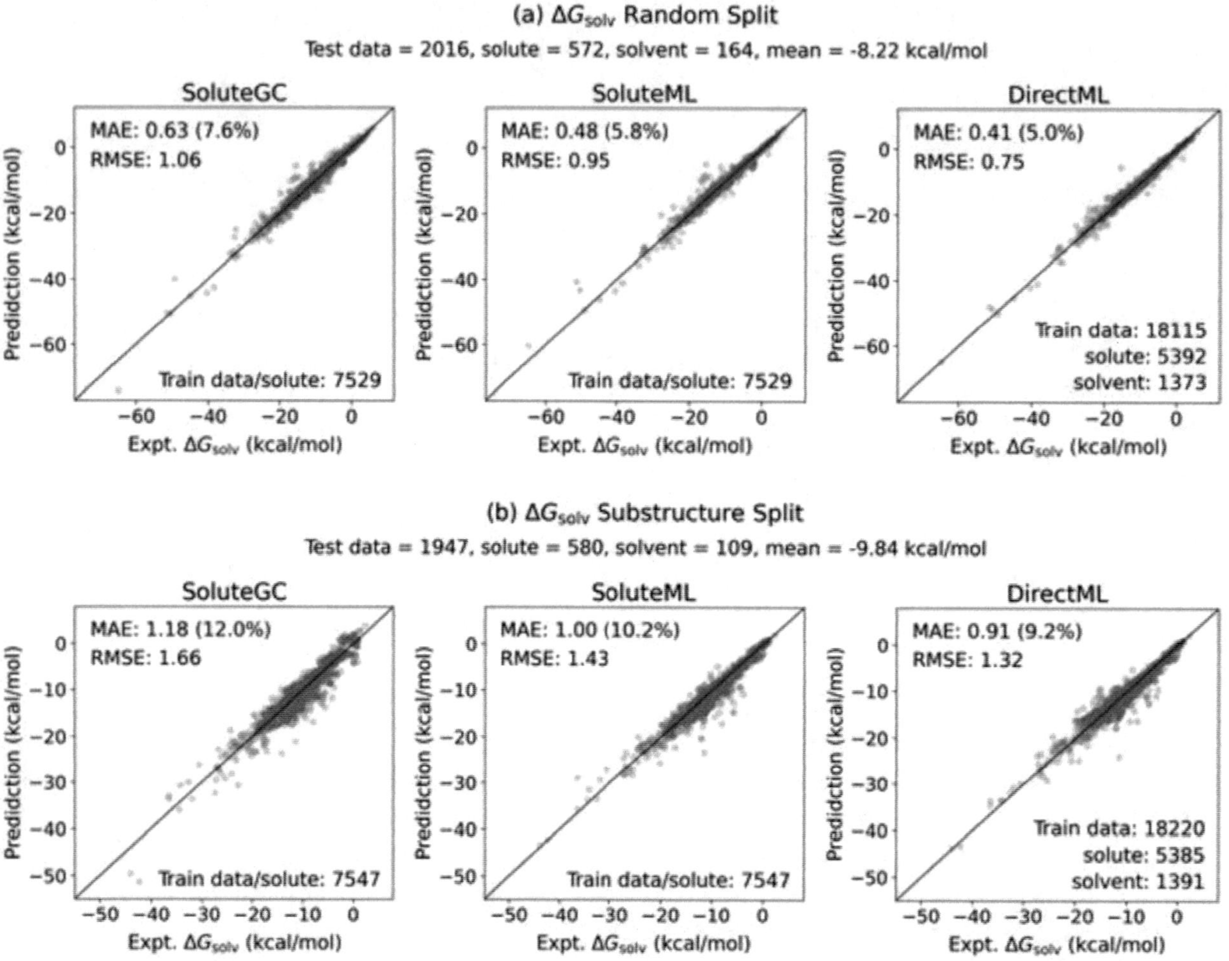

Figure 4: Parity plots for experimental and predicted $\Delta G_{\text{solv}}(298\,\text{K})$ on the 10 % test sets. The plots only show the overlapping test data in each split type. The three different models are tested on pairs of unseen solutes and trained solvents for the random and substructure solute splits. The MAE and RMSE are in kcal/mol, and the PMAE is given in parenthesis. Information on the training and test data are given on the figures together with the overall errors.

unseen solutes in Figure 5. Again, all solvents in the overlapping test data appear in the training set of the DirectML model, and hence, the figure reflects the results on pairs of unseen solutes and trained solvents. The DirectML model is trained using much fewer data that contain only about 1500 solutes due to the limited availability of data on solvation enthalpies. Nevertheless, the DirectML model achieves a similar accuracy as the SoluteML model on a random split and outperforms the SoluteML model on the substructure split.

The MAE of each substructure used for the substructure splits is compared for the three models in Figure 6. The SMARTS strings and drawings of the substructures can be found in Section 2.2 of the Supporting Information. The three models have errors of similar magnitudes

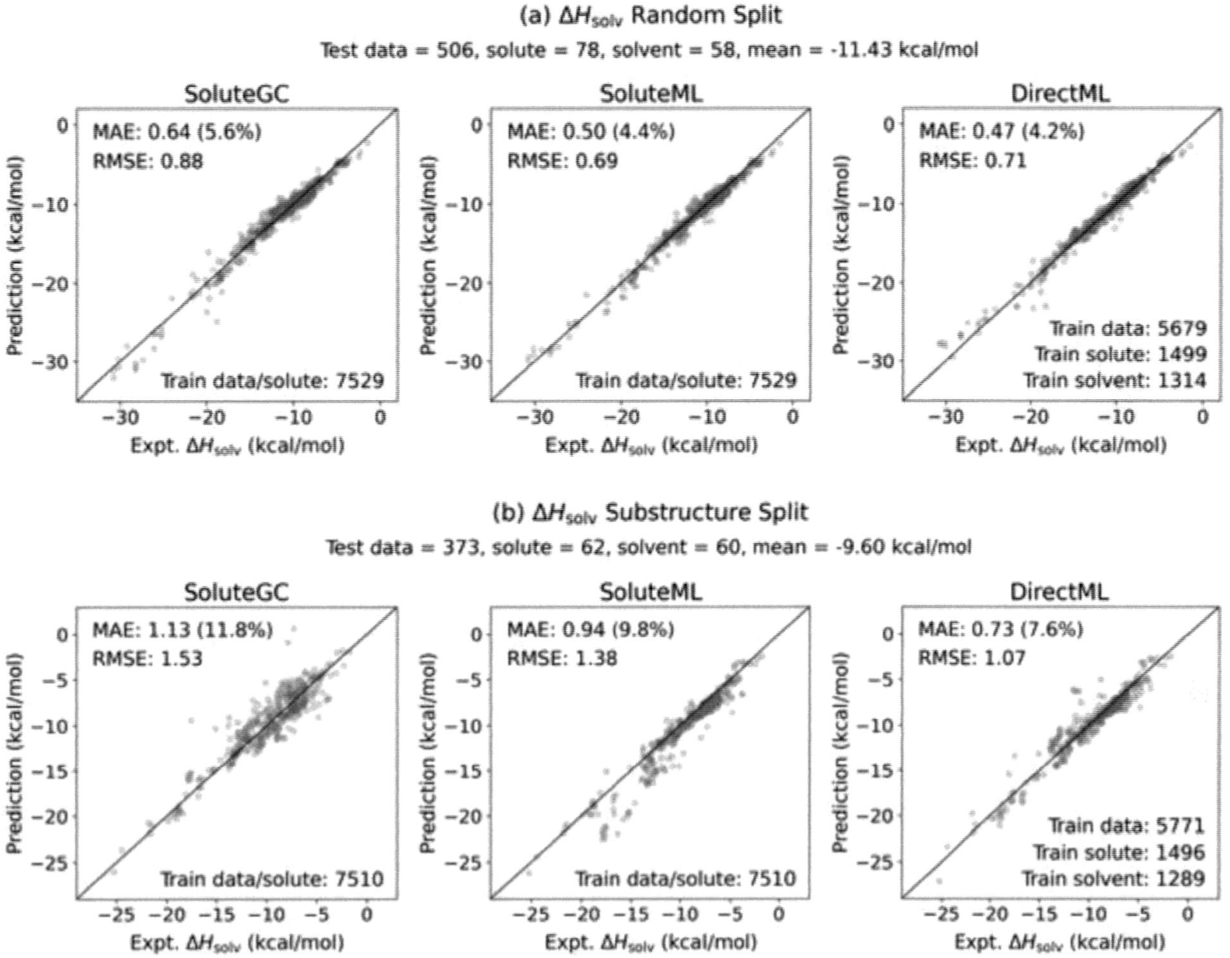

Figure 5: Parity plots for experimental and predicted $\Delta H_{\mathrm{solv}}(298\,\mathrm{K})$ on the 10 % test sets. The plots only show the overlapping test data in each split type. The three different models are tested on pairs of unseen solutes and trained solvents for the random and substructure splits. The MAE and RMSE are in kcal/mol, and the PMAE is given in parenthesis. Information on the training and test data are given on the figures together with the overall errors.

for many substructures, but they have very different levels of accuracy for some substructures. For example, for the $\Delta G_{\mathrm{solv}}(298\,\mathrm{K})$ predictions, the solutes containing methanesulfonamide functional group (SMARTS: NS(=O)(=O)C) are the main outliers of the SoluteGC and SoluteML models whereas the solutes with cyclopentene scaffold (C1=CCCC1) give the highest error for the DirectML model. The difference between the key outliers is more pronounced for the $\Delta H_{\mathrm{solv}}(298\,\mathrm{K})$ predictions; here, the SoluteML model has much higher MAE than the other two models for the solutes containing adamantane scaffold (C1C2CC3CC1CC(C2)C3) while it has much lower error for imidazole scaffold (c1c[n&H1]cn1), which is the main outlier of the other two models.

In the random splits, the common outliers of the three models are cyclic or polycyclic so-

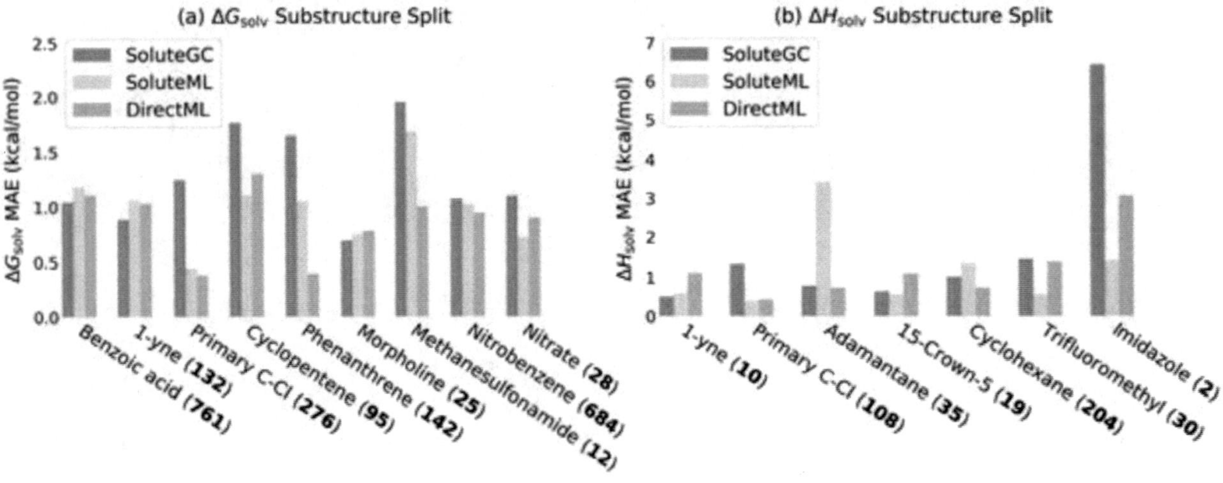

Figure 6: The MAE of each substructure for the prediction of $\Delta G_{\text{solv}}(298\,\text{K})$ and $\Delta H_{\text{solv}}(298\,\text{K})$ in the substructure split. Only the overlapping test data of the three models in each split type are compared in the plots. The number of solute data containing the corresponding substructure in each test set is given in parenthesis in bold.

lute compounds such as guanabenz, galantamine, methotrexate, and colchicine for ΔG_{solv} predictions and salicylamide, pyrene, 1,4-diphenylbenzene, and benzo-15-crown-5 for ΔH_{solv} predictions. However, the majority of the main outliers are found to be different for each model.

The solvent-wise errors of the three models are compared for the 30 most frequently appearing solvents in each test set as shown in Figure 7. Note that the figure shows the results on the unseen solutes in the trained solvents. For the majority of the solvents for $\Delta G_{\text{solv}}(298\,\text{K})$ prediction, the DirectML model performs similar to or better than the SoluteML model, and the SoluteGC model gives the highest errors. In the random split set of $\Delta G_{\text{solv}}(298\,\text{K})$, the SoluteML model performs better than the DirectML model for non-polar solvents such as alkanes, carbon tetrachloride (CCl4), isooctane, and p-xylene. However, the DirectML model gives the best predictions for the same non-polar solvents in the substructure split set. Similarly, the SoluteGC and SoluteML models have much higher errors for dimethylformamide (DMF), ethyl acetate, and ethyl ether than the DirectML model in the random split, but their performance is similar to the DirectML model in the substructure split for the same solvents (with exception of the SoluteML model for DMF). No clear correlation between the model performance and the types of solvents is found from the results as there are no specific types of solvents that a certain model always underperforms or outperforms. It is also noted that the three models have very similar error scales for most of the solvents.

Similar to the prediction of $\Delta G_{\text{solv}}(298\,\text{K})$, the DirectML model performs better than the other two models for the majority of the solvents in the prediction of $\Delta H_{\text{solv}}(298\,\text{K})$ in Figure 7. Some exceptions include hexadecane, ethanol, dichloromethane, and tert-butanol for which the SoluteML model gives much lower errors than the DirectML model in the

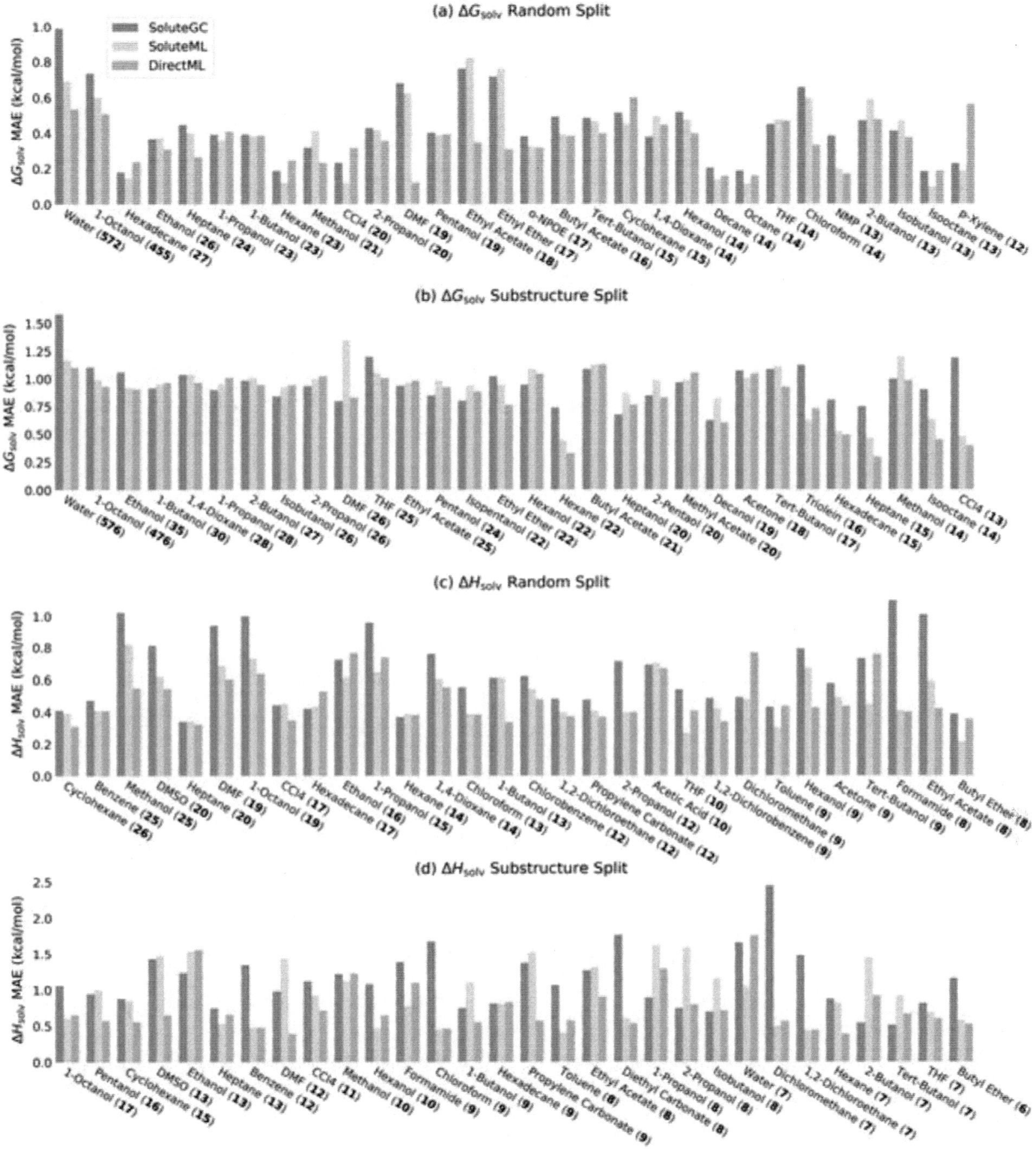

Figure 7: The MAE for the prediction of $\Delta G_{\text{solv}}(298\,\text{K})$ and $\Delta H_{\text{solv}}(298\,\text{K})$ on the 30 most frequently appearing solvents in each test set. The performance of the different models is compared only using the overlapping test data. The number of solvent data in each test set is given in parenthesis in bold. All solvents are included in the training sets of the DirectML models. (CCl4: carbon tetrachloride, DMF: dimethylformamide, o-NPOE: 2-nitrophenyl octyl ether, THF: tetrahydrofuran, NMP: N-methyl-2-pyrrolidone, DMSO: dimethyl sulfoxide)

random split, but the SoluteML and DirectML models have similar prediction accuracy for the same solvents in the substructure split. The three models have different levels of accuracy for many solvents for the $\Delta H_{\text{solv}}(298\,\text{K})$ prediction in the substructure split compared to the $\Delta G_{\text{solv}}(298\,\text{K})$ prediction.

Overall, the DirectML model outperforms the SoluteML model despite having fewer solute data. The SoluteGC model gives the highest error but also contains the most useful information that relates physical contributions to the solvation properties to chemical substructures in the solutes. The underperformance of the SoluteML and the SoluteGC models are most likely related to the approximation of using the linear relationships for the calculation of energy-related solvation properties. For the SoluteGC model, an additional error comes from limiting most of the groups to the nearest neighboring atom interaction. The graph convolutional neural networks in the ML methods, on the other hand, allow the information of each atom to propagate into further-distanced atoms within the molecule and hence can include the non-nearest atom interaction.

Yet, as can be concluded from Figures 6 and 7, there is not one model that outperforms the others on all substructures and all solvents. Based on the observation that the three methods have different levels of accuracy for various solute substructures and solvents and have different outliers, we expect that using the average predictions of the different models may be able to give better predictions by suppressing the large errors from outliers. To test this, we compare the average predictions of the SoluteML and DirectML models (2 model average) and of the SoluteGC, SoluteML, and DirectML models (3 model average) with the experimental data. The resulting parity plots and error summary are shown in Figure 8 and Table 5. The average predictions indeed lead to slight improvements for most test sets compared to the single predictions from the DirectML model. More significant improvements are observed for some of the main outliers of the DirectML model where the other models had better performance.

In summary, even though the DirectML model performs better in general, the three models can offer different levels of information on the predicted values, and, when combined together, they can provide even more accurate predictions of $\Delta G_{\text{solv}}(298\,\text{K})$ and $\Delta H_{\text{solv}}(298\,\text{K})$. The three models as a whole can be used to identify errors or outliers in any of the models. However, users of these models should still be cautious even when using the average values of the three models since it is possible that the particular solute-solvent pair of greatest interest to a user would be one that is significantly mis-predicted by all three models. We also caution that while the models work very well overall, this is not a guarantee that all the model parameters have been well-determined. In particular for the SoluteGC model, as is typical with large linear models, some linear combinations of group contribution values may have not been very well determined or tested by our data set. Therefore, it would be inadvisable to combine these group values with other group values determined some other way, without validating the newly formed combined model on independent data.

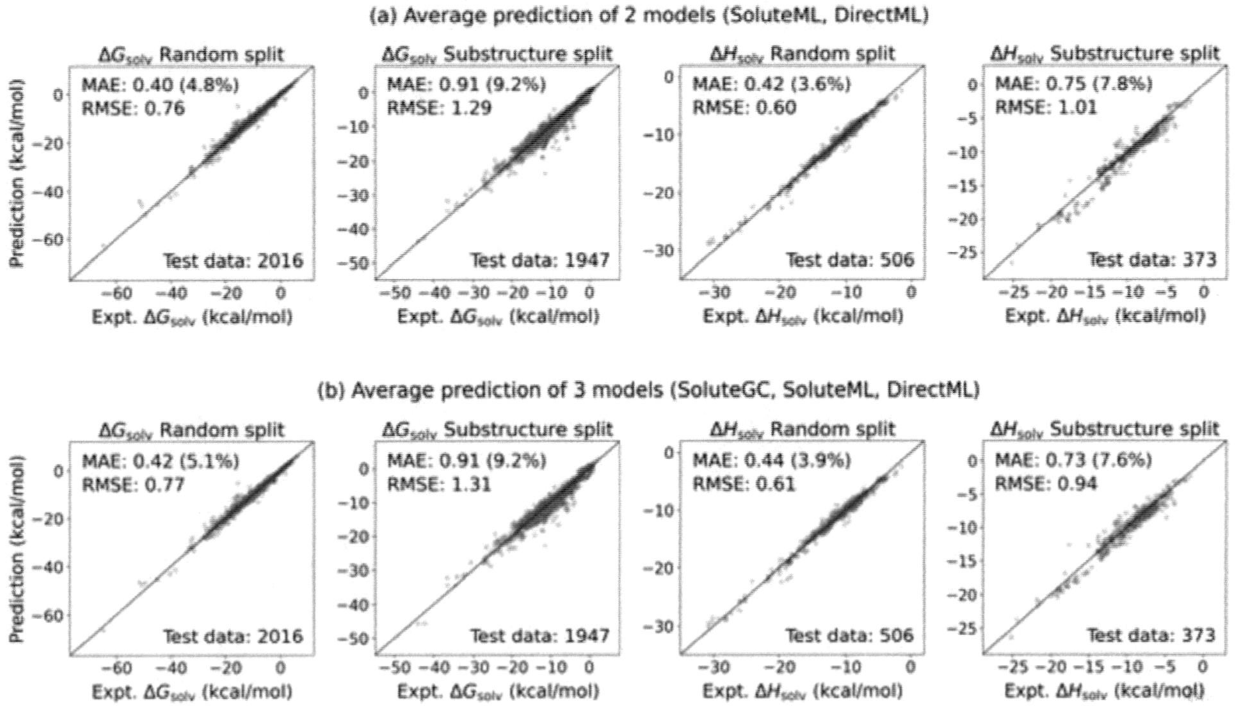

Figure 8: Parity plots between the experimental data and average predictions of the two different models (SoluteML, DirectML) and the three different models (SoluteGC, SoluteML, DirectML) on the overlapping test data. The parity plots are for $\Delta G_{\text{solv}}(298\,\text{K})$ and $\Delta H_{\text{solv}}(298\,\text{K})$ predictions tested on pairs of unseen solutes and trained solvents for the random and substructure splits. The MAE and RMSE are in kcal/mol, and the PMAE is given in parenthesis. The number of test data are also given on the figures along with the test errors. The plots only show the overlapping test data in each split type, and therefore the test data are identical to those in Figures 4 and 5.

Table 5: Test set error summary of the SoluteGC, SoluteML, and DirectML models and the average predictions of multiple models on unseen solute and trained solvent pairs. "Avg. 2Models" represents the average predictions of the SoluteML and DirectML models, and "Avg. 3Models" represents the average predictions of the SoluteGC, SoluteML, and DirectML models. The MAE and RMSE are reported in kcal/mol. The lowest errors in each test set are marked in bold. The table shows the results on the overlapping test data, and therefore the test data are identical to those in Figures 4 and 5.

Model	ΔG_{solv} Random		ΔG_{solv} Substr.		ΔH_{solv} Random		ΔH_{solv} Substr.	
	MAE	RMSE	MAE	RMSE	MAE	RMSE	MAE	RMSE
SoluteGC	0.63	1.06	1.18	1.66	0.64	0.88	1.13	1.53
SoluteML	0.48	0.95	1.00	1.43	0.50	0.69	0.93	1.37
DirectML	0.41	**0.75**	**0.91**	1.32	0.47	0.71	**0.73**	1.07
Avg. 2Models	**0.40**	0.76	**0.91**	**1.29**	**0.42**	**0.60**	0.75	1.01
Avg. 3Models	0.42	0.77	**0.91**	1.31	0.44	0.61	**0.73**	**0.94**

4.2.2 Performance on Unseen Solutes and Unseen Solvents

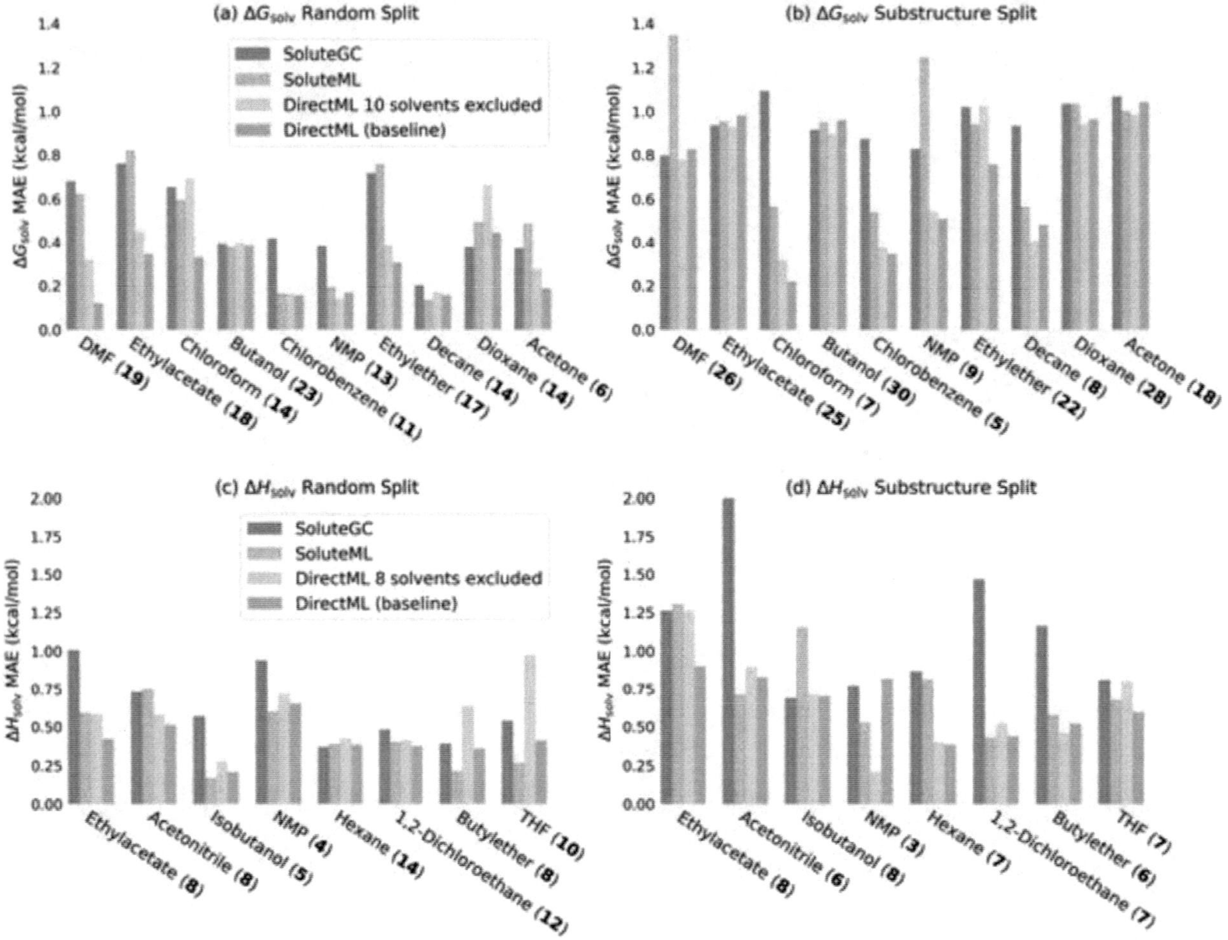

Figure 9: The MAE for the prediction of $\Delta G_{\mathrm{solv}}(298\,\mathrm{K})$ and $\Delta H_{\mathrm{solv}}(298\,\mathrm{K})$ on the solvents excluded from the DirectML training and validation sets. The performance of the different models is compared using only the overlapping test data. The number of solvent data in the test set is given in parenthesis in bold. (DMF: dimethylformamide, NMP: N-methyl-2-pyrrolidone, THF: tetrahydrofuran)

To test the predictive performance of the DirectML model on pairs of unseen solutes and unseen solvents, 10 solvents and 8 solvents are excluded from the training and validation sets of dGsolvDB3 and dHsolvDB, respectively. New DirectML models are trained and validated on these reduced data sets, and their performance on the excluded solvents are compared with the baseline DirectML models trained using all solvents. The SoluteGC and SoluteML models from the earlier section are also compared to see whether the new DirectML models can still outperform them on the unseen solvents. The results on the random and substructure splits for $\Delta G_{\mathrm{solv}}(298\,\mathrm{K})$ and $\Delta H_{\mathrm{solv}}(298\,\mathrm{K})$ are presented in Figure 9.

For the majority of the out-of-sample solvents for $\Delta G_{\mathrm{solv}}(298\,\mathrm{K})$ prediction, the DirectML models trained on the reduced data set still outperform the SoluteGC and SoluteML models.

Compared to the baseline case, the new DirectML model has on average a higher error for the random split as expected, but achieves comparable or even better performance on some solvents for the substructure split. Most likely, this is because the prediction error of the substructure split is predominantly caused by the lack of solute substructures in the training set rather than missing solvent information. Moreover, the training set still includes solvents with similar structures to the excluded solvents. For example, even though decane is excluded, the training set still contains a series of other alkanes. Hence, for the substructure split, it is expected that the DirectML model still provides a similar performance even if solvents are excluded from the training set.

Similar results are observed for the prediction of $\Delta H_{\mathrm{solv}}(298\,\mathrm{K})$; the DirectML model has higher errors than the baseline model for all out-of-sample solvents in the random split and has similar or even lower errors for some solvents in the substructure split. Even though the DirectML model for the prediction of $\Delta H_{\mathrm{solv}}(298\,\mathrm{K})$ is trained using much fewer data points (142 solvents when not including the solvents that are only present in self-solvation data) compared to the $\Delta G_{\mathrm{solv}}(298\,\mathrm{K})$ model trained on 303 solvents, the error increase for the excluded solvents is on average similar to that of the $\Delta G_{\mathrm{solv}}(298\,\mathrm{K})$ prediction.

Overall, the prediction error of the DirectML on out-of-sample solvents increases in the random splits on average by 0.1 - 0.2 kcal/mol and by 0.4 - 0.5 kcal/mol at most. On the contrary, a lack of solvent information does not affect the predictive performance of the DirectML model much in the substructure splits where the error is primarily caused by a lack of solute substructures in the training sets. The DirectML model can still provide better or similar predictions for most of the out-of-sample solvents compared to the SoluteML model, which uses empirical solvent parameters. However, caution should be made when applying the DirectML model to a solvent with a foreign structure or characteristic that is vastly different from the solvents in the training set as the error is usually much larger for a compound with a unique substructure.

4.3 Comparison to Existing Methods

Finally, the prediction errors of our models on unseen solutes that are presented in Section 4.2 are summarized in Table 6. In the table, the prediction errors of our models are reported using the entire 10 % test set data of each model while smaller common test sets were used in the previous sections for model comparison. All DirectML models in the table are trained, validated, and tested using the dGsolvDB3 data set for the solvation free energy prediction.

The prediction errors of the selected existing models from Section 3.3 are also included in Table 6. Lim and Jung[43] reported results on multiple split types for their MLSolvA model, but only the results on solute clustering are included in the table as this is most similar to the random and substructure-based solute splits used throughout our work. Similarly, for the transfer learning model by Vermeire and Green,[44] the results on element-based solute splits are used for comparison. To the best of our knowledge, there is no ML model that can predict solvation enthalpies in a variety of solvents. One model reported by Jaquis et al.[102] is used for comparison, but it only considers ethanol as a solvent. Note that the size and the nature of the training and test sets used for each method varies, and therefore the relative ranking of these methods could change when evaluated on different data sets. The parity plots comparing the experimental data with the in-house calculations using the existing methods, namely the COSMO-RS and the UFZ-LSER GC methods, are provided in the Supporting Information Sections 4.3.1 and 4.3.2.

For both $\Delta G_{\text{solv}}(298\,\text{K})$ and $\Delta H_{\text{solv}}(298\,\text{K})$ predictions, our ML models achieve a similar or even better accuracy than the QM methods on random split but have higher errors on the substructure split. MLSolvA gives lower error than the DirectML and SoluteML models for the solute substructure splits. Nevertheless, it is difficult to draw a solid conclusion without knowing what substructures are found in MLSolvA's solute clustering. The transfer learning model by Vermeire and Green[44] yields relatively low error for the element-based solute splits, which are considered more challenging than the substructure-based splits. Their transfer learning model, however, is pre-trained on 1 million quantum calculations while our models are limited to the use of experimental data. For the solvation enthalpy prediction, our models are more accurate than the ML model by Jaquis et al.,[102] even though their model is limited to ethanol as a solvent.

The performance of the SoluteGC model on random solute split is similar to or better than that of the UFZ-LSER GC model evaluated on all solvation free energy and enthalpy data. Nevertheless, the reported errors of the UFZ-LSER in Table 6 are not true test errors since the training set of the UFZ-LSER is unknown. A considerable portion of its training set solutes are expected to overlap with the test set solutes used in the work, and hence we are unable to test the performance of the UFZ-LSER model on any solute splits.

Because of the paucity of experimental data, often all or nearly all data are used for training with no or a very small independent test set as it is done for our final models. The resulting final DirectML and SoluteGC models achieve very low training and testing errors similar to the expected experimental error bars as it can be seen from Table S7 in the Supporting Information. The final SoluteGC model has larger training errors, likely because the true solvation energies are not exactly linear in the functional groups. All of our final models can be accessed through various ways described in Section 6.

Table 6: Comparison of various QM, ML, and GC methods for solvation energy and enthalpy prediction at 298 K. The MAE and RMSE are in kcal/mol, and the number of each test set data (N_{test}) is presented. Our model results are written in bold. The results here for each of our models use its own full test set. In the comparisons in the text, a smaller common test set was used for all the models.

target	method	method type	test set split type	MAE	RMSE	N_{test}	Ref.
ΔG_{solv}	SMD/IEF-PCM/ M05-2X/6-31G*	QM	-	0.63	0.86	2346	9 [a]
	COSMO-RS/ BP86/TZVPD-FINE	QM	-	0.46	0.77	14236	[b]
	MLSolvA	ML	K-mean solute cluster CV using scaffold-based split	0.62	1.15	6239	43
	Transfer learning	ML	element-based solute split (O excluded)	0.52	0.91	4684	44
	Transfer learning	ML	element-based solute split (Cl excluded)	0.45	0.63	1124	44
	DirectML	**ML**	**random solute**	**0.40**	**0.73**	**2138**	
	DirectML	**ML**	**substructure solute**	**0.89**	**1.32**	**2033**	
	DirectML	**ML**	**random solute (10 solvents excluded)**	**0.40**	**0.75**	**2138**	[c]
	DirectML	**ML**	**substructure solute (10 solvents excluded)**	**0.89**	**1.32**	**2033**	[c]
	SoluteML	**ML**	**random solute**	**0.48**	**0.95**	**2016**	
	SoluteML	**ML**	**substructure solute**	**1.01**	**1.45**	**1948**	
	UFZ-LSER	GC	evaluated on all ΔG_{solv} data [d]	0.78	1.42	16878	[e]
	SoluteGC	**GC**	**random solute**	**0.63**	**1.06**	**2016**	
	SoluteGC	**GC**	**substructure solute**	**1.18**	**1.66**	**1947**	
ΔH_{solv}	COSMO-RS/ BP86/TZVPD-FINE	QM	-	0.69	1.06	6058	[b]
	Jaquis et al. (only solvent ethanol)	ML	random solute	-	1.58	35	102
	DirectML	**ML**	**random solute**	**0.50**	**0.80**	**643**	
	DirectML	**ML**	**substructure solute**	**0.78**	**1.13**	**551**	
	DirectML	**ML**	**random solute (8 solvents excluded)**	**0.51**	**0.80**	**643**	[c]
	DirectML	**ML**	**substructure solute (8 solvents excluded)**	**0.78**	**1.12**	**551**	[c]
	SoluteML	**ML**	**random solute**	**0.50**	**0.69**	**506**	
	SoluteML	**ML**	**substructure solute**	**0.94**	**1.38**	**373**	
	UFZ-LSER	GC	evaluated on all ΔH_{solv} data [d]	0.59	0.98	4099	[e]
	SoluteGC	**GC**	**random solute**	**0.64**	**0.88**	**506**	
	SoluteGC	**GC**	**substructure solute**	**1.13**	**1.53**	**373**	

[a] The original authors used IEF-PCM protocol implemented in Gaussian03 for the SMD calculations. The errors were reported separately for neutral solutes in aqueous and non-aqueous solvents in the original paper, and we calculated the errors for all solvents based on the reported errors and number of data points. [b] The COSMO-RS calculations are performed in this work using COSMO*therm*[103] and COSMO*base*.[104] [c] These refer to the DirectML models from Section 4.2.2 for which 10 and 8 solvents are excluded from the training

and validation sets of dGsolvDB3 and dHsolvDB, respectively. [d] Note that these are not true test errors since the training set of the UFZ-LSER is unknown and its training set solutes are expected overlap with the test set solutes used in these error calculations. [e] The UFZ-LSER GC calculations are done in this work through the UFZ-LSER database[29] using in-house solvent parameters. The UFZ-LSER calculation is only available for molecular weight less than 1000 g/mol, and therefore, a few solute compounds could not be evaluated for ΔG_{solv} predictions.

5 Conclusions

A group contribution method (SoluteGC) and a machine learning model (SoluteML) are constructed for the Abraham solute parameter prediction that are used to estimate solvation free energy and enthalpy via the LSERs. Additionally, a machine learning model (DirectML) is developed for the direct prediction of the solvation free energy and enthalpy for a given solvent-solute pair. The predictive performance of the three models is evaluated on common test sets of the solvation free energy and enthalpy for out-of-sample solute compounds prepared using a random and substructure-based split. The results show that the DirectML model is superior to the SoluteGC and SoluteML models for both solvation energy and enthalpy predictions on all data splits. The SoluteML model performs similarly with the mean absolute errors (MAEs) around 0.1 kcal/mol higher than those of the DirectML. The SoluteGC model underperforms with the MAEs about 0.2 - 0.3 kcal/mol higher than those of the DirectML.

It is also found that adding $\log P$ data and a substantial amount of $\log K_{\mathrm{w}}$ data to the solvation free energy data set improves the performance of the DirectML model mainly for the substructure solute split. Although including these data introduces additional data uncertainty and causes the solvent data to be biased toward water and 1-octanol, the information gained from the additional data can compensate for these drawbacks and gives overall better results.

The present models and some other recently developed models[43,44] trained with large data sets are all accurate enough that they give $\Delta G_{\mathrm{solv}}(298\,\mathrm{K})$ predictions close to the aleatoric uncertainties for random-split test data sets. Each model's predictions for solutes and solvents that are very different from those in its training data are less reliable, and hence an averaging or consensus-of-models approach is recommended. Here we provide a convenient set of 3 rather different models for $\Delta G_{\mathrm{solv}}(298\,\mathrm{K})$ and also provide 3 models for $\Delta H_{\mathrm{solv}}(298\,\mathrm{K})$. Together these models provide good inputs for predicting the solvation thermodynamics of a very large range of solutes and solvents at temperatures up to each solvent's critical point.[5]

Finally, we present our compiled solute parameter (SoluteDB), solvation free energy (dGsolvDBx), and solvation enthalpy (dHsolvDB) databases and provide public access to our final prediction models through a simple web-based tool, software package, and GitHub. A web-based tool is designed for quick calculations and others are more suited for bulk, automated calculations.

6 Data and Software Availability

The in-house Abraham solute parameter database (SoluteDB) along with the in-house $\log K_{\mathrm{w}}$ and $\log P_{\mathrm{ow}}$ data are provided as a part of the Supplementary Information. The publicly available $\Delta G_{\mathrm{solv}}(298\,\mathrm{K})$ and $\Delta H_{\mathrm{solv}}(298\,\mathrm{K})$ data are also provided in the Supplementary Information. This excludes the $\Delta G_{\mathrm{solv}}(298\,\mathrm{K})$ data from the Minnesota solvation database, which is not open-source. For each entry of a solute-solvent pair, the list of all individual data points, mean value, and standard deviation are tabulated in our databases. Although we limit our work to the solute compounds containing H, C, N, O, S, P, F, Cl, Br, or I atoms, the open-source solute parameter, $\Delta G_{\mathrm{solv}}(298\,\mathrm{K})$, and $\Delta H_{\mathrm{solv}}(298\,\mathrm{K})$ data that contain the elements out of our scope are also provided as Supplementary Information.

The final SoluteGC, SoluteML, and DirectML models that are constructed using all data are made publicly available through GitHub, conda software package, and web-based search tool. The web-based tool is available on `https://rmg.mit.edu/database/solvation/search/` and is the simplest way to search for solute parameters, solvation free energies, and solvation enthalpies. Temperature-dependent ΔG_{solv} calculation based on our earlier work[5] is also available through the web-based tool for the solvents whose temperature-dependent densities can be computed by a free fluid modeling software CoolProp.[108] For bulk queries, one can download the source code using GitHub or install a conda package. The SoluteGC model can be accessed by installing the source code from RMG-Py and RMG-database git repositories (`https://github.com/ReactionMechanismGenerator`) with a sample code located at `https://github.com/ReactionMechanismGenerator/RMG-Py/blob/master/ipython/estimate_solvation_thermo_and_search_available_solvents.ipynb`. The SoluteML and DirectML models can be downloaded as a conda package from `https://anaconda.org/fhvermei/chemprop_solvation`. The source code for the ML models can be found from chemprop_solvation git repository (`https://github.com/fhvermei/chemprop_solvation`).

Acknowledgement

We gratefully acknowledge Eni S.p.A. for supporting this research. We also acknowledge additional financial support from the Machine Learning for Pharmaceutical Discovery and Synthesis Consortium (MLPDS). Yunsie Chung acknowledges financial support from National Science Foundation (subawarded by the MolSSI under fellowship agreement number 479590-19825A). Florence Vermeire acknowledges financial support from the Belgian American Educational Foundation (BAEF). We thank Jonathan Zheng for his help with revising the code for the web-based tool.

References

(1) Thompson, J. D.; Cramer, C. J.; Truhlar, D. G. Predicting aqueous solubilities from aqueous free energies of solvation and experimental or calculated vapor pressures of pure substances. *J. Chem. Phys.* **2003**, *119*, 1661–1670.

(2) Li, C.-J.; Chan, T.-H. *Comprehensive Organic Reactions in Aqueous Media*, 2nd ed.; John Wiley & Sons, Inc.: Hoboken, NJ, USA, 2007.

(3) Sixt, M.; Koudous, I.; Strube, J. Process design for integration of extraction, purification and formulation with alternative solvent concepts. *C. R. Chim.* **2016**, *19*, 733–748.

(4) Ruelle, P. The n-octanol and n-hexane/water partition coefficient of environmentally relevant chemicals predicted from the mobile order and disorder (MOD) thermodynamics. *Chemosphere* **2000**, *40*, 457–512.

(5) Chung, Y.; Gillis, R. J.; Green, W. H. Temperature-dependent vapor–liquid equilibria and solvation free energy estimation from minimal data. *AIChE J.* **2020**, *66*, e16976.

(6) Hildebrand, J. H. A History of Solution Theory. *Annu. Rev. Phys. Chem.* **1981**, *32*, 1–24.

(7) Cramer, C. J.; Truhlar, D. G. Molecular Orbital Theory Calculations of Aqueous Solvation Effects on Chemical Equilibria. *J. Am. Chem. Soc.* **1991**, *113*, 8552–8554.

(8) Marenich, A. V.; Olson, R. M.; Kelly, C. P.; Cramer, C. J.; Truhlar, D. G. Self-consistent reaction field model for aqueous and nonaqueous solutions based on accurate polarized partial charges. *J. Chem. Theory Comput.* **2007**, *3*, 2011–2033.

(9) Marenich, A. V.; Cramer, C. J.; Truhlar, D. G. Universal solvation model based on solute electron density and on a continuum model of the solvent defined by the bulk dielectric constant and atomic surface tensions. *J. Phys. Chem. B* **2009**, *113*, 6378–6396.

(10) Klamt, A. Conductor-like screening model for real solvents: A new approach to the quantitative calculation of solvation phenomena. *J. Phys. Chem.* **1995**, *99*, 2224–2235.

(11) Klamt, A.; Jonas, V.; Bürger, T.; Lohrenz, J. C. Refinement and parametrization of COSMO-RS. *J. Phys. Chem. A* **1998**, *102*, 5074–5085.

(12) Klamt, A.; Eckert, F. COSMO-RS: a novel and efficient method for the a priori prediction of thermophysical data of liquids. *Fluid Phase Equilib.* **2000**, *172*, 43–72.

(13) Abraham, M. H.; Ibrahim, A.; Zissimos, A. M. Determination of sets of solute descriptors from chromatographic measurements. *J. Chromatogr. A* **2004**, *1037*, 29–47.

(14) Abraham, M. H.; Acree, W. E. Correlation and prediction of partition coefficients between the gas phase and water, and the solvents dodecane and undecane. *New J. Chem.* **2004**, *28*, 1538–1543.

(15) Kamlet, M. J.; Taft, R. W. The solvatochromic comparison method. I. The .beta.-scale of solvent hydrogen-bond acceptor (HBA) basicities. *J. Am. Chem. Soc.* **1976**, *98*, 377–383.

(16) Mintz, C.; Clark, M.; Acree, W. E.; Abraham, M. H. Enthalpy of Solvation Correlations for Gaseous Solutes Dissolved in Water and in 1-Octanol Based on the Abraham Model. *J. Chem. Inf. Model.* **2007**, *47*, 115–121.

(17) Jalan, A.; Ashcraft, R. W.; West, R. H.; Green, W. H. Predicting solvation energies for kinetic modeling. *Annu. Rep. Prog. Chem., Sect. C: Phys. Chem.* **2010**, *106*, 211–258.

(18) Bradley, J. C.; Abraham, M. H.; Acree, W. E.; Lang, A. S. Predicting Abraham model solvent coefficients. *Chem. Cent. J.* **2015**, *9*, 1–10.

(19) Abraham, M. H.; Acree, W. E. Estimation of heat capacities of gases, liquids and solids, and heat capacities of vaporization and of sublimation of organic chemicals at 298.15 K. *J. Mol. Liq.* **2020**, *317*, 113969.

(20) Abraham, M. H.; Acree, W. E. Estimation of enthalpies of sublimation of organic, organometallic and inorganic compounds. *Fluid Phase Equilib.* **2020**, *515*, 112575.

(21) Snelgrove, D. W.; Lusztyk, J.; Banks, J. T.; Mulder, P.; Ingold, K. U. Kinetic solvent effects on hydrogen-atom abstractions: Reliable, quantitative predictions via a single empirical equation. *J. Am. Chem. Soc.* **2001**, *123*, 469–477.

(22) Havelec, P.; Ševčík, J. G. Extended additivity model of parameter log(L16). *J. Phys. Chem. Ref. Data* **1996**, *25*, 1483–1493.

(23) Svozil, D.; Ševčík, J. G.; Kvasnička, V. Neural network prediction of the solvatochromic polarity/polarizability parameter πH2. *J. Chem. Inf. Comput. Sci.* **1997**, *37*, 338–342.

(24) Platts, J. A.; Butina, D.; Abraham, M. H.; Hersey, A. Estimation of molecular linear free energy relation descriptors using a group contribution approach. *J. Chem. Inf. Comput. Sci.* **1999**, *39*, 835–845.

(25) Platts, J. A.; Abraham, M. H.; Butina, D.; Hersey, A. Estimation of Molecular Linear Free Energy Relationship Descriptors by a Group Contribution Approach. 2. Prediction of Partition Coefficients. *J. Chem. Inf. Comput. Sci.* **2000**, *40*, 71–80.

(26) GHAFOURIAN, T.; DEARDEN, J. C. The Use of Atomic Charges and Orbital Energies as Hydrogen-bonding-donor Parameters for QSAR Studies: Comparison of MNDO, AM1 and PM3 Methods. *J. Pharm. Pharmacol.* **2000**, *52*, 603–610.

(27) Zissimos, A. M.; Abraham, M. H.; Klamt, A.; Eckert, F.; Wood, J. A comparison between the two general sets of linear free energy descriptors of Abraham and Klamt. *J. Chem. Inf. Comput. Sci.* **2002**, *42*, 1320–1331.

(28) Arey, J. S.; Green, W. H.; Gschwend, P. M. The electrostatic origin of Abraham's solute polarity parameter. *J. Phys. Chem. B* **2005**, *109*, 7564–7573.

(29) Ulrich, N.; Endo, S.; Brown, T.; Watanabe, N.; Bronner, G.; Abraham, M.; Goss, K.-U. UFZ-LSER database v 3.2 [Internet]. 2017; `http://www.ufz.de/lserd`.

(30) Liang, Y.; Torralba-Sanchez, T. L.; Di Toro, D. M. Estimating system parameters for solvent-water and plant cuticle-water using quantum chemically estimated Abraham solute parameters. *Environ. Sci.: Process. Impacts* **2018**, *20*, 813–821.

(31) ACD/Labs, ACD/ADME Suite 5.0, version 12.0. `https://www.acdlabs.com/`.

(32) Stenzel, A.; Goss, K. U.; Endo, S. Prediction of partition coefficients for complex environmental contaminants: Validation of COSMOtherm, ABSOLV, and SPARC. *Environ. Toxicol. Chem.* **2014**, *33*, 1537–1543.

(33) Brown, T. N.; Arnot, J. A.; Wania, F. Iterative fragment selection: A group contribution approach to predicting fish biotransformation half-lives. *Environ. Sci. Technol.* **2012**, *46*, 8253–8260.

(34) Brown, T. N. Predicting hexadecane-air equilibrium partition coefficients (L) using a group contribution approach constructed from high quality data. *SAR QSAR Environ. Res.* **2014**, *25*, 51–71.

(35) Mobley, D. L.; Guthrie, J. P. FreeSolv: A database of experimental and calculated hydration free energies, with input files. *J. Comput. Aided Mol. Des.* **2014**, *28*, 711–720.

(36) Wu, Z.; Ramsundar, B.; Feinberg, E. N.; Gomes, J.; Geniesse, C.; Pappu, A. S.; Leswing, K.; Pande, V. MoleculeNet: A benchmark for molecular machine learning. *Chem. Sci.* **2018**, *9*, 513–530.

(37) Yang, K.; Swanson, K.; Jin, W.; Coley, C.; Eiden, P.; Gao, H.; Guzman-Perez, A.; Hopper, T.; Kelley, B.; Mathea, M.; Palmer, A.; Settels, V.; Jaakkola, T.; Jensen, K.; Barzilay, R. Analyzing Learned Molecular Representations for Property Prediction. *J. Chem. Inf. Model.* **2019**, *59*, 3370–3388.

(38) Marenich, A. V.; Kelly, C. P.; Thompson, J. D.; Hawkins, G. D.; Chambers, C. C.; Giesen, D. J.; Winget, P.; Cramer, C. J.; Truhlar, D. G. Minnesota Solvation Database – version 2012, University of Minnesota, Minneapolis. 2012.

(39) Hutchinson, S. T.; Kobayashi, R. Solvent-Specific Featurization for Predicting Free Energies of Solvation through Machine Learning. *J. Chem. Inf. Model.* **2019**, *59*, 1338–1346.

(40) Ramsundar, B.; Eastman, P.; Walters, P.; Pande, V.; Leswing, K.; Wu, Z. *Deep Learning for the Life Sciences*; O'Reilly Media, 2019.

(41) Hille, C.; Ringe, S.; Deimel, M.; Kunkel, C.; Acree, W. E.; Reuter, K.; Oberhofer, H. Generalized molecular solvation in non-aqueous solutions by a single parameter implicit solvation scheme. *J. Chem. Phys.* **2019**, *150*, 041710.

(42) Pathak, Y.; Mehta, S.; Priyakumar, U. D. Learning Atomic Interactions through Solvation Free Energy Prediction Using Graph Neural Networks. *J. Chem. Inf. Model.* **2021**, *61*, 689–698.

(43) Lim, H.; Jung, Y. J. MLSolvA: solvation free energy prediction from pairwise atomistic interactions by machine learning. *J. Cheminformatics* **2021**, *13*, 1–10.

(44) Vermeire, F. H.; Green, W. H. Transfer learning for solvation free energies: From quantum chemistry to experiments. *Chem. Eng. J.* **2021**, *418*, 129307.

(45) Weininger, D. SMILES, a Chemical Language and Information System: 1: Introduction to Methodology and Encoding Rules. *J. Chem. Inf. Comput. Sci.* **1988**, *28*, 31–36.

(46) Heller, S. R.; McNaught, A.; Pletnev, I.; Stein, S.; Tchekhovskoi, D. InChI, the IUPAC International Chemical Identifier. *J. Cheminf.* **2015**, *7*, 23.

(47) Swain, M. PubChemPy. 2013; `https://pypi.org/project/PubChemPy/1.0/`.

(48) Swain, M. CIRpy. 2016; `https://pypi.org/project/CIRpy/`.

(49) Landrum, G. RDKit: Open-Source Cheminformatics. 2006; `http://www.rdkit.org/`.

(50) Grinberg Dana, A.; Ranasinghe, D.; Wu, H.; Grambow, C.; Dong, X.; Johnson, M.; Goldman, M.; Liu, M.; Green, W. H. ARC - Automated Rate Calculator, version 1.1.0. https://github.com/ReactionMechanismGenerator/ARC, 2019.

(51) Moine, E.; Privat, R.; Sirjean, B.; Jaubert, J. N. Estimation of solvation quantities from experimental thermodynamic data: Development of the comprehensive CompSol databank for pure and mixed solutes. *J. Phys. Chem. Ref. Data* **2017**, *46*.

(52) Abraham, M. H.; Zissimos, A. M.; Acree, W. E., Jr. Partition of solutes from the gas phase and from water to wet and dry di-n-butyl ether: a linear free energy relationship analysis. *Phys. Chem. Chem. Phys.* **2001**, *3*, 3732–3736.

(53) Abraham, M. H.; Zissimos, A. M.; Acree, W. E., Jr. Partition of solutes into wet and dry ethers; an LFER analysis. *New J. Chem.* **2003**, *27*, 1041–1044.

(54) Abraham, M. H.; Acree, W. E., Jr. Comparative analysis of solvation and selectivity in room temperature ionic liquids using the Abraham linear free energy relationship. *Green Chem.* **2006**, *8*, 906–915.

(55) Acree William E, J.; Abraham, M. H. The analysis of solvation in ionic liquids and organic solvents using the Abraham linear free energy relationship. *J. Chem. Technol. Biotechnol.* **2006**, *81*, 1441–1446.

(56) Abraham, M. H.; Enomoto, K.; Clarke, E. D.; Rosés, M.; Ràfols, C.; Fuguet, E. Henry's Law constants or air to water partition coefficients for 1,3,5-triazines by an LFER method. *J. Environ. Monit.* **2007**, *9*, 234–239.

(57) Sprunger, L. M.; Proctor, A.; Acree, W. E.; Abraham, M. H.; Benjelloun-Dakhama, N. Correlation and prediction of partition coefficient between the gas phase and water, and the solvents dry methyl acetate, dry and wet ethyl acetate, and dry and wet butyl acetate. *Fluid Phase Equilib.* **2008**, *270*, 30–44.

(58) Sprunger, L. M.; Gibbs, J.; Acree, W. E.; Abraham, M. H. Correlation and prediction of partition coefficients for solute transfer to 1,2-dichloroethane from both water and from the gas phase. *Fluid Phase Equilib.* **2008**, *273*, 78–86.

(59) Sprunger, L. M.; Achi, S. S.; Pointer, R.; Blake-Taylor, B. H.; Acree, W. E.; Abraham, M. H. Development of Abraham model correlations for solvation characteristics of linear alcohols. *Fluid Phase Equilib.* **2009**, *286*, 170–174.

(60) Sprunger, L. M.; Achi, S. S.; Acree, W. E.; Abraham, M. H.; Leo, A. J.; Hoekman, D. Correlation and prediction of solute transfer to chloroalkanes from both water and the gas phase. *Fluid Phase Equilib.* **2009**, *281*, 144–162.

(61) Abraham, M. H.; Acree, W. E.; Leo, A. J.; Hoekman, D. The partition of compounds from water and from air into wet and dry ketones. *New J. Chem.* **2009**, *33*, 568–573.

(62) Abraham, M. H.; Acree Jr, W. E.; Cometto-Muñiz, J. E. Partition of compounds from water and from air into amides. *New J. Chem.* **2009**, *33*, 2034–2043.

(63) Abraham, M. H.; Acree, W. E., Jr.; Leo, A. J.; Hoekman, D. Partition of compounds from water and from air into the wet and dry monohalobenzenes. *New J. Chem.* **2009**, *33*, 1685–1692.

(64) Sprunger, L. M.; Achi, S. S.; Pointer, R.; Acree, W. E.; Abraham, M. H. Development of Abraham model correlations for solvation characteristics of secondary and branched alcohols. *Fluid Phase Equilib.* **2010**, *288*, 121–127.

(65) Grubbs, L. M.; Saifullah, M.; De La Rosa, N. E.; Ye, S.; Achi, S. S.; Acree, W. E.; Abraham, M. H. Mathematical correlations for describing solute transfer into functionalized alkane solvents containing hydroxyl, ether, ester or ketone solvents. *Fluid Phase Equilib.* **2010**, *298*, 48–53.

(66) Abraham, M. H.; Smith, R. E.; Luchtefeld, R.; Boorem, A. J.; Luo, R.; Acree, W. E. Prediction of Solubility of Drugs and Other Compounds in Organic Solvents. *J. Pharm. Sci.* **2010**, *99*, 1500–1515.

(67) Ye, S.; Saifullah, M.; Grubbs, L. M.; McMillan-Wiggins, M. C.; Acosta, P.; Mejorado, D.; Flores, I.; Acree, W. E.; Abraham, M. H. Determination of the Abraham model solute descriptors for 3,5-dinitro-2-methylbenzoic acid from measured solubility data in organic solvents. *Phys. Chem. Liq.* **2011**, *49*, 821–829.

(68) Stephens, T. W.; De La Rosa, N. E.; Saifullah, M.; Ye, S.; Chou, V.; Quay, A. N.; Acree, W. E.; Abraham, M. H. Abraham model correlations for solute partitioning into o-xylene, m-xylene and p-xylene from both water and the gas phase. *Fluid Phase Equilib.* **2011**, *308*, 64–71.

(69) Saifullah, M.; Ye, S.; Grubbs, L. M.; De La Rosa, N. E.; Acree, W. E.; Abraham, M. H. Abraham Model Correlations for Transfer of Neutral Molecules to Tetrahydrofuran and to 1,4-Dioxane, and for Transfer of Ions to Tetrahydrofuran. *J. Solution Chem.* **2011**, *40*, 2082–2094.

(70) Abraham, M. H.; Acree, W. E. The transfer of neutral molecules, ions and ionic species from water to benzonitrile; comparison with nitrobenzene. *Thermochim. Acta* **2011**, *526*, 22–28.

(71) Holley, K.; Acree, W. E.; Abraham, M. H. Determination of Abraham model solute descriptors for 2-ethylanthraquinone based on measured solubility ratios. *Phys. Chem. Liq.* **2011**, *49*, 355–365.

(72) Stephens, T. W.; Wilson, A.; Dabadge, N.; Tian, A.; Zimmerman, M.; Hensley, H. J.; Acree, W. E. J.; Abraham, M. H. Correlation of solute partitioning into isooctane from water and from the gas phase based on updated Abraham equations. *Global J. Phys. Chem.* **2012**, *3*, 1–16.

(73) Stephens, T. W.; Quay, A. N.; Chou, V.; Loera, M.; Shen, C.; Wilson, A.; Acree, W. E.; Abraham, M. H. Correlation of solute transfer into alkane solvents from water and from the gas phase with updated Abraham model equations. 2012; `https://digital.library.unt.edu/ark:/67531/metadc152452/`.

(74) Stephens, T. W.; Loera, M.; Calderas, M.; Diaz, R.; Montney, N.; Acree, W. E.; Abraham, M. H. Determination of Abraham model solute descriptors for benzoin based on measured solubility ratios. *Phys. Chem. Liq.* **2012**, *50*, 254–265.

(75) Abraham, M. H.; Acree, W. E. Linear free-energy relationships for water/hexadec-1-ene and water/deca-1,9-diene partitions, and for permeation through lipid bilayers; comparison of permeation systems. *New J. Chem.* **2012**, *36*, 1798–1806.

(76) Abraham, M. H.; Gola, J. R. M.; Gil-Lostes, J.; Acree, W. E.; Cometto-Muñiz, J. E. Determination of solvation descriptors for terpene hydrocarbons from chromatographic measurements. *J. Chromatogr. A* **2013**, *1293*, 133–141.

(77) Abraham, M. H.; Acree, W. E. Physicochemical and biochemical properties for the dialkyl phthalates. *Chemosphere* **2015**, *119*, 871–880.

(78) Bradley, J.-C.; Abraham, M. H.; Acree, W. E.; Lang, A. S. I. D.; Beck, S. N.; Bulger, D. A.; Clark, E. A.; Condron, L. N.; Costa, S. T.; Curtin, E. M.; Kurtu, S. B.; Mangir, M. I.; McBride, M. J. Determination of Abraham model solute descriptors for the monomeric and dimeric forms of trans-cinnamic acid using measured solubilities from the Open Notebook Science Challenge. *Chem. Cent. J.* **2015**, *9*, 11.

(79) Brumfield, M.; Wadawadigi, A.; Kuprasertkul, N.; Mehta, S.; Stephens, T. W.; Barrera, M.; De La Rosa, J.; Kennemer, L.; Meza, J.; Acree, W. E.; Abraham, M. H. Determination of Abraham model solute descriptors for three dichloronitrobenzenes from measured solubilities in organic solvents. *Phys. Chem. Liq.* **2015**, *53*, 163–173.

(80) Abraham, M. H.; Acree, W. E. Equations for water-triolein partition coefficients for neutral species; comparison with other water-solvent partitions, and environmental and toxicological processes. *Chemosphere* **2016**, *154*, 48–54.

(81) Abraham, M. H.; Acree, W. E. Descriptors for Pentane-2,4-dione and Its Derivatives. *J. Solution Chem.* **2017**, *46*, 1625–1638.

(82) Abraham, M. H.; Acree, W. E.; Liu, X. Partition of Neutral Molecules and Ions from Water to o-Nitrophenyl Octyl Ether and of Neutral Molecules from the Gas Phase to o-Nitrophenyl Octyl Ether. *J. Solution Chem.* **2018**, *47*, 293–307.

(83) Abraham, M. H.; Acree, W. E.; Cometto-Muñiz, J. E. Descriptors for terpene esters from chromatographic and partition measurements: Estimation of human odor detection thresholds. *J. Chromatogr. A* **2020**, *1609*.

(84) Sushko, I. et al. Online chemical modeling environment (OCHEM): Web platform for data storage, model development and publishing of chemical information. *J. Comput. Aided Mol. Des.* **2011**, *25*, 533–554.

(85) Wishart, D. S. et al. DrugBank 5.0: A major update to the DrugBank database for 2018. *Nucleic Acids Res.* **2018**, *46*, D1074–D1082.

(86) Syracuse Research Corporation. Physical/Chemical Property Database (PHYSPROP); SRC Environmental Science Center: Syracuse, NY. 1994.

(87) Abraham, M. H.; Acree, W. E. The solubility of liquid and solid compounds in dry octan-1-ol. *Chemosphere* **2014**, *103*, 26–34.

(88) Acree, W.; Lang, A. Acree Enthalpy of Solvation Dataset. *Figshare* **2015**,

(89) Systems, D. C. I. SMARTS - A Language for Describing Molecular Patterns. `https://www.daylight.com/dayhtml/doc/theory/theory.smarts.html`.

(90) Gao, C. W.; Allen, J. W.; Green, W. H.; West, R. H. Reaction Mechanism Generator: Automatic construction of chemical kinetic mechanisms. *Comput. Phys. Commun.* **2016**, *203*, 212–225.

(91) Liu, M.; Grinberg Dana, A.; Johnson, M. S.; Goldman, M. J.; Jocher, A.; Payne, A. M.; Grambow, C. A.; Han, K.; Yee, N. W.; Mazeau, E. J.; Blondal, K.; West, R. H.; Goldsmith, C. F.; Green, W. H. Reaction Mechanism Generator v3.0: Advances in Automatic Mechanism Generation. *J. Chem. Inf. Model.* **2021**, *61*, 2686–2696.

(92) Benson, S. W.; Golden, D. M.; Haugen, G. R.; Shaw, R.; Cruickshank, F. R.; Rodgers, A. S.; O'neal, H. E.; Walsh, R. Additivity rules for the estimation of thermochemical properties. *Chem. Rev.* **1969**, *69*, 279–324.

(93) Han, K.; Jamal, A.; Grambow, C. A.; Buras, Z. J.; Green, W. H. An Extended Group Additivity Method for Polycyclic Thermochemistry Estimation. *Int. J. Chem. Kinet.* **2018**, *50*, 294–303.

(94) Green, W. H.; West, R. H. RMG - Reaction Mechanism Generator. 2021; `https://rmg.mit.edu/`.

(95) Lay, T. H.; Bozzelli, J. W.; Dean, A. M.; Ritter, E. R. Hydrogen atom bond increments for calculation of thermodynamic properties of hydrocarbon radical species. *J. Phys. Chem.* **1995**, *99*, 14514–14527.

(96) Walker, P. J.; Haslam, A. J. A New Predictive Group-Contribution Ideal-Heat-Capacity Model and Its Influence on Second-Derivative Properties Calculated Using a Free-Energy Equation of State. *J. Chem. Eng. Data* **2020**, *65*, 5809–5829.

(97) Pedregosa, F. et al. Scikit-learn: Machine Learning in Python. *J. Mach. Learn. Res.* **2011**, *12*, 2825–2830.

(98) Hoerl, A. Application of ridge analysis to regression problems. *Chem. Eng. Prog.* **1962**, *58*, 54–59.

(99) Distributed Asynchronous Hyperparameter Optimization in Python. `https://github.com/hyperopt/hyperopt`.

(100) Bergstra, J.; Yamins, D.; Cox, D. D. Making a Science of Model Search: Hyperparameter Optimization in Hundreds of Dimensions for Vision Architectures. Proceedings of the 30th International Conference on Machine Learning. Atlanta, Georgia, 2013; pp I–115 to I–23.

(101) Glorot, X.; Bengio, Y. Understanding the difficulty of training deep feedforward neural networks. *J. Mach. Learn. Res.* **2010**, *9*, 249–256.

(102) Jaquis, B. J.; Li, A.; Monnier, N. D.; Sisk, R. G.; Acree, W. E.; Lang, A. S. Using Machine Learning to Predict Enthalpy of Solvation. *J. Solution Chem.* **2019**, 564–573.

(103) Dassault Systèmes, BIOVIA COSMOtherm, Release 2020. `http://www.3ds.com`.

(104) Dassault Systèmes, BIOVIA COSMObase, Release 2020. `http://www.3ds.com`.

(105) Thompson, J. D.; Cramer, C. J.; Truhlar, D. G. New universal solvation model and comparison of the accuracy of the SM5.42R, SM5.43R, C-PCM, D-PCM, and IEF-PCM continuum solvation models for aqueous and organic solvation free energies and for vapor pressures. *J. Phys. Chem. A* **2004**, *108*, 6532–6542.

(106) Nicholls, A.; Mobley, D. L.; Guthrie, J. P.; Chodera, J. D.; Bayly, C. I.; Cooper, M. D.; Pande, V. S. Predicting small-molecule solvation free energies: An informal blind test for computational chemistry. *J. Med. Chem.* **2008**, *51*, 769–779.

(107) Guthrie, J. P. A blind challenge for computational solvation free energies: Introduction and overview. *J. Phys. Chem. B* **2009**, *113*, 4501–4507.

(108) Bell, I. H.; Wronski, J.; Quoilin, S.; Lemort, V. Pure and pseudo-pure fluid thermophysical property evaluation and the open-source thermophysical property library coolprop. *Ind. Eng. Chem. Res.* **2014**, *53*, 2498–2508.

632638 *Electrosprayed Scalable 3D Graphene-CNT Electrodes for Li-Ion and Fuel Cell Applications*

Abstract:

Lithium-sulfur batteries (LSB) and advanced lithium-ion batteries (LIB) demonstrate promise as the next generation energy storage and conversion (ESC) technology especially as it pertains to wearable technology and electric vehicles. Although LIB dominate the battery market (>90%) due its reliability, long cycle life, and market maturity, alternative means of renewable energy must be discovered in order to reduce the cost of materials, dendritic issues at the solid-electrolyte interphase (SEI) layer, and limited charge capacity (~250 mAH h g^{-1}). The next generation of future batteries is the lithium-sulfur battery due its inexpensive cost ($0.25/kg sulfur), high theoretical energy density (2600 Wh kg^{-1}), and safety. How LIB and LSB systems incorporate into military applications such as drones and battery backs is what the interests the Army. However, LSB face many challenges before commercialization into wearable technology and electric vehicles due its fast capacity degradation, polysulfide "shuttling" effect, and short cycle life (<300 cycles). Carbon nanomaterials constitute an ideal miniaturized composite platform for integration with other solid-state electrode materials. Air-controlled electrospray has been recognized as an effective means of fabricating cathodes using a convective air jet flow that accelerates the drying and deposition process. This technique provides a uniform and well-dispersed coating onto the current collector while using low-cost commercial carbon nanomaterials in a matter of minutes. Currently, self-assembly is conducting using the bottom–up synthesis of 3D macrostructures using graphene and carbon nanotubes (CNTs) as building blocks. We demonstrate a novel, all-aqueous, and scalable 3D platform materials design process for the synthesis of carbon nanocomposite aerogels to serve as porous ultralightweight electrodes for the next generation energy storage and conversion applications. Specifically, the negatively charged surface functional groups on graphene oxide and oxidized CNTs will be the sites for electrostatic coordination of positively charged noble metal cations and polyelectrolytes from aqueous solutions. This enables the incorporation of carbon nanomaterials with any polyelectrolyte and noble metal nanostructures with precise connection of each of the individual components while maintaining their unique properties in both LIB and LSB systems. Through a scalable air-controlled electrospray approach, we can develop hierarchical porous 3D nanocomposite cathodes comprised of carbon nanomaterials with any precious or noble metal nanostructures (Pt, Pd, Au, Ru, Ag, and Ru) and polyelectrolyte that demonstrate a synergistic network between the active materials, carbon materials, and metal nanostructures. Electrochemical techniques such as cyclic voltammetry, linear sweep voltammetry, galvanostatic charge-discharge, and impedance spectroscopy are used to determine the electrical and ionic conductivity, electrocatalytic activity, and supercapacitor performance of the aerogels as lightweight high-power and energy density electrodes.

624456 Modeling the Brønsted Acidity of Lanthanum Exchanged Faujasite

Abstract:

Lanthanum stabilizes faujasite in fluid catalytic cracking, which increases catalyst material lifetimes. Despite its extensive use for decades, many unknowns remain on how lanthanum increases faujasite hydrothermal stability. The hydrothermal stability of zeolites decreases by dealumination, which occurs via reactions that involve water molecules. Brønsted acid sites give desired catalytic properties to zeolites, however they reduce hydrothermal stability by attracting water molecules to aluminum tetrahedra and promoting dealumination. The energetically favored binding sites of water molecules and lanthanum ions in hydrogen exchanged faujasite were determined with density functional theory (DFT) calculations and ab initio molecular dynamics (AIMD) simulations. Potentials of mean force of proton transfer were obtained with constrained AIMD simulations to quantify the Brønsted acidity of faujasite, with and without lanthanum ions. Results indicate that water and lanthanum ions are energetically favored to bind in the same sites in faujasite, and that lanthanum ions increase the acidity of Brønsted sites adjacent to, and not adjacent to, bound lanthanum ions in faujasite. Implications toward hydrothermal stability are discussed, since lanthanum ions change the protonation state, and therefore hydrophilicity, of aluminum tetrahedra in faujasite.

629492 Charge Transport Mechanism in Sodium Selenide (Na_2Se) for Sodium-Selenium Batteries: A First Principles Study

Abstract:

Sodium−selenium (Na−Se) batteries are recognized as promising energy storage systems because of their high theoretical volumetric capacity (~ 3250 mAh/cm^3), which is of particular importance for portable devices and electric vehicles (EVs) with the limited battery packing space. Sodium selenide (Na_2Se) is the final discharge product for the Na−Se batteries and is expected to limit the overall battery conductivity because of the insulating nature. To effectively utilize the active materials, it is necessary to have a better understanding of the charge transport mechanism in Na_2Se. Based on first-principles simulations, we first evaluate the contribution of potential charge carriers (ionic and electronic defects) to the conductivity in Na_2Se by calculating the formation energies and diffusion barriers. Negatively charged sodium vacancies (V_{Na}^-) and positive sodium interstitials (Na_i^+) are found to be the defects with the lowest formation energy (0.80 eV) in Na_2Se, followed by hole polarons (h_p^+) with a formation energy of 1.95 eV. Our simulations demonstrate that while the V_{Na}^- has a higher diffusion barrier than the h_p^+ (0.25 eV *vs.* 0.11 eV), its contribution to the conductivity is the most dominant, on the order of 10^{-15} S/cm, due to the high concentration. On the other hand, the electronic conductivity mediated by the h_p^+ is predicted to be negligible because of its low concentration. Once the vacancy−polaron ($V_{Na}^- - h_p^+$) complex forms, the energy cost to break the pair is predicted to be 0.78 eV, implying that the charge transport could be hindered due to its charge neutrality. Our calculations also illustrate that compressive strain (which is potentially induced by surrounding host materials) could reduce both ionic and electrical conductivity of Na_2Se. Lastly, we will discuss a possible strategy to modify the ionic and electronic conductivities in Na_2Se by controlling the strain.

624018 Nitrogen-Rich Free-Standing Carbon Cathode for Improved Lithium-Sulfur Battery Performance

Abstract:

A study on the production of free-standing carbon cathode is an attractive method to improve the low active material ratio, which is a problem with existing Li-S batteries. This work introduces a free-standing carbon material with superior porosity properties for enhanced Li-S battery performance. The free-standing carbon cathode synthesized in this study shows outstanding physical stability by showing superior elastic properties based on a porous structure. Additionally, the free-standing cathode made through electrospinning of $NaBH_4$ and polyacrylonitrile (PAN) has a high nitrogen atom equivalent to about 20%. This in turn inhibits the shuttle phenomenon of polar lithium polysulfides, which is highly soluble in electrolyte, based on strong chemical adsorption ability between, and at the same times improves conductivity to deliver high specific capacity at high current densities. As a result, the free-standing electrode with 2.54 mg cm^{-2} sulfur loading exhibited a capacity of about 650 mAh g^{-1} at 5.0 C. By maximizing the advantages of the free-standing electrode, the areal loading of the active material has been increased to 6.11 mg cm^{-2}, which delivers an areal capacity of 3.1 mAh cm^{-2} at 500th.

624433 Simulating the Glass Transition of Steam Cracked Tar

Abstract:

Molecular modeling tools, like Molecular Dynamics, provide a unique perspective on microscopic properties and processes. While there has been a lot of effort using Molecular Dynamics to simulate oil, there has been a reoccurring concern: "Do the molecules being simulated accurately represent the oil's composition?" While there has been significant progress towards determining the structure of various components in crude, there are still challenges in determining the whole composition accurately. In addition, there are practical concerns given that Molecular Dynamics can only simulate a relatively small sampling of the great diversity of molecules that exist in a typical crude. Significant progress has been made to mitigate these problems, with a great example being the development of "Digital Oil"[1]. Using a variety of analytical techniques, representative molecules are generated to fit available experimental data. The number of molecules used is chosen by the modeler based on how many molecules they wish to simulate and how many are required to sufficiently fit the data.

This route for producing Digital Oil is very promising. However, this method—like any Molecular Dynamics based approach—is still limited by the fact that a small subset of molecules must be chosen from the whole crude. The goal of the work being presented here is to provide a proof of concept for an approach that captures the sensitivity of the predicted property to the particular molecules being selected. This is useful when there is uncertainty regarding how many structures are actually required to precisely calculate a property of interest.

In this study, we investigated a model system involving different solubility fractions of steam cracked tar with different glass transition temperatures. Using a variety of analytical and compositional modeling techniques, a composition of well over 2,500 different molecules was generated for each sample. In an automated workflow, force field parameters were generated for each of these molecules, and then molecules were randomly selected to generate multiple simulations for each sample allowing confidence intervals to be determined. Calculated glass transition results were then compared to experimental data generated by differential scanning calorimetry. There was a good match between calculated and experimental results for most samples, with the few outliers likely being caused by flux contamination that was not included in the composition model. These positive results indicate that this method is a promising tool to virtually probe the properties of crude-based materials. This may be especially helpful for developing novel processes where limited experimental data is available, such as creating higher value carbon-based products from petroleum rather than fuel.

References

[1] M. Iwase; S. Sugiyama; Y. Liang; Y. Masuda; M. Morimoto; T. Matsuoka; E. S. Boek; R. Ueda; K. Nakagawa, *Energy Fuels,* **32** (2018) *2781.*

627766 Multiscale Modeling and Characterization of Radical-Initiated Modification of Molten Polyolefins

Abstract:

Radical grafting mechanisms provide powerful and frequently-used methods for the functionalization of polyolefin chains which are relatively devoid of functional handles, having found applications as structural adhesives, high performance elastomers, and reactive copolymer intermediate, etc. Here, by investigating the reaction kinetics through a hybrid quantum calculation procedure, a comprehensive mechanistic view of the complex free-radical mediated synthesis of functional polyolefins is achieved, which includes grafting of vinyl silane monomers, inter-and intramolecular H-transfer as well as chain scission. Different radical species are found to be in kinetic equilibrium with their relative concentration depending on both the thermal stability and the abundance of specific carbon moiety. Our results clearly show that the overall mechanism is dominated by grafting of single vinyl silane monomer via intramolecular H-transfer to hydrocarbon substrates, rather than forming localized or oligomeric grafts, and this occurs at the expense of polymer crosslinking due to the termination of radicals via combination.

NMR+GC/MS spectroscopy of model compounds were also used to further elucidate the dominant reaction pathways for the initiation and the termination mechanisms. On the basis of the intrinsic kinetic dataset discerned from both computational chemistry and model compound studies, quantitative relationships between the product properties and the reaction conditions are therefore revealed with the intension to establish a polymer topology based kinetic model. To evaluate the quality of the resulting kinetic model for radical grafting reactions, it is compared to a factorial design of melt processing experiments on industrial grade polyolefin samples, including comparisons of the degree of grafting via FTIR, the molar mass measured via high temperature GPC, and the linear rheology of the polymer melts. The results show strong correlation between model predictions and the experimental results across a wide range of parameter space, suggesting the predictive ability of this approach for complex chemical networks.

SARS-CoV-2 is the virus responsible for the 2020 global pandemic. Although multiple vaccines have been approved for use against the virus throughout the world, the threat of disease remains largely unsolved for individuals with underlying medical conditions and people residing in regions where vaccines and the necessary drug treatments are not available. A potential solution to this problem is the use of phytochemicals (Chemical compounds obtained from plants) that inhibit viral activity through the binding of certain SARS-CoV-2 proteins. The phytochemicals selected in this study were obtained from the USDA Phytochemical and Ethnobotanical Database, and many of them have documented anti-viral properties and are generally regarded as safe. To identify potential therapeutic phytochemicals, we used the Rosetta software, a computational biology tool with ligand-protein docking capabilities, to screen a phytochemical library against selected viral proteins. From this analysis, we identified 35 potential binders for various SARS-CoV-2 proteins. (e.g., Non-structural proteins and spike proteins) In order to obtain high-quality docking data, we protonated ligands to body pH, generated 100 conformers for each ligand, and evaluated the performance of different energy scoring functions. To validate our docking results, we compared our data with known inhibitors' data. We got a 66% successful rate on identifying them. To speed up the virtual screening process, we developed a ligand-based virtual screening that takes place before Rosetta Docking. (The structure-based screening) This ligand-based screening uses the results from docking simulation and molecules clustering to identify lead clusters that mostly contain computationally identified binders. Then, classification algorithms were applied to classify new molecules to clusters, and we predict those been classified to lead clusters as potential binders. Overall, this study will provide useful information on plant-based drug development for COVID-19 treatment. The models and high throughput screening protocol established may be used in a wide range of drug discovery scenarios.

621588 The Influence of Glycosylation on the Binding of Sars-Cov-2 Spike

Abstract:

Glycosylation is an important posttranslational modification of proteins and plays a crucial role in protein efficacy. It also needs to be understood for optimization of recombinant protein expression. We developed a methodology for modeling glycoproteins with different glycosylation patterns. We use this toolbox to study differently glycosylated 3D structures of the receptor binding domain (RBD) of the spike protein of Sars-Cov-2 in atomistic detail starting from known crystal structures . Computational methods for modeling glycoprotein structures are employed to understand how changes in site-specific oligosaccharide composition and structure influence 3D glycoprotein structure and binding to targets including the Ace2 receptor which again is studied using different glycosylation patterns. We observe e.g. that some ACE2 glycans directly interact with the spike fragment. We also find correlation between glycosylation and the conformation of an ACE2-Fc construct which is a therapeutic candidate. Particularly we find that glycosylation of the fusion protein impacts the average distance between the two major subdomains. Finally we determine the free energy of binding between the RBD and Ace2.

629018 Investigation of Finite-Size Effects of Rutile Titanium Dioxide Clusters

Abstract:

Supported solid catalysts are useful as functional materials and are used in many industrial processes. Nanoparticles of active elements are dispersed on the surface of carrier in supported solid catalysts, yielding a high surface-area to volume ratio. Titanium dioxide is of paramount technological importance and it is widely used for photo-catalytic processes and photo-electrochemistry. It is also one of the most well-studied oxide materials. In this work, we use density functional theory (DFT) calculations to investigate and de-convolute the geometric and electronic finite-size effects of rutile titania clusters. We determine the size of titania clusters at which the surface energy converges to the infinite slab limit. Surface energies are calculated for rutile titania clusters ranging from 24 to 384 atoms. We observe that surface energies of the clusters are very close to the infinite slab limit at 384 atoms. Finite-size effects are observed below this critical size which are due to geometric and electronic structure effects. The geometric defects are due to the presence of edge, corner and sub-surface atoms that have unique geometric configurations and are classified using machine-learning fingerprints. This approach enables identification of the length scale where electronic finite size effects are significant, suggesting that particles with a minimum dimension > 1 nm can be accurately modeled using the slab approximation. We also utilize hybrid functionals to analyze the optical properties of the particles, revealing that defect states play a key role in determining the bandgap of nanoparticles, and that optical properties converge much more slowly than surface properties.

621410 Teaching Process Data Analytics and Machine Learning

Abstract:

Teaching chemical engineers to be effective in data analytics is as important today than in any time in the history of the discipline, and advances in sensor technologies, machine learning, and associated software have enabled applications that were not formerly possible. This abstract describes experiences with teaching process data analytics and machine learning over the last few years, including in (1) a new chemical engineering elective course in which the students come from chemical engineering, mechanical engineering, and engineering management, (2) a professional education course cotaught by chemical and mechanical engineering faculty whose learners primarily come with chemistry and chemical engineering backgrounds and many years of industrial experience, and (3) an undergraduate chemical engineering concentration in process data analytics with courses coming from a mixture of multiple disciplines including chemical engineering, computer science, economics, management. The overview will describe the challenges in teaching data science to chemical engineers and scientists and engineers working in the chemical and biotechnology industries, and strategies for overcoming those challenges.

622120 Teaching Big Data Science to Undergraduate Students in the University of Iowa

Abstract:

In this presentation, we will provide an overview of our course development and our experiences in teaching big data science and analytical skills to the undergraduate students in chemical engineering in the University of Iowa. These courses include: (a) Machine Learning (ML) and Artificial Intelligence (AI) course; (b) statistical analysis course; and (c) thermodynamic course. They are designed to be offered to students from freshman to junior years. We found that in general, the chemical engineering students are not getting used to use programming to analyze the data, in part because many courses they are taking focus more on the lab experiment while the analysis of the lab results is elementary and often done in Excel or by calculator. However, the students are willing to learn new analytical skills, and provided with well-designed teaching materials and problems, they can grasp the big data skills quickly. While the outcome of our experience has been encouraging, we also found there is clear lag in the field of chemical engineering to train our next generation with big data tools. This lag is part is due to the institutional tradition that emphasizes on the lab experiments and often deterministic way of thinking (with less consideration of probabilities), and in part due to the slowness in re-design the curriculum to reflect new development and ease access of ML and AI. We will provide examples to support the findings above, share the lessons we learned, and provide thoughts for next steps.

623878 Textile: Tutorials in Experimentalist Interactive Learning

Abstract:

Background:

Confocal microscopy is a common preclinical imaging modality to study a range of tissue features. Fluorescent image microscopy can provide insights into cell density, morphology, and cell-cell relationships. In our lab, we use immunofluorescent staining to investigate changes in cell and microenvironment features in response to development, injury, and treatment in the brain. We have developed image processing pipelines that threshold, segment, and quantify cell features. More recently, we have incorporated data science-enabled analysis and techniques into our imaging workflow. While the lab is comprised of researchers who are wet-lab experimentalists and data-driven analysts, only the second group has formal data science training. Wet-lab experimentalists can significantly benefit from having data science literacy to apply data science methodologies to their imaging datasets. However, there is no formal training regime to train wet-lab experimentalists in lab-specific data science methodologies. To address this training gap, we developed Tutorials in EXperimentalisT Interactive LEarning (TEXTILE) for data-science enabled fluorescent image processing.

TEXTILE is a module-based, semi-linear educational system developed with laboratory-based learning in mind. Modules combine data science fundamentals with specific biological imaging processing techniques. For example, we pair a data science module about data management with an image processing module on quantitative cell morphology characterization. The modular format allows us to train students on managing experimental data and pair that with quantifying features of cells that are relevant to their specific project of interest. Additionally, we designed TEXTILE such that the learning process is adaptable to starting skill level. Learning paths are based on entry knowledge but allow students to pick and choose their modules based on interest or previous experience. For example, someone with ample coding knowledge can skip the introductory Python modules and begin with the experimental image processing modules. Finally, we designed TEXTILE for a variety of degree levels. As a laboratory at a large public university, we engage with undergraduate and high school students, in addition to master's thesis and doctoral thesis students. We accomplished degree-level accessibility through the semi-linear system that assumes no prior knowledge in our research area. Since we already use TEXTILE to train high school students participating in laboratory research that have no previous knowledge of the subject topic, we aim to include more local high school students and expand TEXTILE for science education and outreach as well as in-laboratory education.

Methods:

TEXTILE is composed of topic-specific modules. Each module is designed around a data science topic, lasts an hour and a half, and contains four sections: the pre-module activity, main module, post-module reflection, and feedback form (Figure 1a). The instructor sends the pre-module activity to students for self-driven learning a week before teaching the main module. The pre-module's purpose is to prime students for the module's main lesson with a high-level activity. For example, in our data management module, students analyze a grocery shopping trip and decide where to store specific categories of food and perform a space evaluation on how much room in each storage location the food will take up.

The main module aims to teach the bulk material in digestible lectures broken up by interactive sessions. The main module's detailed structure consists of six parts: introduction presentation, interactive session one, main presentation workshop, interactive session two, concluding presentation, and a reflection session (Figure 1b). The introduction presentation introduces the module's topic, outlines the structure the main module will follow, and leads into the first interactive session. Interactive session one guides the student through an exploration of the module material. For coding-based modules, instructors lead interactive sessions with Jupyter Notebook. During interactive session one, the instructor walks students through the code's basics as students code along. After the students have written the code, the instructor begins the main presentation, which has two purposes: (1) to increase the students' connection between the pre-module activity and the module introduction and (2) to dive deeper into the technical aspects of the first interactive session. Once students understand the technical elements of the main module, they begin interactive session two, where they apply technical details of the lesson to further explore the code written in the earlier interactive session. After students have an opportunity to explore, the instructor gives the concluding presentation, which includes wrapping up the main takeaways, emphasizing the module's motivation, and thanking students for their participation.

Finally, the main module ends with a prompt for students to reflect on what they have learned and how their opinions about the module's topic have changed since the pre-module activity. The post-module reflection is completed individually by each student a day or two after the main module. The purpose is to provide a student-led review activity that encourages them to explore content from the main module on their own time to solidify concepts and build confidence. After the whole module is completed, students are sent a feedback form through Google Forms. With the feedback form, students reflect on their growth and the impact of each component of the module. The feedback is then implemented back into the module for the next cohort of students.

Results:

In the summer of 2020, the first cohort of students went through six TEXTILE modules of paired data science and image processing techniques used within the lab. The first cohort had 15 students: six high school students, four undergraduates, and five graduate students. The first cohort was taught live by a Ph.D. student from the lab's data-driven analysis team. The topics covered included GitHub and version control, introduction to data management, introduction to experimental data management specific to the lab, intro to image processing, thresholding, image segmentation, cellular morphology, and quantifying cell shape features. By the end of the first cohort, all students completed every TEXTILE module and the high school and undergraduate students developed individual projects using protocols learned over the summer. They each created Jupyter Notebooks for a self-led project with a brain cell morphology lit review, data management plan, and image processing pseudo-code. Many students from the summer 2020 cohort chose to continue researching with the lab and are currently developing these projects further.

In fall 2020, a small cohort of two undergraduate students explored the TEXTILE modules asynchronously. The students watched a recording of the main module and participated in asynchronous discussions with previous students and a Ph.D. student-instructor via Slack. We also provided these students with published image data that no one previously analyzed for cell morphology features. The students are currently completing retroactive analysis on this data using techniques learned from various TEXTILE modules.

For summer 2021, TEXTILE is expanded to include three modules on machine learning and two additional modules on experimental image acquisition and manual cell counting, incorporating student feedback from 2020. The main takeaways from student feedback included developing a formalized mentoring process for student projects and teaching via a hybrid live and asynchronous format.

Conclusions:

TEXTILE is a successful data science educational system for teaching data science techniques commonly used in experimental research labs. The TEXTILE system works for protocol training of fellow lab members, onboarding new students, and science education outreach. The combination of interactive sessions with instructor-led presentations gives students ample opportunity to work hands-on with real data from the lab in a guided environment. Students from the first cohort had the opportunity to develop their cell morphology research projects. In contrast, students from the second cohort took the lessons learned from TEXTILE and applied them for retroactive analysis of previously published data. The TEXTILE system is built to be robust and adaptable so that other laboratories can follow the methodology to develop training modules for their own data science techniques.

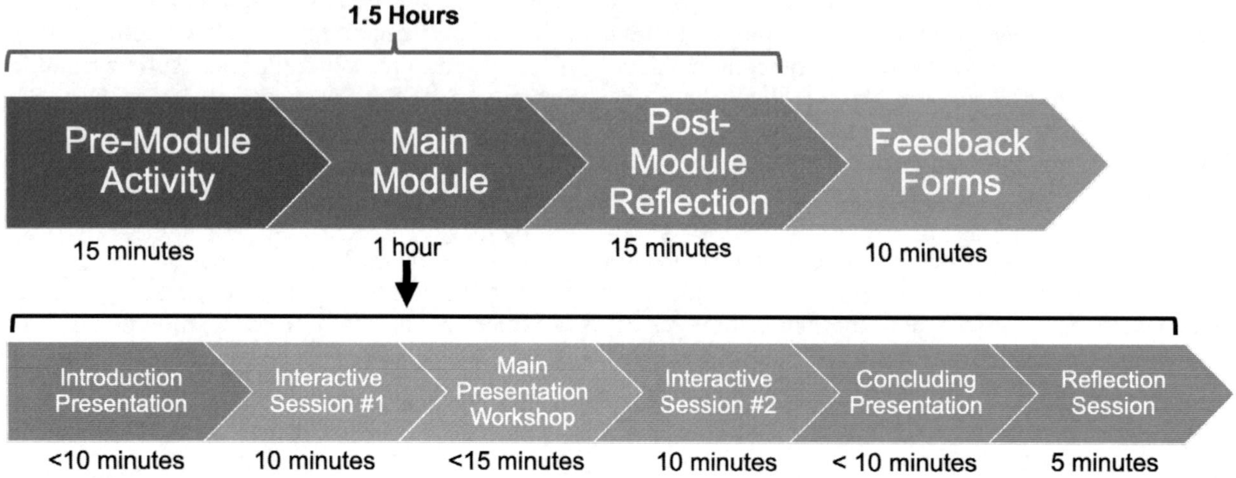

Figure 1. (a) Top row: TEXTILE process composed of four parts – pre-module activity, main modules, post-module reflection, and feedback forms. (b) Bottom Row: The six sections of the main module - introduction presentation, interactive session #1, main presentation workshop, interactive session #2, concluding presentation, reflection session.

622481 Teaching Artificial Intelligence to Chemical Engineers: Experiences from a 35-Year-Old Course

Abstract:

Given the recent success of artificial intelligence (AI) in a wide variety of applications, it has become quite clear that chemical engineering students need to become familiar with applied AI techniques [1]. There is great interest in the academic community in developing a course to address this important need. I have been teaching such a course to chemical engineering undergraduate and graduate students since 1986 [2], first at Columbia for three years, then at Purdue for about twenty years, and back again at Columbia now. It was mostly on *symbolic AI* in the early years, with an emphasis on *knowledge-based expert system* concepts and techniques, for various applications in chemical engineering such as design, diagnosis, control, optimization, and process safety. From the 1990s onwards, as machine learning techniques started showing promise, my course evolved to include topics such as clustering, neural networks, statistical classifiers, graph-based models, and genetic algorithms. The current version at Columbia has the first one-third of the course on *symbolic AI* and the rest on machine learning-based *numeric AI* concepts and techniques. It is mainly a project-based course where the students develop systems using AI techniques for chemical engineering applications. In the early years, the students programmed in AI languages such as LISP and OPS5, but now using the easier tools such as Python and TensorFlow. In this presentation, I will discuss my experiences in teaching this course, including the syllabus and the projects we have done over the decades.

Reference:

1. Venkatasubramanian, "The Promise of Artificial Intelligence in Chemical Engineering: Is it here, finally?", *Perspective article, AIChE Journal,* 65(2), pp. 466-478, 2019.

- Venkatasubramanian, V. "Artificial Intelligence in Process Engineering: Experiences from a Graduate Course," *Chemical Engineering Education*, Fall Graduate Education Issue, November 1986, pp. 188 – 192.

625508 Determining Optimal Solvents for High-Throughput Processing of Extreme-Strength Covalent Organic Frameworks

Abstract:

Hydrogen-bonded lamellar solids have shown promise as candidates for next-generation extreme strength materials as they combine the advantages of traditional Kevlar and more recent graphene-like materials. Unfortunately, there has been limited success in finding cheap, non-volatile solvents for high-throughput solution-processing due to the above dual interactions. In this study, we utilize quantum mechanical calculations to understand the nature of the interactions between the 2-D polymer and different solvents in both vacuum and non-vacuum environments. Specifically, adsorption configurations, binding energies, and charge transfer between the polymer and solvents will be used to evaluate the strength of interactions and eventually, to predict optimal solvents for these materials.

630839 Computational Modeling of Aluminum Covetics for Improving Manufacturing Yields

Abstract:

Introduction: Manufacturing of a unique covalently bonded graphene-aluminum composite, referred to as aluminum covetics, are inhibited by inhomogeneous distribution of carbon and carbide formation within liquid aluminum; this has resulted in poor manufacturing yields of aluminum covetics. A significant drawback to fabrication is the limited understanding of graphene and graphene-aluminum formation in liquid aluminum. Therefore, the objective of this project is to determine the reaction mechanism pathway for the formation of graphene-aluminum covetics. The ultimate goal is to apply thermodynamic and kinetic properties, ascertained from ab initio molecular dynamic (AIMD) and density functional theory (DFT), to design and fabricate an efficient reactor for improved manufacturing yields of covetics.

Methods: To tackle this challenge, we optimize carbon-aluminum species using a COductor like Screening MOdel (COSMO) in conjunction with DFT, in the presence of an electric field and added charges. Thermochemical analysis is used to calculate thermodynamic properties of the various species. These thermodynamic properties are verified for accuracy by comparing to the thermodynamic properties of aluminum-carbon species formed from explicit AIMD simulations. AIMD simulations are performed with 64 atom aluminum liquid, at 1100°C, and various carbon concentrations (C_1-C_6), with periodic boundary conditions. The two-phase thermodynamic (2PT) method is used to calculate thermochemical properties from AIMD and the radial distribution function (RDF) is used to calculate the bond lengths. The optimized carbon-aluminum species are then used as building blocks to generate larger graphene-aluminum structures. The reaction pathway is then generated from the base case of Al_XC_X to graphene-aluminum covetics using transition state theory to ascertain the intermediate species.

Results: Gibbs energies between AIMD and COSMO-DFT optimized species vary from 0.46% to 3.57%. Our results suggest that COSMO-DFT can accurately mimic AIMD derived thermodynamic properties of aluminum covetics to complement AIMD simulations. Diffusion of carbon-aluminum species increase with increasing external potential, and further increase in the presence of added electrons. The added electrons decrease molecular size of carbon-aluminum species, leading to increased diffusion. Finally, a reaction mechanism pathway and reaction rates are presented from Al_4C_1 to graphene-aluminum covetic formation.

623464 Molecular Aspects of Temperature Swing Solvent Extraction for Brine Desalination Using Imidazole-Based Solvents.

Abstract:

Solvent extraction desalination has shown to be a promising energy-efficient technique when compared to separation processes usually employed. In this context, freshwater from brine is obtained by solubilizing the water in an organic phase followed by a later recovery process induced by a temperature swing. Despite the remarkable impact of the organic solvent's molecular characteristics on this process, few studies have been carried out so far to explicitly ascertain the influence of these characteristics on the solubility of water in the solvent. Here, we assess the viability of four different imidazole-based solvents and compare to an amine compound for solvent extraction desalination of high salinity brines. All-atom molecular simulations of pure solvent and water-solvent bulk mixtures are used to ascertain the influence of the solvents' molecular structures on solvent-solvent and solvent-water interaction. Our results show that water solubilization in imidazole-based solvents is an exothermic process, while for a secondary amine (diisopropylamine) it is an endothermic process. Notably, we explicitly show that imidazole-based solvents can be designed to provide characteristics favorable to brine desalination, namely, dependence on the solubility of water in the solvent within a narrow temperature range, high water solubility, and minimal solvent solubility in the aqueous phase.

629013 Brownian Bridge Method for Studying Rare Events – an Approximation Scheme

Abstract:

A stochastic bridge is a continuous random walk that ends in a given region, stays in a region for its entire path, or reaches one region before another region. This concept has been used in control theory to guide random processes to end in a given state, but has not been used in chemistry although it could have wide ranging implications for polymer physics and molecular simulations. This talk will formulate numerical schemes to generate stochastic bridges for a wide range of chemistry applications where traditional sampling techniques are inefficient due to the intrinsic issues such as trial-and-error and loss of information. Our previous work [1, 2] studied a continuous polymer chain in an external field -- a canonical problem in all polymer field theories. One can demonstrate that a stochastic bridge can sample conformations that end with a given topology or end with a range of final energies, the latter of which is important for rare event sampling.

In this work, an approximation method will be developed to extend the stochastic bridge formalism to more complex systems and higher dimensional problems. In our past work [2], a stochastic bridge was created by calculating a drift that constrained random walks to the correct regions of phase space. This drift was determined by a solution of a Backwards Fokker Planck (BFK) equation, which is infeasible for complicated systems. We will show how to systematically generate approximations to the drift without solving the BFK equation, and how to correct (i.e., re-weight) any errors incurred from this approximation.

Using asymptotic properties of the Backwards Fokker Planck equation for different applications, we will develop approximations for the hitting probability needed to develop a stochastic bridge. We will then develop a novel procedure to reweight the approximations so that one is guaranteed to generate the correct conditional statistics. We accomplish this goal by applying the approximation method to one dimensional Wiener processes (i.e. Ornstein Ulenbeck process and Geometric Brownian motion). Next, the approach will be generalized for high dimensional systems. The main outcome of this talk will be to show that one can now generate a stochastic bridge without knowing the solution to a high dimensional Fokker Planck equation, which will allow one to apply this methodology to a wide range of problems. We outline many applications that will be served by this development.

[1] S. Krishnaswami, D. Ramkrishna, and J. M. Caruthers, "Statistical-mechanically exact simulation of polymer conformation in an external field," The Journal of Chemical Physics, vol.107, no. 15, pp. 5929–5944, 1997.

[2] S. Wang, D. Ramkrishna, and V. Narsimhan, "Exact sampling of polymer conformations using Brownian bridges," The Journal of Chemical Physics, vol. 153, no. 3, p. 034901, 2020.

630467 Development of Two Sequential Kmc Models to Describe Ligand Exchange and Charge Transport for $CsPbBr_3$ Quantum Dots with Improved Stability and Charge Carrier Mobility

Abstract:

Inorganic halide perovskite quantum dots (QDs) such as $CsPbBr_3$ are in the spotlight as materials for semiconductors, photodetectors, LEDs, and solar cells due to their relatively easy synthesis method and excellent optoelectronic properties [1]. However, QDs have a fatal disadvantage where the efficiency is easily deteriorated by light, high temperature, and moisture. Hence, the stability of QDs must be increased for its widespread implementation. Ligands located on the outer side of the QDs are commonly utilized to increase the stability of QDs because ligands physically protect the QDs by preventing dispersion and aggregation [2]. During the QD synthesis, long-chain ligands (e.g., oleic acid) act as natural surfactants to facilitate synthesis and stabilize precursors. However, long-chain ligands may cap the surface and hinder efficient electron transport by increasing the gap between the QDs (i.e, working as an insulator) [3]. Therefore, the ligand-exchange procedure with short-chain ligands such as 1,2-ethanedithiol (EDT) or Mercaptopropanoic acid (MPA) is essential to reduce the gap and thereby increase conductivity. Furthermore, the type of ligands alters electronic structure of QDs confined exciton state, which significantly affects the surface, electrical, and optical properties of QDs [4]. While several experimental studies have been recently carried out, there is a lack of the fundamental understanding of the ligand exchange process.

First, we proposed a kinetic Monte Carlo (kMC) model to enhance our understanding of the ligand-exchange phenomenon. The developed kMC model describes the surface of QDs in the solution using a two-dimensional lattice of size N [5]. Four microscopic events were considered in the kMC simulation to describe the ligand-exchange process: (i) attachment of long-chain ligand, (ii) detachment of long-chain ligand, (iii) attachment of short-chain ligand, and (iv) detachment of short-chain ligand. Parameters such as the ligand concentration, the size of QDs, and the solution concentration were available from the relevant experimental data [6,7]. The activation energy of the attachment and detachment rate was derived from the difference in bandgap energy values before and after ligand exchange. The kMC model enables the prediction of ligand exchange ratio and morphology on the surface of QDs.

However, the direct measurement of surface characteristics of QDs in the solution (i.e., surface morphology and exchanged ligand ratio) is not readily available in the literature [8]. Therefore, a model (i.e., a charge-transport kMC model) was developed to quantify the state of ligand exchange. Since the activation energy values for the transport of exciton and charge carrier are affected by the QDs' surface characteristics [9], the simulation results of the charge-transport kMC model can reasonably explain the ligand-exchange phenomena. Similar to the ligand-exchange kMC model, four events were considered in the kMC simulation for charge transport: (i) charge carrier generation by light absorption, (ii) diffusion by hopping, (iii) decay, and (iv) recombination. Additionally, the charge-transport rates of the QD core and short-chain ligands were calculated from Marcus formula. The long-chain ligand rate was computed from the Miller-Abrahams formula that is widely used in the charge-transport kMC of organic solar cells. The developed model that integrates the ligand-exchange kMC model and the charge-transport kMC model was validated by the Current-Voltage curve data. [10]

In conclusion, our study addressed the knowledge gap present in the interpretation and modeling of the ligand-exchange phenomenon. Specifically, a first-principled ligand-exchange kMC model and a charge-transport model were integrated to simulate the ligand-exchange phenomenon. Subsequently, this model was validated against a system of (long-chain) oleic acid ligands with (short-chain) MPA ligands. The developed model provided a foundation for optimizing the QD surface passivation, by enhancing stability and charge mobility.

Literature Cited

[1] J. Zhang *et al.*, "Low-Dimensional Halide Perovskites and Their Advanced Optoelectronic Applications," *Nano-Micro Lett.*, vol. 9, no. 3, p. 36, 2017, doi: 10.1007/s40820-017-0137-5.

[2] A. Heuer-Jungemann *et al.*, "The role of ligands in the chemical synthesis and applications of inorganic nanoparticles," *Chem. Rev.*, vol. 119, no. 8, pp. 4819–4880, 2019, doi: 10.1021/acs.chemrev.8b00733.

[3] M. J. Smith, C. H. Lin, S. Yu, and V. V Tsukruk, "Composite Structures with Emissive Quantum Dots for Light Enhancement," *Adv. Opt. Mater.*, vol. 1801072, no. 4, pp. 1–35, 2019, doi: 10.1002/adom.201801072.

[4] G. H. Carey, A. L. Abdelhady, Z. Ning, S. M. Thon, O. M. Bakr, and E. H. Sargent, "Colloidal Quantum Dot Solar Cells," *Chem. Rev.*, 2015, doi: 10.1021/acs.chemrev.5b00063.

[5] N. Sitapure, R. Epps, M. Abolhasani, and J. S.-I. Kwon, "Multiscale modeling and optimal operation of millifluidic synthesis of perovskite quantum dots: Towards size-controlled continuous manufacturing," *Chem. Eng. J.*, vol. 413, p. 127905, 2021, doi: https://doi.org/10.1016/j.cej.2020.127905.

[6] R. W. Epps, K. C. Felton, C. W. Coley, and M. Abolhasani, "Automated microfluidic platform for systematic studies of colloidal perovskite nanocrystals: towards continuous nano-manufacturing," *Lab Chip*, vol. 17, no. 23, pp. 4040–4047, 2017.

[7] S. R, H. V Vishaka, K. J, and R. G. Balakrishna, "Green to Blue Light Emitting CsPbBr3 Perovskite by Ligand Exchange and its Encapsulation by TiO2 for Tandem Effect in Photovoltaic Applications," *ACS Appl. Nano Mater.*, vol. 3, no. 6, pp. 6089–6098, Jun. 2020, doi: 10.1021/acsanm.0c01293.

[8] R. Maugi, P. Hauer, J. Bowen, E. Ashman, E. Hunsicker, and M. Platt, "A methodology for characterising nanoparticle size and shape using nanopores," *Nanoscale*, vol. 12, no. 1, pp. 262–270, 2020.

[9] C. R. Kagan, E. Lifshitz, E. H. Sargent, and D. V Talapin, "Building devices from colloidal quantum dots," *Science (80-.).*, vol. 353, no. 6302, 2016.

[10] J. Li *et al.*, "50-Fold EQE improvement up to 6.27% of solution-processed all-inorganic perovskite CsPbBr3 QLEDs via surface ligand density control," *Adv. Mater.*, vol. 29, no. 5, p. 1603885, 2017.

628251 Mechanism of Inorganic Phosphate Release Reaction in Actin Subunits

Abstract:

Actin filaments and their dynamic response to environmental perturbations impart cells with varied functionalities such as their ability to migrate, divide, shape change, transport cargo, etc. The ATP bound within actin subunits in a filament undergoes a series of reactions called ATP hydrolysis and inorganic phosphate (Pi) release. These reactions play a crucial role in dynamically remodeling the higher-order structures of actin filaments by modulating their interactions with other binding proteins. Although a lot of studies have focused on unraveling the key steps driving the ATP hydrolysis reaction, a lot is not known about the mechanism of the Pi release. In this regard, here, we investigate the mechanism of Pi release reaction and correlate it with actin subunit conformational fluctuations. To do so, we adopt Transition Tempered Metadynamics (TTMetaD), an enhanced free energy sampling technique, to efficiently sample the underlying free-energy landscape and identify the minimum free energy path driving the release of inorganic phosphate. Comprehensive knowledge of the mechanism of phosphate release from actin subunits would help in elucidating the potential binding modes of several regulatory proteins, which play important roles in the remodeling of actin networks. Furthermore, it would enable the development of rigorous bottom-up coarse-grained models of actin filaments that can be used to simulate actin networks efficiently.

631567 Mesoscale Simulation Approach for Dynamics and Assembly of Deformable Objects

Abstract:

In dilute suspensions, surfactant micelles and emulsion droplets often take spherical shapes. But at high enough concentrations, contact among the micelles or droplets leads to nonspherical shapes. In this limit, the dynamics and assembly of the suspension depend more on the interfaces between objects than the bulk objects themselves. This has similarities with foams and biological tissues in which the interfaces between deformable objects (gas bubbles or cells) play a critical role. We have adapted a mesoscale approach (called vertex models) in order to apply it to the dynamics and assembly of small deformable objects where thermal fluctuations are important. In particular, we have quantified the phase diagram of assembly and kinetics of assembly when a system is quenched into an ordered region of the phase diagram. A unique feature of 3D materials is that multiple ordered states are possible. We have quantified the meta-stability of these ordered states and found a "diffusionless transformation" between ordered states. These transformations are well-known in metallic systems, but have only more recently been found in experiments with soft materials. Exchange of material and volume between objects also plays an important role in determining the ideal packing structures.

631005 Entropy Compartmentalization of Open Host-Guest Colloidal Clathrates with Rotating Guest Particles

Abstract:

Clathrates are open crystals in which atoms and molecules are arranged in a hierarchy of polyhedral cages. In this study, we report the entropy-driven self-assembly of new open-caged host-guest colloidal clathrate crystals. In Monte Carlo simulation, hard truncated triangular bipyramids (TBPs) self-assemble into five clathrates where guests and hosts are the same species. Crystallization occurs via entropy compartmentalization, where the entropy of the system partitions into a low entropy (host particle) subsystem and a high entropy (guest particle) subsystem with rotating guests. The type of clathrate depends on the amount of truncation, which determines the guest/cavity size ratio, while the dynamical behavior of the guests depends on the shape and symmetry of the guest and cavity. When TBPs are mixed with another particle shape, the traditional host-guest structure self-assembles, again driven by entropy maximization via subsystem compartmentalization.

626887 Effect of Particle Anisotropy and Rotational Symmetry on the Kinetics of Disorder-to-Order Phase Transitions

Abstract:

It is well-known that the phase behavior of polyhedral colloidal nanoparticles is influenced by their anisotropy and rotational symmetry.[1,2] In this study Metropolis Monte Carlo simulations are conducted to understand the effect of anisotropy and rotational symmetry on the kinetics of disorder to order transitions in athermal systems. Previous simulation studies have explored the effect on transition kinetics in particles with high rotational symmetry and minimal anisotropy.[3–5] This study thus primarily focuses on the effect of particle anisotropy and is exemplified through simulations of a system with limited rotational symmetry, namely, a family of gyrobifastigia (GBF) whose aspect ratio (AR) is varied keeping rotational symmetry unaffected.

A regular GBF (AR=√3) has a disordered phase and a crystal phase and no mesophases in-between. The disorder-to-order transition for GBF undergoes nucleation and growth with very large free energy barriers for low-to-moderate degrees of supersaturation, requiring biased simulations to overcome such barriers. This transition is precluded by glassy jamming at high degrees of supersaturation. In addition, the nuclei are aspherical and faceted. To elucidate this unique nucleus shape, direct measurement of the interfacial free energy was conducted for different crystal planes using the cleaving walls method. These measurements were used to perform a Wulff construction which predicts a nucleus shape consistent with our previously reported observations made through nucleus-size pinning.[6]

The unique kinetics for GBF is then studied for selected aspect ratios by varying the height of its triangles along the long axis while maintaining a square base. While reducing the GBF aspect ratio (up to AR=1) maintains a kinetic behavior akin to that of the base case (AR=√3), an elongation of the GBF (up to AR=2√3-1) lowers the nucleation barrier enough such that the ordering transition can be spontaneously observed in unbiased simulations. This trend is related to the change in interfacial free energy and nucleus geometry. A further increase in aspect ratio (i.e. to AR=3) stabilizes an orientationally ordered but translationally disordered mesophase which is also spontaneously formed in unbiased simulations (see Image). This nematic-like mesophase transitions to the crystalline phase upon compression. Translational and orientational order parameters allow us to follow these transitions and elucidate their mechanisms. These findings reveal that both phase behavior and the kinetics of phase transition of polyhedral nanoparticles can be engineered by controlling their aspect ratio, making otherwise inaccessible phases accessible. While the effect of AR on phase and kinetic behavior has been studied before through simulations of axisymmetric particle shapes (like spherocylinders[7,8]) hence having high rotation symmetry around the main particle axis, GBF shapes, lacking such symmetry, and having pronounced facet-facet interactions and complex ABCD crystal packing, introduce additional constraints on the entropic degrees of freedom leading to more complex phase ordering mechanisms.

Complementary studies were also performed on particle shapes with high rotational symmetry and minimal anisotropy. Such systems tend to form plastic 'rotator' mesophases that have particles arranged on a lattice but are not orientationally aligned, wherein the rotator lattice need not fully match that of the ensuing crystalline phase. To illustrate the effect of this incongruence, simulations were conducted for two particle shapes from the truncated cubes family with truncation parameter s=0.527 (TC52) and s=0.572 (TC57). TC52 has a dissimilar lattice geometry between its rotator and crystal while TC57 has an identical lattice. Our results illustrate how such differences set the stage for qualitatively distinct mechanisms of the diffusionless mesophase-to-crystal transition, the extend of cooperativity, and the appearance of novel intermediates.

(1) Agarwal, U.; Escobedo, F. A. Mesophase Behaviour of Polyhedral Particles. *Nat. Mater.* **2011**, *10* (3), 230–235.

(2) Damasceno, P. F.; Engel, M.; Glotzer, S. C. Predictive Self-Assembly of Polyhedra into Complex Structures. *Science (80-.).* **2012**, *337* (6093), 453–457.

(3) Thapar, V.; Escobedo, F. A. Localized Orientational Order Chaperones the Nucleation of Rotator Phases in Hard Polyhedral Particles. *Phys. Rev. Lett.* **2014**, *112* (4), 048301.

(4) Sharma, A. K.; Thapar, V.; Escobedo, F. A. Solid-Phase Nucleation Free-Energy Barriers in Truncated Cubes: Interplay of Localized Orientational Order and Facet Alignment. *Soft Matter* **2018**, *14* (11), 1996–2005.

(5) Karas, A. S.; Dshemuchadse, J.; Van Anders, G.; Glotzer, S. C. Phase Behavior and Design Rules for Plastic Colloidal Crystals of Hard Polyhedra: Via Consideration of Directional Entropic Forces. *Soft Matter* **2019**, *15* (27), 5380–5389.

(6) Sharma, A. K.; Escobedo, F. A. Nucleus-Size Pinning for Determination of Nucleation Free-Energy Barriers and Nucleus Geometry. *J. Chem. Phys.* **2018**, *148* (18), 184104.

(7) Schilling, T.; Frenkel, D. Self-Poisoning of Crystal Nuclei in Hard-Rod Liquids. *J. Phys. Condens. Matter* **2004**, *16* (19), 085505.

(8) Ni, R.; Belli, S.; Van Roij, R.; Dijkstra, M. Glassy Dynamics, Spinodal Fluctuations, and the Kinetic Limit of Nucleation in Suspensions of Colloidal Hard Rods. *Phys. Rev. Lett.* **2010**, *105* (8), 7–10.

630751 Two-Dimensional Self-Assembly of "Ionic" Colloidal Crystals

Abstract:

Two-dimensional crystals have promising electronic and optical applications. In order to design new functional 2D materials, a greater understanding of the phase behavior of soft particles in a 2D plane must be obtained. We investigate binary systems of particles in two dimensions using molecular dynamics simulations. We simulate "ionic" colloidal interactions, where unlike particles interact *via* attractive isotropic potentials and like particles interact *via* repulsive isotropic potentials. The resulting particle systems form crystalline assemblies across a range of compositions and particle sizes, producing compositionally ordered structures with various symmetries and complexities. We report the self-assembly phase diagram of these particles in 2D monolayers and discuss the observed ordered structures, as well as how their structural identification can be streamlined or automated. This work will inform the controlled assembly of 2D colloidal crystals and direct future assembly studies toward favorable areas of phase space.

622172 Self-Assembly Templates for Two-Dimensional Colloidal Crystals Derived from Symmetry

Abstract:

We investigate self-assembling rings that can template the organization of an arbitrary colloid into any desired periodic symmetry in two Euclidean dimensions. By viewing this as a tiling problem, we illustrate how the shape and chemical patterning of these rings may be derived from symmetry, rather than through iterative inverse-design methods. Moreover, we illustrate how these features are explicitly reflected by their orbifold symbol, which provides a natural language to express them. We performed molecular dynamics simulations to observe the self-assembly of these rings and found 5 different characteristics which could be easily rationalized on the basis of their orbifold symbol. These include systems which (1) undergo chiral phase separation, (2) are addressably complex, (3) exhibit self-limiting growth into clusters, (4) may form a nematic phase, and (5) those from symmetry groups which allow one to select rings which exhibit different self-assembly behaviors. We discuss how the curvature of a ring's edges plays an integral role in achieving correct self-assembly, and we illustrate how to derive these ring shapes. This provides a method for patterning colloidal systems at interfaces without explicitly programming this information onto the colloid itself.

632672 Free Energy Contributions to Template-Assisted Self-Assembly of Sub-10 Nm Particles from Steered Molecular Dynamics Simulations

Abstract:

Directed self-assembly of nanoparticles (DSA-n) over a templated surface has extensive applications in miniaturization of nanodevices, including semiconducting, plasmonic,and photovoltaic devices. The pre-designed templated surface is able to selectively capture nanoparticles, induce interaction-based orientation and even organize them into a complex structure using a bottom-up fabrication process. Taking advantage of the great potential in template-assisted DSA-n, many studies have utilized a thin liquid film sliding over a templated surface to realize the assembly process, which allows high yield of nanoparticle deposition with specific ordering or alignment. In DSA-n with sub-10 nm particles, the thin liquid film provides a restriction of long-range molecular interactions from multiple nanoparticle layers; that in turns facilitates the positioning of nanoparticles in DSA-n. Although the fabrication strategy of template-assisted DSA-n is able to achieve ordered arrays of nanoparticles over large areas of the templated surface, the yield of self-assembly is sensitive to the nanoparticle density among other parameters. This variation of yield at different nanoparticle density indicates that the density is one of the structural determinants of the efficiency of DSA-n, and plays a key role in the energetics of the complex self-assembly process. In this process, computational modeling has come illustrative to aid with explaining the contributions to the associated free energy.

Calculation of free energy change between two states, in the liquid film and on the templated surface, for a self-assembled nanoparticle will be immensely aided in locating the optimal yield. Various molecular-simulation-based approaches such as probability ratio method (PRM) in molecular dynamics (MD) studies aimed at estimating the free energy have been reported. PRM tracks the positions of neighboring molecules as a function of time, which are transformed into a probability distribution function with respect to the states of tracked nanoparticle. The probability distribution function is then used to calculate the relative free energy. Nevertheless, reports indicate that PRM may suffer from insufficient sampling, in addition to the unreliable estimates of the probability density distribution across the surface. An alternative method to the PRM is the thermodynamic integration (TI) in which the position-dependent potential of mean force, as a representative of the free energy, acting on the molecule of interest at different positions is integrated over the distances of constraint forces. Both PRM and TI are equilibrium methods that require the simulation environment to be at equilibrium during the transition of tracked molecule. This means the simulation needs to regain an equilibrium state after each motion of tracked molecule. As such, aforementioned techniques seem not suitable for measuring free energy contributions to the DSA-n, considering that the process is continuously out of equilibrium. The lengthy equilibration processes can therefore lead to a high computational cost.

Enlightened by the Jarzynski's equality in statistical mechanics, the steered molecular dynamics (SMD) extends the realm of free energy calculation to non-equilibrium simulations at a reasonable computational cost. SMD simulations have been widely applied in revealing mechanical process of protein binding, folding, and stretching. Recently some studies have also attempted to apply SMD simulations to calculate the free energy by employing Jarzynski's equality. Jarzynski's equality is a relation that explains the free energy between two equilibrium states, ΔE12, in terms of the external work done through a non-equilibrium process between above-mentioned two states, W12. The work is obtained from an ensemble of finite-time measurements of external work performed on the system during the transition, i.e.

$\exp(-\beta\Delta E12) = \langle\exp(-\beta W12)\rangle$ (1)

where $\beta = (kBT)-1$, kB is the Boltzmann constant, T is the temperature, and $\langle . \rangle$ represents averaging over independent simulations. Applying Jarzynski's equality, SMD simulations enable free energy calculations from non-equilibrium processes. To obtain an accurate estimate of free energy, one needs to collect enough samples as the average of exponential term in Equation 1 requires a large number of independent SMD simulations.

The purpose of this work is to utilize Jarzynski's equality to calculate the free energy contributions to the DSA-n into templated surfaces from SMD simulations. For SMD simulations, we use many-body dissipative particle dynamics (MDPD). As coarse- gained molecular dynamics (MD) simulations, MDPD simulations are capable of combining both microscopic and mesoscopic interactions while capturing the properties of fluid interfaces at much larger scales than traditional MD simulations, such as those in evaporation-mediated DSA-n. Without loss of generality, we focus on directed self-assembly of a spherical nanoparticle into a circular nanocavity, which is etched out of a substrate with an otherwise flat surface. The self-assembly is observed in two cases: from (1) a stagnant bulk thin liquid film, and (2) a thin liquid film with flow field and a receding interface to the nanocavity. In the second case, to generate the flow field and receding interface in the thin liquid film, we model a hydrophilic surface with a downward moving piston. The transition of states is induced by attaching one end of a spring to a tracer nanoparticle while pulling the other end towards the nanocavity in the substrate, mimicking the DSA-n as shown in Figure 1. Free energy variation is calculated as a function of pulling distance. All MDPD simulations are performed using LAMMPS package at a constant temperature in a manner that satisfy the Markov property and detailed balance. The initial states, which depends on the positions and velocities of the nanoparticles, liquid, and the tracer particle are sampled from the canonical ensemble corresponding to the Hamiltonian of the particle system.

Our results suggest that overall the free energy contribution to DSA-n decreases with the nanoparticle density until a critical density after which the free energy increases. Moreover, we find that the free energy is minimum at this critical density whether the liquid film is stagnant or moving. The results indicate that at low nanoparticle density, it is entropically less favorable for the nanoparticle to be deposited in the nanocavity; however, at the critical density, we find that there is no change in the entropy as the nanoparticle leaves the bulk liquid and enters the nanocavity. Once the nanoparticle density goes beyond the critical density, there will be an energetic barrier for the nanoparticle deposition due to a favorable interaction with other nanoparticles in the bulk, which increases the free energy of DSA-n.

Figure 1. Panel A presents a top view of a liquid film going over a templated surface used in SMD simulations. Side view of SMD simulations at 3 stages (from B to D) show the progression of a tracer nanoparticle deposition guided by a spring while the liquid film is receding downward.

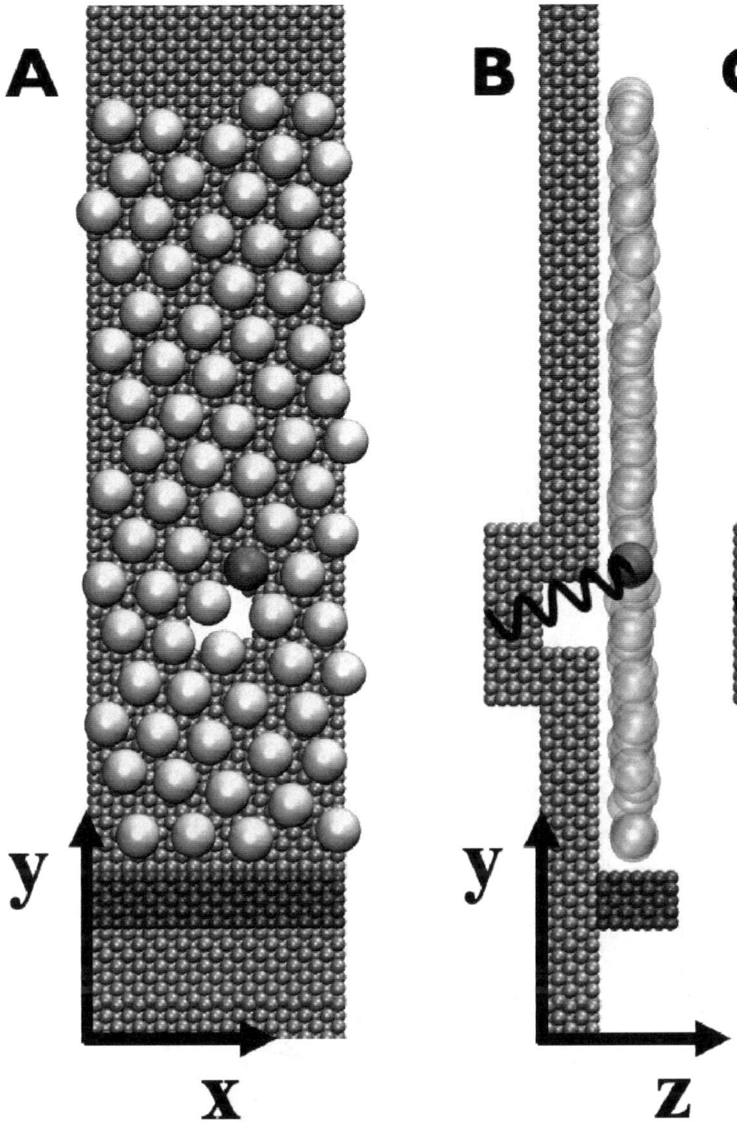
A
B
C
D
y
x
y
z

632748 Parallel Bias Method for Free Energy Study of Colloidal Clusters

Abstract:

Controlling the assembly of colloidal particles into specific structures hasbeen a long-term goal of the soft materials community. Much can be learnedabout the process of self-assembly by examining the early-stage assembly of par-ticles into clusters. Free energy calculations help to determine how one mighttarget specific clusters relative to other isomers. Such computational studiesbenefit from sampling along with a variety of collective variables (CV) within agiven system, though increasing complexity limits the extent of sampling whichcan be achieved. Here, we explore the parallel bias metadynamics technique asan alternative to high-dimensional sampling to extract free energy landscapesand stability criteria for small colloidal clusters with multiple metastable states.We anticipate this strategy to be extensible to free energy calculations in com-plex self-assembling systems, particularly when coupled with emerging data-driven CV discovery and sampling methods.

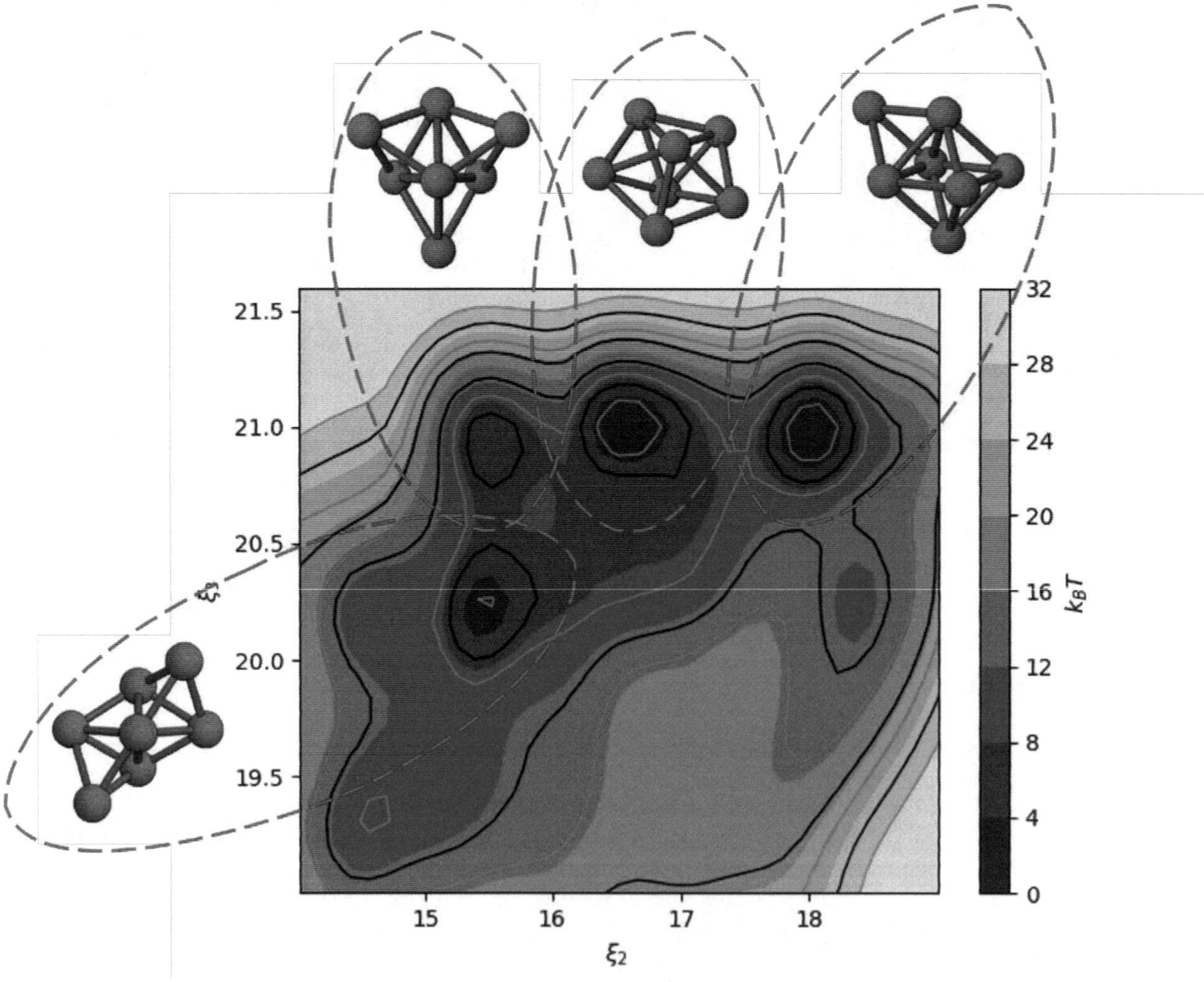

623578 Understanding Sequence-Dependent DNA Dynamics through Self-Associative Machine Learning and Temperature-Jump Spectroscopy

Abstract:

Despite rapid advances in DNA nanotechnology and a robust understanding of the associated thermodynamics, the sequence-dependent mechanisms of DNA hybridization are not fully understood. In this work, we investigate these dynamics by performing equilibrium coarse-grained simulations of oligonucleotide sequences with varied G:C placement. We employ recently developed State-Free Reversible VAMPnets to directly learn the slowest dynamical modes of each sequence and to optimize Markov State Models construction. Furthermore, we perform elevated temperature simulations to recapitulate temperature-jump IR spectroscopy and FTIR data collected on the oligonucleotides. For repetitive sequences, we find a spectrum of slow dynamics associated with out-of-register base pairing and kinetically relevant transitions between these states. In contrast, G:C pairs near the center of the duplex induce more rapid fraying dynamics. In both cases, hybridization/dissociation mechanisms deviate from an "all-or-nothing" model. Our computational predictions are in excellent accord with experiment and elucidate new insights into sequence-dependent dynamics, metastable structures, and relative timescales. These results can inform a generalized model between DNA sequence and hybridization mechanism with application towards nanotechnologies such as DNA-PAINT and barcoding.

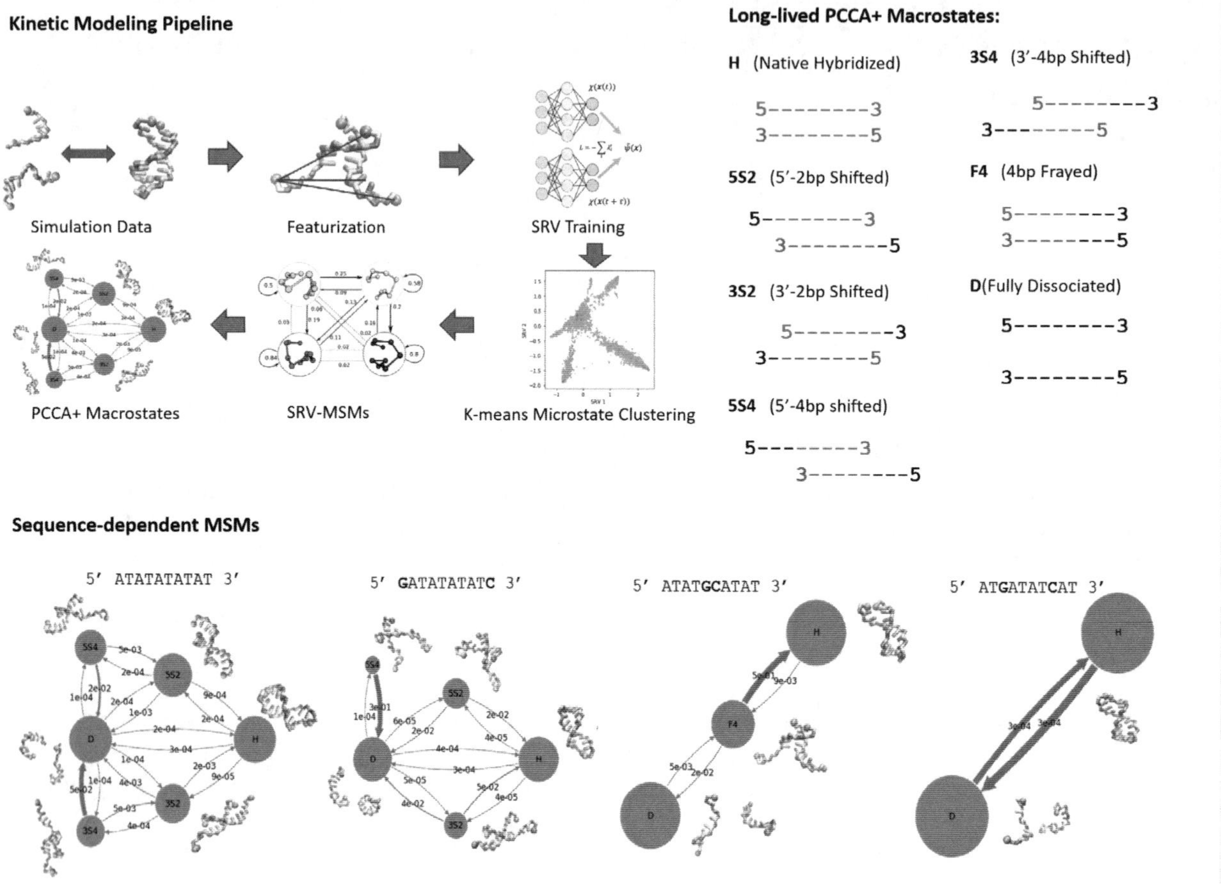
Kinetic Modeling Pipeline
Simulation Data
Featurization
SRV Training
PCCA+ Macrostates
SRV-MSMs
K-means Microstate Clustering
Long-lived PCCA+ Macrostates:
H (Native Hybridized)
3S4 (3'-4bp Shifted)
5S2 (5'-2bp Shifted)
F4 (4bp Frayed)
3S2 (3'-2bp Shifted)
D(Fully Dissociated)
5S4 (5'-4bp shifted)
Sequence-dependent MSMs
5' ATATATATAT 3'
5' GATATATATC 3'
5' ATATGCATAT 3'
5' ATGATATCAT 3'

632088 Hierarchical Self-Assembly of Thermo-Responsive Nanomaterials from Lipidated Disordered Proteins

Abstract:

We investigate hybrid functional biomaterials by combining biocompatible building blocks—lipids and peptides—with distinct physicochemical and functional properties. The work involved combining complementary expertise in biosynthesis as well as computational and experimental characterization of lipidated proteins to elucidate the design principles for lipid-modified intrinsically disordered protein polymers. The simulations were performed using atomistic and coarse grained molecular dynamics to determine the design rules for programming the thermo-response and hierarchical assembly of lipidated-peptides at physiologically relevant temperatures. The development of coarse-grained force fields required integrated and iterative feedback from in silico and experimental characterization techniques with data analytics. Our results show development on new hybrid materials that are promising candidates for applications ranging from healthcare to nano- and biotechnology.

630817 Exploring the Role of Lipid Tail Length on the Structure of Self-Assembled Stratum Corneum Lipids

Abstract:

The barrier function of the human epidermis is primarily located in the stratum corneum (SC), which is composed of corneocytes surrounded by a dense, lamellar lipid matrix made up of ceramides (CERs), cholesterol, and free fatty acids of various lengths. While, experimentalists are able to reproduce the lipid organization of intact SC with model systems using synthetic lipids,[1] the role of each lipid in determining the molecular structure of the SC, and hence the barrier properties of skin, is not well understood due to the complex nature of the SC lipid matrix. To address this need, we have used molecular dynamics (MD) simulations to study the microscopic behavior of the SC lipids.[2,3] MD simulations of SC systems using atomistic models are typically performed with pre-assembled bilayer structures. However such simulations can suffer from initialization bias due to low lipid mobility and further, simple bilayers are not representative of the complex multilayer structure of the SC.[4] While self-assembly eliminates any bias associated with assuming a starting structure, the long timescales and large system sizes needed to study multilamellar structures make performing self-assembly using atomistic models infeasible. Thus, we employ computationally efficient coarse-grained (CG) models that have been parameterized using the Multi-State Iterative Boltzmann Inversion (MS-IBI) method[2] based on simulation data from fully atomistic models.[3] Here, we examine mixtures of lesser-studied ceramides, including CERs AP, NP, AS, and NS that form the short periodicity phase (SPP) of the SC lipid matrix. We examine how free fatty acid chain length and CER acyl chain length affects lipid packing and structural properties in multilamellar skin lipid systems in order to compare to experimental work. The simulations show the effect of interdigitation of the lipid tails leading to changes in tail confirmation and lamellar repeat distances, as well as on the stability of the SPP. Through a molecular-level understanding of the packing and formation of these regions, we are able to provide molecular level explanations of the experimental results.[5]

References:

[1]J. A. Bouwstra and M. Ponec, "The skin barrier in healthy and diseased state," *Biochimica et Biophysica Acta - Biomembranes*. 2006, doi: 10.1016/j.bbamem.2006.06.021.[2]T. C. Moore, C. R. Iacovella, and C. McCabe, "Development of a Coarse-Grained Water Forcefield via Multistate Iterative Boltzmann Inversion," *Found. Mol. Model. Simulation. Mol. Model. Simul.*, pp. 37–52, Sep. 2016, doi: 10.1007/978-981-10-1128-3_3.[3]T. C. Moore, C. R. Iacovella, R. Hartkamp, A. L. Bunge, and C. McCabe, "A Coarse-Grained Model of Stratum Corneum Lipids: Free Fatty Acids and Ceramide NS," *J. Phys. Chem. B*, vol. 120, no. 37, pp. 9944–9958, 2016, doi: 10.1021/acs.jpcb.6b08046. [4]I. Iwai *et al.*, "The human akin barrier is organized as stacked bilayers of fully extended ceramides with cholesterol molecules associated with the ceramide sphingoid moiety," *J. Invest. Dermatol.*, vol. 132, no. 9, pp. 2215–2225, 2012, doi: 10.1038/jid.2012.43. [5]D. Kessner, A. Ruettinger, M. A. Kiselev, S. Wartewig, and R. H. H. Neubert, "Properties of ceramides and their impact on the stratum corneum structure: A review - Part 2: Stratum corneum lipid model systems," *Skin Pharmacology and Physiology*. 2008, doi: 10.1159/000112956.

623179 Molecular Simulation of Adsorption Hysteresis of n-Alkanes in Nanoporous Materials

Abstract:

Adsorption of alkanes is crucial to a wide variety of applications ranging from adsorption cooling[1] to separations[2] to storage of natural gas.[3] Previous studies have shown that porous materials such as zeolites and metal-organic frameworks (MOFs) can be useful for all these applications.[4,5] Despite the progress in these applications, capillary condensation and hysteresis, which may occur in such systems, are still not well understood. In addition, although hysteresis has been investigated for decades, most works, except for a few simulation studies,[6,7] have focused on simple molecules such as argon, nitrogen, and methane.[8,9]

Here we present a computational study of hysteresis of C_1 to C_6 n-alkanes in selected MOFs from the ToBaCCo 1.0 MOF database of Colon et al.[10] As complementary model systems, slit pores of varying widths were also studied. Two complementary simulation approaches were employed. First, we used grand canonical Monte Carlo (GCMC) and obtained the hysteresis loops (if any) by performing GCMC simulations sequentially "up and down" the isotherm, where the final configuration from a given pressure is used as the initial configuration for the simulation at the next pressure. In the second approach, we used canonical ensemble molecular dynamics (MD) simulations at fixed loadings combined with Widom insertions[11] that employ configurational-bias Monte Carlo (CBMC)[12,13] to determine the fugacities. Note the complementary nature of the approaches: GCMC sets the fugacity and calculates the loading, while MD plus Widom insertions sets the loading and calculates the fugacity. We show that the MD plus Widom approach is able to simulate the van der Waals (vdW) loop for an isotherm while GCMC cannot. GCMC simulations generate hysteresis loops, but they may suffer from sampling inefficiencies: at high loading, GCMC insertion and deletion moves have low acceptance rates due to the high density in the pores. A canonical simulation avoids these inefficiencies by fixing the number of molecules in the simulation box. Thus, a vdW loop serves as an additional proof that hysteresis observed in GCMC has a physical origin and is not simply due to poor sampling during the simulation. Utilizing these two methods, we were able to identify MOFs that exhibit hysteresis for n-butane and n-hexane at room temperature. The presence of hysteresis in these systems may affect their use in various applications.

We also tested the applicability of ideas from the principle of corresponding states as well as the Dubinin and Raduschekevich (DR) theory to n-alkane adsorption in MOFs. Following the concept of corresponding states, we simulated methane adsorption at the same reduced temperatures as for n-butane and n-hexane at 298 K, and we were able to observe hysteresis and vdW loops also for methane. Using the inverse of saturation loading as affinity coefficients for methane through n-hexane, we were able to collapse the isotherms from GCMC of these molecules onto a characteristic curve,[14] although there is some scatter. We were also able to analyze the effect of size and flexibility of long chain molecules on adsorption and hysteresis comparing to small molecules: smaller molecules have smaller hysteresis loops and form plateaus before capillary phase transition, while larger and more complicated adsorbates broaden the hysteresis loops in terms of fugacity and loading ranges. These results indicate the usefulness of traditional adsorption and thermodynamic theories for novel sorbents such as MOFs and suggest future work to refine these theories.

References:

(1) Chen, H.; Chen, Z.; Farha, O. K.; Snurr, R. Q. High Propane and Isobutane Adsorption Cooling Capacities in Zirconium-Based Metal–Organic Frameworks Predicted by Molecular Simulations. *ACS Sustainable Chem. Eng.* **2019**, *7* (22), 18242–18246. https://doi.org/10.1021/acssuschemeng.9b05368.

(2) Sholl, D. S.; Lively, R. P. Seven Chemical Separations to Change the World. *Nature News* **2016**, *532*

(7600), 435. https://doi.org/10.1038/532435a.

(3) Zhang, H.; Deria, P.; K. Farha, O.; T. Hupp, J.; Q. Snurr, R. A Thermodynamic Tank Model for Studying the Effect of Higher Hydrocarbons on Natural Gas Storage in Metal–Organic Frameworks. *Energy & Environmental Science* **2015**, *8* (5), 1501–1510. https://doi.org/10.1039/C5EE00808E.

(4) Yaghi, O. M.; O'Keeffe, M.; Ockwig, N. W.; Chae, H. K.; Eddaoudi, M.; Kim, J. Reticular Synthesis and the Design of New Materials. *Nature* **2003**, *423* (6941), 705–714. https://doi.org/10.1038/nature01650.

(5) R. Long, J.; M. Yaghi, O. The Pervasive Chemistry of Metal–Organic Frameworks. *Chemical Society Reviews* **2009**, *38* (5), 1213–1214. https://doi.org/10.1039/B903811F.

(6) Jin, Z. Bubble/Dew Point and Hysteresis of Hydrocarbons in Nanopores from Molecular Perspective. *Fluid Phase Equilibria* **2018**, *458*, 177–185. https://doi.org/10.1016/j.fluid.2017.11.022.

(7) Rasmussen, C. J.; Vishnyakov, A.; Neimark, A. V. Calculation of Chemical Potentials of Chain Molecules by the Incremental Gauge Cell Method. *J. Chem. Phys.* **2011**, *135* (21), 214109. https://doi.org/10.1063/1.3657438.

(8) Neimark, A. V.; Vishnyakov, A. Gauge Cell Method for Simulation Studies of Phase Transitions in Confined Systems. *Phys. Rev. E* **2000**, *62* (4), 4611–4622. https://doi.org/10.1103/PhysRevE.62.4611.

(9) Zeng, Y.; Prasetyo, L.; Tan, S. J.; Fan, C.; Do, D. D.; Nicholson, D. On the Hysteresis of Adsorption and Desorption of Simple Gases in Open End and Closed End Pores. *Chemical Engineering Science* **2017**, *158*, 462–479. https://doi.org/10.1016/j.ces.2016.10.048.

(10) Colón, Y. J.; Gómez-Gualdrón, D. A.; Snurr, R. Q. Topologically Guided, Automated Construction of Metal–Organic Frameworks and Their Evaluation for Energy-Related Applications. *Crystal Growth & Design* **2017**, *17* (11), 5801–5810. https://doi.org/10.1021/acs.cgd.7b00848.

(11) Widom, B. Some Topics in the Theory of Fluids. *J. Chem. Phys.* **1963**, *39* (11), 2808–2812. https://doi.org/10.1063/1.1734110.

(12) Siepmann, J. I.; Frenkel, D. Configurational Bias Monte Carlo: A New Sampling Scheme for Flexible Chains. *Molecular Physics* **1992**, *75* (1), 59–70. https://doi.org/10.1080/00268979200100061.

(13) Dubbeldam, D.; Calero, S.; Ellis, D. E.; Snurr, R. Q. RASPA: Molecular Simulation Software for Adsorption and Diffusion in Flexible Nanoporous Materials. *Molecular Simulation* **2016**, *42* (2), 81–101. https://doi.org/10.1080/08927022.2015.1010082.

(14) Severson, B. L.; Snurr, R. Q. Monte Carlo Simulation of N-Alkane Adsorption Isotherms in Carbon Slit Pores. *J. Chem. Phys.* **2007**, *126* (13), 134708. https://doi.org/10.1063/1.2713097.

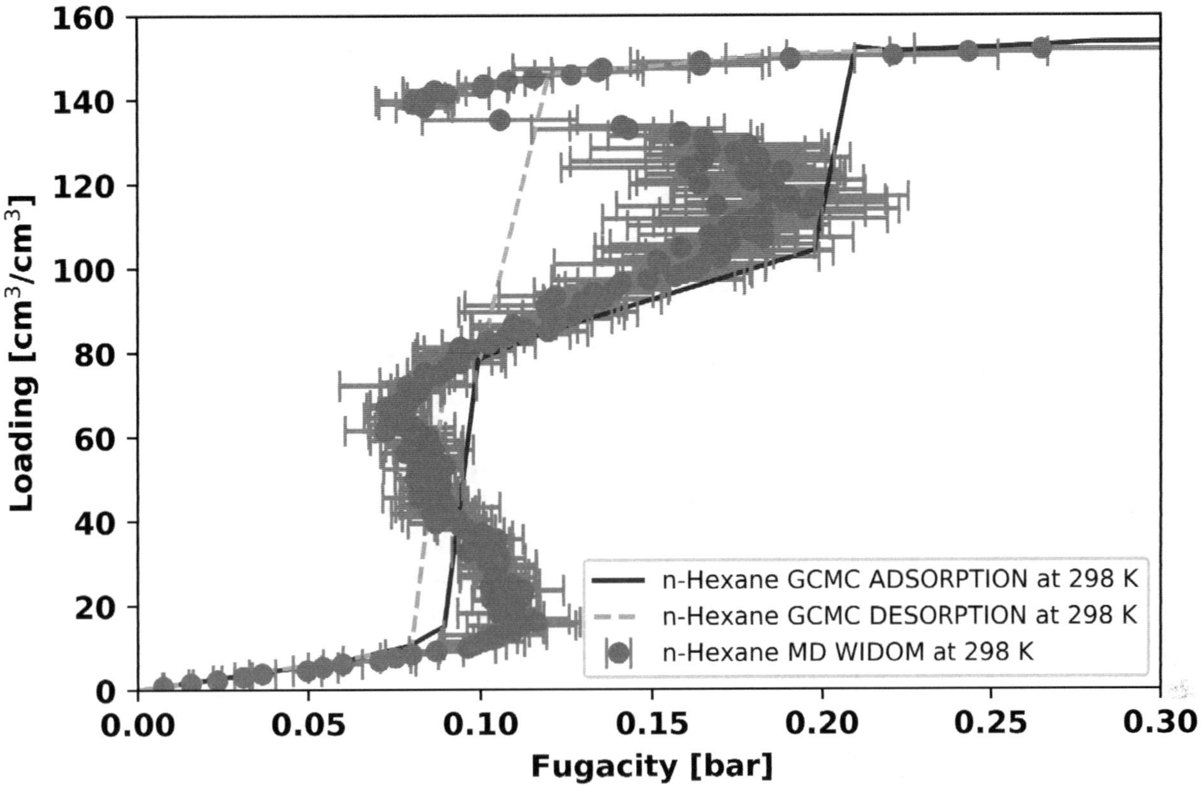

Loading [cm³/cm³]
Fugacity [bar]
0
20
40
60
80
100
120
140
160
0.00
0.05
0.10
0.15
0.20
0.25
0.30
n-Hexane GCMC ADSORPTION at 298 K
n-Hexane GCMC DESORPTION at 298 K
n-Hexane MD WIDOM at 298 K

623219 Improving Computational Assessment of Water Adsorption to Enable Large-Scale Screening of Porous Materials for Water Harvesting

Abstract:

Water harvesting using porous materials has recently drawn considerable scientific attention for its potential in producing usable water. This technology works on the principle of temperature and pressure swing adsorption (T/PSA). In this application, water is adsorbed into nanoporous materials at high relative humidity and/or low temperature while being desorbed at low relative humidity and/or high temperature. One of the most important performance metrics of a nanoporous material for such an application is its deliverable capacity—the difference between how much the material can adsorb (take in) at the adsorption condition and desorb (release) at the desorption condition. A key step to the development of such technologies is the selection of optimal adsorbent materials. In this case, for example, an adsorbent with a high deliverable capacity for the desired adsorption and desorption conditions is preferred. However, owing to the rapid advancement of materials synthesis in the last two decades, there are thousands of available materials. High-throughput computational approaches can therefore play an important role in facilitating the search for optimal adsorbents.

The widely-used approach to study water adsorption in materials is the grand canonical Monte Carlo (GCMC) method. However, a drawback of this method is that it favors short-term reduction of energy over a long-term reduction. This can be particularly problematic for water adsorption studies because water may form clusters during adsorption due to their strong hydrogen-bonding networks. This causes water adsorption simulations to converge very slowly, leading to uncertainties in the outcome of the simulations as well as long computational wall-time. In this study, we demonstrated that this drawback can be mitigated using techniques from a class of methods known as flat histogram methods. These methods force (bias) the system to sample all loading macrostates with equal probability. A variant of this technique—the NVT + W (NVT simulations and Widom sampling) has been used in this work. An important advantage of flat histogram methods is that the macrostate probability distributions can be easily reweighted to compute adsorption uptake at other temperatures and pressures. Thus, a single macrostate probability distribution can yield adsorption isotherms at a range of temperatures. We applied and validated this approach to study CO_2 adsorption in the metal-organic framework (MOF) structure IRMOF-1, as well as demonstrated its potential in studying water adsorption in MOF-806. While the probability distribution yielded by the NVT + W method is quite useful in predicting adsorption behavior, it could be computationally expensive as it requires up to hundreds of calculations per structure or tens of calculations with appropriate interpolation approaches.

Fortuitously, to screen materials for a particular application, the knowledge of the entire probability landscape (entire isotherm) is not necessary. When computing the deliverable capacity, the only properties of interest are the loading values at the adsorption and desorption conditions. Thus, we further proposed an approach for large-scale screening of materials—the so-called C-map method, which can estimate the equilibrium loading values at a specific temperature (T) and pressure (P) much more efficiently. The method exploits the fact that, for a given value of T, P, and loading (N), the ratio of the probability of deletion to that of insertion indicates the position of that loading macrostate (N) relative to the closest local probability maximum. A ratio of unity implies that the loading (N) is a local probability maximum or minimum. A ratio of < 1 indicates that the local probability maximum occurs at a larger loading than N. Conversely, a ratio > 1 indicates that the local probability maximum occurs at a smaller loading than N. Thus, for given adsorption and desorption conditions, the range of local probability maxima can be found with only a few NVT + W calculations. This knowledge leads to the prediction of a range of adsorption loadings, desorption loadings, and deliverable capacities for the simulated structures. The C-map method was subsequently applied to screen a large number of topologically diverse MOF structures for water harvesting. We identified a number of promising MOFs that can provide deliverable capacities larger than that of the state-of-the-art materials, as well as shed light on their atomic-level adsorption mechanism. Structure-property relationships for water adsorption in MOFs were also established, offering insights into the future rational design of materials with improved harvesting performance. We anticipate that this study can facilitate the search for optimal materials to harvest atmospheric water at a given geographical location.

625318 Predictive Modeling of Microporous Polymeric Adsorbents of Complex Sorbates

Abstract:

While the use of molecular simulations is standard practice in many research areas, studies focusing on microporous polymeric adsorbents are often limited to a few amorphous frameworks, a handful of adsorbates, and typically restricted to rigid framework methods. This talk will highlight our recent work using molecular simulations to assess microporous polymer structure rearrangements and adsorption performance across dilute, swollen, and plasticized loading conditions. Despite the prevalence of species diversity in industrial gas mixtures, a broader understanding of the effect that chemical identity has on polymeric adsorbent restructuring is lacking. In light of this, we report a screening study to assess 15 polymers of intrinsic microporosity (PIMs) in combination with 23 diverse adsorbate species, which results in 345 distinct single-component isotherms adsorption isotherms and a database of over 240,000 polymer conformations. The simulations were conducted using a combined Monte Carlo and molecular dynamics approach that allows for the analysis of sorption-induced changes to microporous structural features. Using this data, we have developed correlations that can provide rapid approximations of polymer swelling and fractional free volume dilation with appreciable accuracy. These approximations are a function of readily obtainable adsorbate properties (critical temperature and molar mass) and adsorbent features (surface area and pore volume), which is indicative of their potential to be broadly applied in both computational and experimental efforts. Additionally, the separation selectivity of the 253 unique binary pairs at dilute loading conditions is reported and high-performing adsorbent systems are identified. Overall, our study reports the largest dataset of simulated adsorption for amorphous polymeric materials to date with framework rearrangement accounted for. This provides key first steps toward achieving screening studies of amorphous polymers that rival the magnitude of ordered adsorbents.

627560 Finding Selective Adsorbent Material for Selenium Oxo-Anion Removal for Water Remediation Using Ab-Initio Study with Hybrid Solvent Method.

Abstract:

Selenium is highly soluble in water as selenate and selenite and is toxic to human and aquatic life even at very low concentration (TWA 0.2 mg/m^3). Adsorption is a promising remediation strategy, however existing metal oxides have low selectivity towards Se and As oxoanions. Here we use first principal calculations to study the facet and material dependent behavior of Se(IV), Se(VI) and S(VI) adsorption. We delineate material properties affecting the adsorption characteristics such as valance electrons, and bond lengths , and surface properties such as interfacial water network, oxo-anion surface configuration, pH, etc. We find that the water network has significant control over adsorption behavior. Since number of H on the surface and oxo-anion directly control the water network, Al_2O_3 is not selective adsorbent towards selenate as compared to sulphate. The adsorption is highest at low pH, as the surface is concentrated with excess H. The facet dependent adsorption is also related to changes in water network over the surface. The adsorbent's electronegativity also affects the adsorption energy of oxo-anion. The hematite binds Se(VI) and Se(IV) more strongly than Al_2O_3 surfaces because the Fe atoms are able to donate electrons to oxoanion-surface bonds . The adsorbent material has stronger effect on oxo-anions in inner sphere configuration. The potential adsorbent material, therefore, needs to bind the oxo-anion in inner sphere configuration, for highly selective adsorption.

631566 Modeling Hydrocarbons Adsorption in Amorphous Nanoporous Carbonaceous Materials

Abstract:

Predicting adsorption in disordered nanoporous carbonaceous materials is important for developing adsorption and membrane separations of hydrocarbons, as well as for oil and gas recovery from shale reservoirs. Many previous molecular simulations studies have typically considered simplified slit pore structures that do not account for the nanoscale specifics of carbonaceous materials. In this work, we use the grand canonical Monte Carlo method to simulate adsorption on amorphous carbonaceous structures, which are created by imitating the conditions of nanoporous carbon formation. We characterize the structures by nitrogen, argon, and carbon dioxide adsorption at standard temperatures. We also simulate adsorption of hydrocarbons on the model structures, and correlate these results with available experimental data. This work provides a new insight into the pore structure and prediction of hydrocarbon adsorption in carbonaceous materials, including the conditions of standard lab experiments, as well as conditions that are not readily accessible in the lab.

629668 Calculation and Analysis of RAST Activity Coefficients for Ethanol/Water Adsorption in All-Silica Zeolites

Abstract:

Understanding mixture adsorption in nanoporous materials such as zeolites is a key piece of information for designing efficient adsorption- or membrane-based separation processes. In practice, the Ideal/Real Adsorbed Solution Theory (IAST/RAST) of Myers and Prausnitiz[1] is often used to predict mixture isotherms from single-component data. However, significant deviations from IAST have been reported for nonideal mixtures[2], while measuring or predicting non-ideal mixture adsorption isotherms remains a challenging task both experimentally and computationally. In this work, we extend our previous study on the adsorption of binary water/ethanol mixtures and apply the RAST formalism to extract the pore-phase activity coefficients in all zeolites catalogued by the Structure Commission of the International Zeolite Association. Grand-canonical Monte Carlo simulations are used to calculate the binary isotherms and the unary isotherms of ethanol. To obtain the unary isotherms of water that typically require long equilibration periods, we adopt an alternative, computationally less expensive method by fitting a Langmuir-Freundlich isotherm for each zeolite to its Henry's constant (computed from Widom insertion), saturation loading (estimated from pore volume), and pressure at half loading (computed from an iterative GCMC procedure). With the three isotherms, RAST activity coefficients are extracted and expressed in terms of the Margules activity model. Comparing ethanol-selective zeolites of different pore dimensions, we observe that activity coefficients are framework and topology dependent, and generally increase with pore sizes. The activity coefficients are higher than those in bulk solutions for medium- and large-pore zeolites (e.g., MFI, BEA, FAU zeolites) and lower for small-pore zeolites (e.g., LTA zeolite). We examine how these observations generalize to the adsorption of other aqueous mixtures and anticipate that these results will provide new insights into how confinement modifies bulk nonidealities.

[1] Myers & Prausnitz. AIChE J. 11, 121 (1965).

[2] Bai, Tsapatsis & Siepmann. Langmuir 28, 15566 (2012).

624331 Modeling the Adsorption Behavior in Irmofs Using Monte Carlo Simulations

Abstract:

Metal-organic frameworks are porous materials built from inorganic nodes connected by organic linkers that are increasingly being used for applications in gas adsorption and storage owing to their high porosities and crystallinity. Isoreticular MOFs, a series of MOFs existing in the same framework topology but differing in the substituent groups on the linker installed during formation of the framework or through post-synthetic modifications, have shown promise to tailor the guest-host interactions, thereby altering the adsorption behavior of the guest molecules in the MOFs.[1] The large number of MOFs resulting from this tunability can be investigated computationally for evaluating their effectiveness for a desired application. The structural flexibility of these materials due to the relatively weaker coordination links between the inorganic and organic units, however, allows for structural deformations of the framework in the presence of guest molecules, thereby making it challenging to model the adsorption behavior of these materials.[2] Conventional grand canonical Monte Carlo (GCMC) adsorption simulations treat the framework as rigid, thereby limiting their ability to model the flexibility of MOFs in the presence of guest molecules. Recent efforts to address this challenge either employ complex hybrid (Monte Carlo + molecular dynamics) simulation protocols[2] or use GCMC simulations in the "osmotic" statistical ensemble which may require *apriori* knowledge of the different deformed configurations of the framework.[3,4]

In this work, we have employed force-field-based Monte Carlo simulations in the $N_{guest}N_{host}PT$-Gibbs ensemble to investigate the adsorption of N,N-dimethylformamide (DMF), *para*-xylene, and *ortho*-xylene in the classic IRMOF-1 (MOF-5) and its linker-functionalized isoreticular analogue, IRMOF-3, which are both modelled as flexible MOFs. In this ensemble, only guest molecules are allowed to transfer between reservoir phase and MOF phase, the volume of the reservoir phase can change only isotropically, whereas changes in individual elements of the **H** matrix allow for both volume and shape fluctuations for the MOF phase. MOF-5 is built from inorganic Zn_4O nodes connected by 1,4-benzenedicarboxylate (BDC) linkers while the inorganic nodes in IRMOF-3 are coordinated by 2-amino-1,4-benzenedicarboxylate (NH_2-BDC) linkers. We observe lower onset pressures for sorbate uptake in IRMOF-3 as compared to MOF-5 owing to favorable interactions of the sorbates with the linker amine groups in IRMOF-3 as evidenced from their respective radial and spatial distribution plots. The isotherms for both *p*-xylene and *o*-xylene adsorption in MOF-5 and IRMOF-3 at 295 K indicate a "pore-filling" adsorption mechanism in these MOFs with negligible framework deformation. Remarkably, these isoreticular MOFs exhibit different DMF adsorption characteristics. While the isotherm for DMF adsorption in MOF-5 at 295 K shows a steep increase in the DMF uptake corresponding to a "pore-filling" adsorption mechanism, the DMF adsorption isotherm in IRMOF-3 at 295 K shows a gradual increase in the DMF uptake which is significantly influenced by the framework deformation. Indeed, at $P/P_o = 0.001$, we observe an approximately 30% reduction in the unit cell volume and a change in the unit cell shape from cubic to triclinic in order to maximize the DMF uptake at this intermediate pressure. Moreover, our TraPPE-based "flexible" force field employed for the IRMOFs in this work along with a new TraPPE force field proposed for DMF is able to capture the previously reported dynamic binding of DMF to the Zn^{2+} ions of the node thereby leading to an increased coordination number of the Zn^{2+} ions in the MOF. Interestingly, we also observe a bimodal distribution of the sorbate loading in individual pores of the MOF, thereby pointing to the existence of two types of pores in the MOF. This behavior is attributed to preferred linker orientations identified in the MOF resulting in adjacent pores of differing pore volume and hence, different sorbate loadings.

References:

(1) Babarao, R.; Jiang, J. Molecular Screening of Metal–Organic Frameworks for CO_2 Storage. *Langmuir* **2008**, *24* (12), 6270–6278. https://doi.org/10.1021/la800369s.

(2) Rogge, S. M. J.; Goeminne, R.; Demuynck, R.; Gutiérrez-Sevillano, J. J.; Vandenbrande, S.; Vanduyfhuys, L.; Waroquier, M.; Verstraelen, T.; Van Speybroeck, V. Modeling Gas Adsorption in Flexible Metal–Organic Frameworks via Hybrid Monte Carlo/Molecular Dynamics Schemes. *Adv. Theory Simul.* **2019**, *2* (4), 1800177. https://doi.org/10.1002/adts.201800177.

(3) Jeffroy, M.; Fuchs, A. H.; Boutin, A. Structural Changes in Nanoporous Solids Due to Fluid Adsorption: Thermodynamic Analysis and Monte Carlo Simulations. *Chem. Commun.* **2008**, No. 28, 3275. https://doi.org/10.1039/b805117h.

(4) Zhao, X. S.; Siepmann, J. I.; Xu, W.; Kiang, Y.-H.; Sheth, A. R.; Karaborni, S. Exploring the Formation of Multiple Layer Hydrates for a Complex Pharmaceutical Compound. *J. Phys. Chem. B* **2009**, *113* (17), 5929–5937. https://doi.org/10.1021/jp808164t.

629511 Molecular Modeling and Adsorption Characterization of Micro-Mesoporous Kerogen Nanostructures

Abstract:

The aim of this work is a better understanding of structure and adsorption properties of organic porosity in kerogen, as it is related to predicting storage capacities of oil and gas, and hydrocarbon recovery. We hypothesize that the pore structure of kerogen can be represented as a network of mesopores distributed in the microporous matrix composed of packed kerogen molecules of given density. We generate 3D molecular models of kerogen micro-mesoporous structure with mesopores of different size and shape, and use GCMC simulations to calculate adsorption isotherms of typical adsorbates used for pore structure characterization. The kerogen matrix is generated by equilibrating the packing of kerogen molecules of given density and microporosity around mesopores. The system energy is minimized followed by MD simulations in NPT ensemble while decreasing the temperature. Adsorption isotherms of Ar, N_2 and CO_2 are generated in the 3D kerogen models of different levels of maturity, pore sizes and pore geometries, and compared with typical experimental isotherms on reference carbonaceous materials of different degree of graphitization.

635246 Synergistic Integration of Coarse-Grained Models, PRISM Theory, and Molecular Simulations for Design and Characterization of Macromolecular Materials

Abstract:

The Jayaraman research group members' expertise lies in development of coarse-grained models and computational approaches involving liquid state theory, molecular simulation, and machine learning for designing and characterizing macromolecular materials. Specifically, in this talk, I will present a few examples from my group's recent work that demonstrate how we use these computational approaches together with experimental results from collaborators to obtain molecular-level insight into structure and phase behavior in polymer nanocomposites and solutions.

635247 Computational Exploration of High-Chi Block Oligomers

Abstract:

Using molecular dynamics simulations and transferable force fields, we explore linear di- and triblock and miktoarm amphiphiles (or high-chi block oligomers) comprising incompatible sugar-based (A) and hydrocarbon (B) blocks that can self-assemble into ordered nanostructures with sub-2 nm domains. The ordered morphologies include lamellae, (hexagonally) perforated lamellae, hexagonally packed cylinders, hexagonal close-packed spheres, and also various three-dimensional network phase. The self-assembly of these amphiphiles with strong directional interactions and high stiffness bears some similarities, but also some differences, to those formed by block polymers.

635251 Molecular and Data-Centric Modeling of Nanoparticle Interactions with Biological Interfaces

Abstract:

The properties of soft materials typically emerge from the interplay of molecular interactions, thermal fluctuations, and entropy. These factors are difficult to interrogate experimentally or predict from chemical structure alone. Molecular-scale simulations of soft materials also face challenges associated with modeling long timescale processes and navigating large design spaces. In this talk, I will describe our efforts to address these challenges by utilizing molecular dynamics simulations, enhanced sampling

techniques, and data-centric analysis to guide the design of nanoparticles (NPs) functionalized with organic, small-molecule ligands. NPs are versatile soft materials for biomedical applications, including targeted delivery, biosensing, and photothermal therapy, because ligands can be selected to tailor interactions at the nano-bio interface. However, subtle differences in NP composition (e.g., ligand selection and core size) can trigger large changes in macroscopic behavior (e.g., cellular uptake) that

cannot be predicted a priori. I will discuss our efforts to model interactions between small (less than 10 nm in diameter) NPs and model cell membranes. These simulations, and complementary experiments, reveal how systematic variations to ligand properties influence the minimum free energy pathways underlying membrane adsorption. I will also discuss the parameterization of quantitative structure-activity relationship models for NP cellular uptake using features obtained from high-throughput simulations. This work highlights our approach to derive chemically specific design guidelines for NPs with desired nano-bio interactions.

635248 Mechanisms of Ion Transport in Polymeric Ionic Liquids

Abstract:

Polymeric ionic liquids (PILs) are an emerging class of materials which combines the attractive properties of ionic liquids with the sequence complexity and mechanical characteristics of macromolecules. While significant advances have occurred in the context of synthesis and characterization of such materials, comparatively less understanding exists on the mechanisms underlying ion transport in such materials. In this talk, I discuss some recent developments in the context of PILs relating to the issue of ion transport in such materials. We focus on the mechanisms of ion transport in such materials, the influence of counterions, and whether such materials do live upto the promise of high transference numbers.

635305 Elucidating Chemical Evolution in Shocked Materials

Abstract:

Understanding evolution of organic molecular materials (OMMs) subject to extreme temperature and pressure conditions (e.g., 1000s of K and 10s of GPa) is crucial to fields spanning astrobiology to nanomaterial fabrication. However, a confluence of challenges leaves a clear picture of this phenomena elusive. Experimentally, these conditions are often realized by subjecting samples to shockwaves via an external driver or detonation. Ensuing material evolution is rapid (e.g., occurring over 10^{-9} – 10^{-6}s timescales) and often highly multiscaled (e.g., where reactivity, phase separation, and material strength are determined on approximately $<10^{-9}$, 10^{-7} , and 10^{-6} m scales, respectively); as a result, direct experimental determination of properties as fundamental as temperature is intractable.

Atomistic simulations can be a powerful tool for shock experiment interpretation by providing an atomistically-resolved view into OMM evolution. However, characteristic problem scales approaching a μm and μs preclude use of highly predictive first principles-based simulation approaches (e.g., Kohn-Sham density functional theory), and existing molecular mechanics-based interatomic models are generally not designed for such high temperature and pressure conditions. Recently, the ChIMES machine-learned reactive interatomic model and development framework was developed to overcome these challenges. In this presentation, first-of-their-kind simulations using these models will be presented, and implications for our understanding OMMs evolution under extreme conditions are discussed.

This work is performed under the auspices of the U.S. Department of Energy by Lawrence Livermore National Laboratory under Contract DE-AC52-07NA27344.

631383 Efficient Measurement of Anharmonic Mechanical Properties of Crystals Using Normal-Mode Mapping

Abstract:

Statistical thermodynamics provides a connection between mechanical properties and free energy derivatives with respect to some external strain. For example, the stress tensors and elastic constants are expressed as first and second derivatives of the free energy, respectively. For crystalline systems, the harmonic model provides a (simple) first-order approximation to these properties, which has increasingly accurate results as the temperature decreases, but fails to describe the system at high temperatures due to anharmonic effects.

Measuring anharmonic effects, however, is not as straightforward as it requires running molecular simulation to generate several configurations. Standard expressions of crystalline properties (e.g., virial pressure) obtain the anharmonic contribution on an average-basis level, by a naïve subtraction of the harmonic contribution from the total. This introduces large noise in the (small) anharmonic effects, due to its contamination with large fluctuations from the harmonic behavior that exists for each configuration.

To alleviate this problem, we developed a configuration-based formulation that leverages the harmonic character to provide in essence a direct measurement of the anharmonic contributions to the properties. This was possible by applying our previously developed harmonically mapped averaging (HMA) theory [1,2] in the "natural" normal-mode coordinates of the crystalline systems; hence the name NM-HMA. The new formulation provides a significant CPU speedup (relative to conventional methods) in computing several mechanical properties (e.g., stress tensor) to a given precision. One of the advantages of the new formulation does not rely on external parameters like the original HMA method, rather it depends on the harmonic information of the system (e.g., Grüneisen parameters). As with other HMA methods, NM-HMA does not alter the molecular simulation sampling; hence, the properties can be computed by post-processing simulation data (e.g., configurations and forces), and several properties may be measured at once.

[1] S. Moustafa et al., *Phys. Rev. E* **92**, 043303 (2015).
[2] A. Schultz et al., *J. Chem. Theory Comput.* **12**, 1491 (2016).

632946 Diabat Method for Polymorph Free Energies

Abstract:

Polymorphs are different crystal structures for the same molecule or compound. Many molecules can crystallize into several different polymorphs, each with different solubility and growth/dissolution kinetics. Polymorphism is especially important in pharmaceuticals as many drugs are administered as crystalline solids. Relative stability of polymorphs affects solubility which consecutively affects the bioavailability of the solid form. *Ab initio* calculations typically infer stability from static energies or harmonically approximate the entropy. Polymorphs of organic molecular crystals may be separated by small free energy differences and hence it is important to account for the entropy exactly. Classical force field based approaches enable complete free energy calculations with advanced sampling methods. These methods, however, typically use thermodynamic integration and are expensive multi-stage simulations.

In this talk we discuss a method to directly yield the polymorph free energy differences. The framework combines ideas from the well-established lattice-switch Monte Carlo procedure and relationships that have been previously used in the electron transfer literature. We briefly discuss the method and its previous applications to atomic solids. Recently, we generalized and extended the framework to yield relative stability of organic molecular crystals. We demonstrate the method for benzene polymorphs and all five experimentally reported polymorphs of pharmaceutically relevant carbamazepine.

628102 How to Quantify and Avoid Finite Size Effects in Computational Studies of Crystal Nucleation

Abstract:

The thermodynamic, structural and kinetic properties estimated from molecular simulations can deviate considerably from the true values in the thermodynamic limit due to the finite dimensions of the simulation box. Such discrepancies are typically referred to as *finite size effects*. Computational studies of crystal nucleation, in particular, are known to be influenced by system size, as unphysical interactions can arise between crystalline nuclei and their periodic images [1]. However, no rigorous guidelines exist for quantifying such artifacts and *ad hoc* heuristics (such as the 10% rule) have emerged over years in the nucleation community to rule out the existence of finite-size effects. In this work, we systematically investigate the sensitivity of nucleation kinetics and mechanism to system size. We use [2] molecular dynamics simulations and a path sampling technique called jumpy forward flux sampling [3] to compute heterogeneous ice nucleation rates for the coarse-grained monoatomic water (mW) [4] model in the vicinity of square-shaped model structureless [5] ice nucleating particles (INPs) of different sizes. We identify three distinct regimes for the dependence of the computed rate on the INP dimension, L. For small L's, the rate is a strong function of L, due to the artificial spanning of the critical nuclei across the periodic boundary. For intermediate L's, the rate is smaller overall, and the critical nuclei are non-spanning but are 'proximal' in the sense that the supercooled liquid within the inter-image region (i.e., the region along the vector connecting a crystalline nucleus to its closest periodic image) becomes fairly structured. Finally, for sufficiently large L's, critical nuclei are neither spanning nor proximal, yet the rate is a weak function of L. We propose a heuristic suggesting that finite size effects will be minimal if critical nuclei are neither spanning nor proximal and if the intermediary liquid has a region that is structurally indistinguishable from the supercooled liquid under the same conditions. Preliminary analysis of the strength of finite-size effects in homogeneous nucleation in the LJ system will also be discussed.

[1] G. C. Sosso et al., *Chem. Rev.,* **116**: 7078 (**2016**).

[2] Hussain and Haji-Akbari, *J. Chem. Phys.*, **154**: 014108 (**2021**).

[3] Haji-Akbari, *J. Chem. Phys.*, **149**: 072303 (**2018**).

[4] Molinero, Moore, *J. Phys. Chem. B*, **113**: 4008 (**2009**).

[5] Magda, Tirrell, Davis, *J. Chem. Phys.*, **83**: 1888 (**1985**).

629280 Improved Configurational Sampling By the the Introduction of Alchemical Variable in Metadynamics

Abstract:

The expanded ensemble technique, an advanced sampling method that utilizes alchemical states to modify the underlying free energy surface, can improve sampling in systems where the physically relevant collective variables are hard to define by intuitions or experimental observations. However, this approach fails if the alchemical variables are almost orthogonal to the most important slow degrees of freedom in the configurational space. Additionally, if there are free energy basins in the alchemical direction or the free energy barriers exist at all lambda states, the system may also get stuck, leading to slow convergence. To alleviate this issue, we propose to add alchemical variables as an allowed collective variable in metadynamics, augmenting the ability to sample collective variables. In this study, we validate our implementation of alchemical metadynamics in PLUMED with test systems/alchemical processes with varying complexities and dimensions of collective variable space, including the binding and unbinding of CB8-G3 host-guest binding complex. We show that one-dimensional alchemical metadynamics correctly recovers the free energy difference between the coupled state and the uncoupled state. In addition, multi-dimensional alchemical metadynamics has the potential to further facilitate the sampling by introducing configurational collective variables to circumvent barriers. The method can trivially be combined with other metadynamics-based algorithms implemented in PLUMED.

629236 Configurational-Bias Monte Carlo Simulation to Predict the Supramolecular Self-Assemblies of Amphiphiles

Abstract:

Amphiphiles are long-chain molecules that possess a polar head and non-polar tail [1]. This unique chemical property of these molecules enables them to form a wide range of supramolecular self-assembled structure in polar solvents [2]. These structures impart unique properties to the surfactant solutions. Hence, these self-assembled structures have been explored extensively over the past decade [3]. Significant theoretical progress and fundamental understanding of these self-assemblies have been achieved in the advent of simulation methodologies like molecular dynamics (MD) [4]. MD simulations coupled with significant progress in modern supercomputing architectures have paved the path for multiple discoveries in the field of self-assembly [8]. However, MD algorithms are limited in terms of the time-scales they can simulate [7]. MD simulations can simulate only a few microseconds of self-assembly in the most advanced forms while requiring substantial computational resources [5]. This limits MD's application to explore a wide range of self-assembled structures as many self-assemblies take much longer times to form. Moreover, it becomes incredibly challenging to simulate complex self-assembles as numerical solutions tend to diverge in such instances.
Monte Carlo simulations are another class of techniques used for molecular studies at multiple length scales [6]. Specifically, Monte Carlo simulations involve the statistical sampling of energetically stable configurations of molecules in an ensemble. In recent years, many strategies such as the Gibb's ensemble Monte Carlo [7] and grand canonical ensemble Monte Carlo with histogram reweighting methodologies [8] have been used extensively to compute the phase equilibrium of pure molecules with simple chemical structures. To further implement these strategies to complex macromolecules, specific biasing techniques are often utilized. One such biasing technique is the configurational-bias Monte Carlo (CBMC), which involves re-growing a molecule to explore its configurational space. Over the years, to improve the efficiency of CBMC sampling, multiple studies have been proposed [9-12]. However, these studies are specific to particular classes of surfactant molecules and cannot be readily extended to all the amphiphiles in general.
Motivated by the absence of such a generalized framework, we propose a CBMC simulation methodology that can be applied to predict a wide range of supramolecular self-assemblies of the surfactants. Here, we performed fragmentation to explore the bond-angle configurational space and dihedral angle space separately. In the proposed methodology, the amphiphiles are broken into fragments of sizes not greater than three. It was found that with a fragment size of three atoms, sampling of individual fragments enables us to explore the energy variations associated with only the bond-angles. This is because three points are always coplanar, and therefore, no dihedral conformations are explored when sampling individual fragments' configurations. The dihedral space is explored only when the fragments are regrown in the CBMC procedure. This proposed procedure prevents the CBMC algorithm's confinement to a local minimum and improves sampling efficiency. Furthermore, using the proposed CBMC scheme, equilibrium structures of various surfactant self-assemblies were predicted. Finally, the model predictions were validated with experimental data from small-angle X-ray scattering (SAXS).
References.
[1] C. A. Dreiss, "Wormlike micelles: where do we stand? recent developments, linear rheology and scattering techniques," Soft Matter, vol. 3, no. 8, pp. 956–970, 2007.
[2] Markvoort, A. J., Pieterse, K., Steijaert, M. N., Spijker, P., & Hilbers, P. A. J. (2005). The bilayer− vesicle transition is entropy driven. The Journal of Physical Chemistry B, 109(47), 22649-22654.

[3] W. Zou and R. G. Larson, "A mesoscopic simulation method for predicting the rheology of semi-dilute wormlike micellar solutions," Journal of Rheology, vol. 58, no. 3, pp. 681–721, 2014.

[4] Ruiz-Morales, Y., & Romero-Martínez, A. (2018). Coarse-grain molecular dynamics simulations to investigate the bulk viscosity and critical micelle concentration of the ionic surfactant sodium dodecyl sulfate (SDS) in aqueous solution. The Journal of Physical Chemistry B, 122(14), 3931-3943.7

[5] Bai, Y., An, N., Chen, D., Liu, Y. Z., Liu, C. P., Yao, H., ... & Tian, W. (2020). Facile construction of shape-regulated β-cyclodextrin-based supramolecular self-assemblies for drug delivery. Carbohydrate polymers, 231, 115714.

[6] Zhang, J., Clennell, M. B., Dewhurst, D. N., & Liu, K. (2014). Combined Monte Carlo and molecular dynamics simulation of methane adsorption on dry and moist coal. Fuel, 122, 186-197.

[7] Panagiotopoulos, A. Z. (1987). Direct determination of phase coexistence properties of fluids by Monte Carlo simulation in a new ensemble. Molecular Physics, 61(4), 813-826.

[8] Ferrenberg, A. M., & Swendsen, R. H. (1988). New Monte Carlo technique for studying phase transitions. Physical review letters, 61(23), 2635.

[9] De Pablo, J. J., Laso, M., & Suter, U. W. (1992). Simulation of polyethylene above and below the melting point. The Journal of chemical physics, 96(3), 2395-2403.

[10] Pant, P. K., & Theodorou, D. N. (1995). Variable connectivity method for the atomistic Monte Carlo simulation of polydisperse polymer melts. Macromolecules, 28(21), 7224-7234.

[11] Deem, M. W., & Bader, J. S. (1996). A configurational bias Monte Carlo method for linear and cyclic peptides. Molecular Physics, 87(6), 1245-1260.

[12] Zhang, J., Kou, S. C., & Liu, J. S. (2007). Biopolymer structure simulation and optimization via fragment regrowth Monte Carlo. The Journal of chemical physics, 126(22), 06B605.

621692 Bypassing Backmapping By Learning the Noise of Electronically Coarse-Grained Models

Abstract:

Coarse-graining methods for molecular and polymeric materials facilitate the "bottom-up" development of multiscale structural prediction models in soft materials and biological applications. Recently, we introduced Electronic Coarse-Graining (ECG) as a novel coarse-graining strategy for developing electronic structure prediction models utilizing coarse-grained representations of soft materials. Here we extend the ECG methodology within a formalism of Gaussian Processes to learn the all-atom electronic noise distributions associated with the coarse-graining decimation procedure. The ability to learn all-atom noise distributions at the coarse-grained resolution within ECG provides an avenue for stochastically sampling the coarse-grained model's noise distribution for electronic predictions, further enabling the bypassing of traditional multiscale backmapping procedures.

627879 Exploration of the Secondary Structure Peptoid Folding Landscape with Metadynamics

Abstract:

Peptoids, or n-substitued glycines, are increasingly stable but complex and diverse oligomeric structures which have been explored for a number of biomimetic applications including drug molecules, surfactants of catalysts.[1-2] In contrast to their peptide counterparts, on peptoids the sidechain is located on the backbone nitrogen resulting in a flexible omega backbone dihedral that can isomerize into both stable *cis*- and *trans*- backbone conformations. This unique feature of peptoids allows for these structures to span a large design space of chemical and structural functionality through the careful tuning of their side chains. One novel example of this is the N-substituted α-chiral aromatic (S)-N-(1-phenylethyl)glycine (Nspe) oligomers which can produce stable helical structures in both aqueous and non-aqueous solutions.[3] Further, many of these oligomers have been shown to have the capacity to assemble into more complex hierarchical assemblies like microspheres, nanosheets, and dynamic single-walled nanotube structures. However, little is known about the molecular-level mechanisms that drive these assemblies.

Enhanced sampling molecular dynamics approaches provide an opportunity to understand this dynamic behavior at an atomistic level of detail, giving insight into the complex energetic landscape of the elusive peptoid backbone. Parallel biased metadynamics is one such method that can be used to sample highly dimensional systems. Here we present our application of parallel tempering parallel bias metadynamics in the well-tempered ensemble (PTPBMetaD-WTE), which overcomes sampling limitations related to the flexible peptoid backbone, to sample the 12-dimensional energetic landscape for a range of different peptoid chemistries and chain lengths. From this, we can begin to interpret the energetic and entropic mechanisms that give rise to these unique complex assemblies for a number of different peptoid chemistries.

References

[1] Dohm MT, Kapoor R, Barron AE. Peptoids: Bio-Inspired Polymers as Potential Pharmaceuticals. *Curr. Pharm.* **2011,** 47 (17), 2732.

[2] Maayan G, Ward MD, Kirshenbaum K. Folded biomimetic oligomers for enantioselective catalysis. *Proc Natl Acad Sci* **2019**, 106(33):13679-84.

[3] Wu C, Sanborn S, Huang K, Zuckermann RN, Barron AE. Peptoid Oligomers with α-Chiral, Aromatic Side Chains: Sequence Requirements for the Formation of Stable Peptoid Helices. *J. Am. Chem. Soc.* **2001**, 123, 28, 6778-6784.

632537 The Influence of Coverage on Entropy: Changes in Vibrational Frequencies of CO on Pt (111)

Abstract:

Introduction

Significance of coverage-awareness in kinetic modeling of heterogeneous catalysis has been well recognized [1]. In fact, with various models ranging from a simple linear model to a rigorous cluster expansion method [2], coverage-dependent binding energy and coverage-dependent barrier energies are being incorporated in the kinetic modeling. On the other hand, the coverage-dependent entropy, specifically due to the coverage dependency of the adsorbate vibrational frequencies, has not been analyzed in detail when there have been experimental evidences illustrating vibrational frequencies could be significantly affected by the surface coverage [3, 4]. The importance of the vibrational entropy in kinetic modeling is getting more addressed [5, 6] in the field but still, the direct effect of coverage on entropy has not been discussed.

We conducted a study on a model system of CO adsorption on Pt (111) to investigate how coverage affects vibrational frequencies and to quantitatively assess the impact on the adsorption thermodynamics. Furthermore, to improve the adsorption thermodynamics estimation through system-specific coverage-dependent binding energy, a comparison between concepts of average binding energy [7] and incremental binding energy [8] was made and the suitability of each model to different situations was discussed.

Materials and Methods

Global minimum optimization as well as vibrational frequency calculations of CO adsorbates were performed using density functional theory (DFT). A periodic slab model of 3×3×4 Pt was used to model the Pt surface by Quantum ESPRESSO suite [9] with plane wave basis sets and PBE exchange-correlation functional. At each CO coverage, ranging from one adsorbate in the unit cell to a full monolayer (ML) with nine adsorbates, the global minimum energy configuration was discovered by the constrained minima hopping algorithm [8]. Then the vibrational frequency calculations were conducted with the optimized configurations. Free energy, enthalpy, and entropy of adsorption were calculated using the coverage-dependent binding energy and vibrational partition function with the harmonic oscillator approximation.

Results and Discussion

Result of coverage-dependent binding energy is shown first. The average binding energy and the incremental binding energies were calculated by referencing the energy of the empty surface [7] and the energy of the global minimum configuration with one less adsorbate [8], respectively. The two binding energies at different coverages are compared in Figure 1. At low coverages , up to 4/9 ML coverage, the difference was insignificant. However, the differences at higher coverages were considerable. Consequently the choice of which binding energy model to deploy could bring dramatically different adsorption thermodynamics, for instance, at the higher coverages in this system.

In order to calculate coverage-dependent vibrational entropy, frequencies of the six vibrational modes of CO adsorbates were first grouped into four distinct motions: tilting, pendulum, desorption, and stretching motions. Then the frequencies of each group was averaged by the number of CO adsorbates and plotted as a function of coverage in Figure 2. Except for the tilting motion, frequencies of all motions showed an monotonically increasing trend with increasing coverage. But the frequency of the tiling motion showed a parabolic trend: initially decreased until 4/9 ML then increased as the coverage increased. Because the tilting motion is the lowest-frequency mode, its changes affected the overall vibrational entropy the most. Therefore the dip in the tilting motion frequency at the low coverages entropically favors additional adsorptions by increasing the overall vibrational entropy, which is the opposite trend observed in the binding energy. Then at higher coverages, increasing frequencies make adsorption entropically unfavorable.

Significance

Our study provides an insight of how often-neglected coverage dependence on adsorbate vibrational frequencies could result in considerable difference in adsorption thermodynamics.

References

1. Miller, S., Dsilva, C., and Kitchin, J.R. *Catalysis* 24, 83 (2012).

2. Bajpai, A., Frey, K., and Schneider, W.F. *Langmuir* 36, 465 (2020).

3. Ortega, A., Hoffman, F.M., and Bradshaw, A.M. *Surf. Sci.* 119, 79 (1982).

4. Ueba, H. *Surf. Sci.* 188, 421 (1987).

5. Lansford, J. L., Mironenko, A. V., Vlachos, D. G. *Nat Commun* 8, 1842 (2017).

6. Lansford, J. L., Kurdziel, S. J., Vlachos, D., G. *J. Phys. Chem. C.* 125, 13, 7119 (2021).

7. Persson, B. N. J., *Phys. Rev. B* 40, 7115 (1989).

8. Zhang, Y., Sethuraman, V., Michalsky, R., Peterson, A. A. *ACS Catal.* 4, 3742 (2014).

9. Giannozzi, P., Baroni, S., Bonini, N, Calandra, M., Car, R., Cavazzoni, C., Ceresoli, D., Chiarotti, G.L., Cococcioni, M., Dabo, I., Dal Corso, A., Fabris, S., Fratesi, G., de Gironcoli, S., Gebauer, R., Gerstmann, U., Gougoussis, C., Kokalj, A., Lazzeri, M., Martin-Samos, L., Marzari, N., Mauri, F., Mazzarello, R., Paolini, S., Pasquarello, A., Paulatto, L., Sbraccia, C., Scandolo, S., Sclauzero, G., Seitsonen, A.P., Smogunov, A., Umari, P., and Wentzcovitch, R.M. *J. Phys.: Condens. Matter* 21, 39, 395502 (2009)

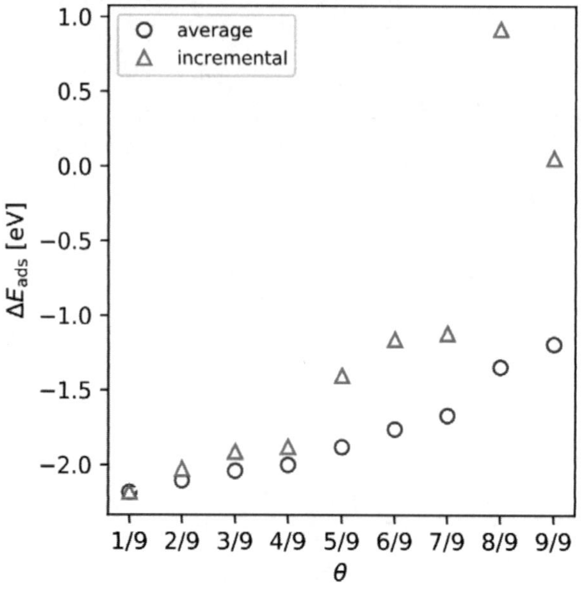

Figure 1: Average binding energy and incremental binding energy of CO adsorption on Pt (111) at different coverages.

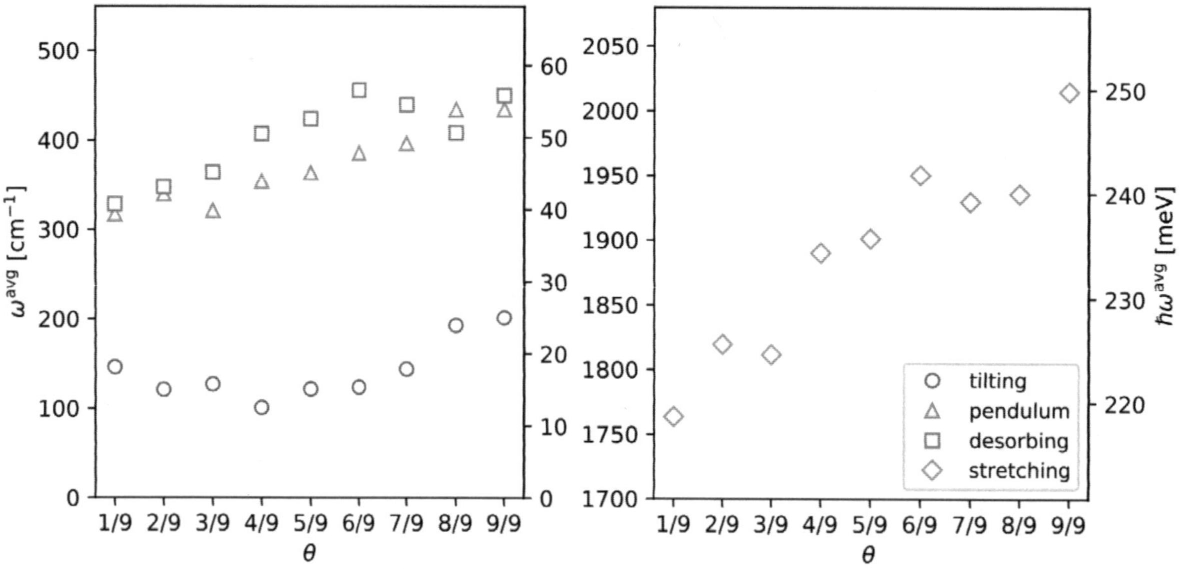

Figure 2: Average vibrational frequencies of four distinct motions of CO adsorbates on Pt (111) as a function of coverage.

631670 Dynamical Evolution of Atomically Dispersed Catalysts: Ab Initio Molecular Dynamics Analysis of Thermal and Adsorbate-Induced Metal Atom Migration

Abstract:

Introduction

Catalysis is a common topic among many areas of industry from petroleum refining to health and food production. Catalysts are used in more than 80% of common methods of production that utilize chemical reactions with the industry itself worth up 33.5 billion annually.[1] Most catalysts consist of rare and expensive metals, such as platinum and palladium, spurring efforts into enhancing atom efficiencies to ease their demand. Naturally, the emerging field of single atom catalysis (SACs), where metal atoms are atomically dispersed on a support interface, is of significant interest.

It has been revealed that these metal atoms often occupy the cation sublattice of the metal oxide supports, but it is not possible to observe the coordination environment, or reactants during a chemical reaction.[1] These metal atoms can often become mobile at low temperatures or when a reaction intermediate becomes adsorbed,[2,3] often forming small clusters of up to four atoms that may be mobile at fairly low temperatures.[4] The growth of these single metal atoms into clusters, known as sintering, often leads to deactivation of these catalysts[5,6] and nucleation of the crystal.[7] Sintering can occur via two mechanisms: by cluster diffusion, and by Ostwald ripening.[1] It has also been observed that support atoms can also diffuse from the bulk to the surface depending on the reactions conditions, as well as diffusion from the surface into the bulk, resulting in a dynamic support interface.[1] For example, oxygen atoms are taken from the surface of the metal oxide interface and metal atoms, such as Fe, can diffuse into the bulk due to under coordination during the oxidation of CO.[1]This results in a surface that changes during the course of the reaction but can be reversed with the addition of oxygen gas to the system.

Parkinson et al. through scanning tunnel microscopy (STM) studies of adsorbate induced migration on Pd/Fe_3O_4 and Pt/Fe_3O_4,[2,3] have shown that the adsorption of CO, H_2, and O_2 can lead to high mobility of the dispersed single atoms which results in the nucleation of small clusters of Pt or Pd. They have determined when the adsorbates are removed from the system, the clusters will redisperse into single atoms on the surface at elevated temperatures.

While microscopy methods demonstrate the importance of probing catalyst stability under reaction conditions, identifying the driving forces for metal atom mobility remains challenging. Our goal is to utilize ab initio molecular dynamics (AIMD) simulations to examine the response of the metal atom to adsorbed intermediates on picosecond timescales. Here, we present our AIMD analysis of metal atom mobility under low-temperature water-gas shift conditions for atomically dispersed Pt on rutile TiO_2.

Methods

All spin polarized calculation are performed using Vienna Ab Initio Simulation Package, VASP 5.4, using the SCAN meta-generalized gradient approximation (meta-GGA).[8] The Atomic Simulation Environment (ASE) is utilized for model construction and geometry relaxation (threshold=0.05 eV/A).[9] All atoms are described using the default projector augmented wave (PAW) potentials available in VASP. The non-spherical contributions related to the gradient of the density in the PAW spheres are included in the calculations with the d-orbitals included in the kinetic energy density mixing. Electronic relaxations are carried out using the preconditioned conjugate gradient algorithm recommended for meta-GGA functionals. The Brillouin Zone is integrated using the Monkhorst-Pack set of k-points (1x1x1). Gaussian smearing ($\sigma = 0.1$ eV) is used along with the cutoff energy of 400eV for DFT and AIMD calculations. We employ Bader and DOS analysis to calculate charges on surface atoms and characterize the binding of the adatoms to the surface.

The AIMD calculations will be carried out in 3 steps:

2. **Temperature ramp**: The DFT-optimized geometries are heated at a rate of 500K/ps from 10K to 500K. The velocities are scaled at every 2.0fs timestep and 40 steps are stored in the Broyden mixer.
3. **Equilibration**: The point of equilibration is identified using the method put forth by Chodera et al.[10] We use the Nosé-Hoover[11] thermostat and calculate the stress tensor at a timestep of 1.0fs. Ten steps are stored in the Broyden mixer.

- **Dynamic Evolution**: Five picosecond trajectories are initiated from the equilibrated structures with a 1fs timestep. Settings for mixing and electronic calculations are identical to the equilibration step.

The quantities measured during simulation are coordination number of Pt to the TiO_2 surface,[12] migration from one DFT site to another,[13] velocities of Pt atom, and geometric analysis such as bond lengths and angles of adsorbates. We carry out hydrogen bonding analysis for the simulations that have potential hydrogen bonds forming between adsorbate and support atoms.

Results

Eight common WGS absorbates were simulated adsorbed to Pt/TiO_2: CO, H, OH, O, CO_2, COOH, CHO, and H_2O. The adsorbate OH was also simulated bound to a Ti atom nearby the single Pt atom because the location of the stabilizing hydroxyl is not reported in prior studies of hydroxyl-stabilized metal atoms.

While we aim to present the results of AIMD simulations for all adsorbates (manuscript in preparation), here in Figure 1, we illustrate the dynamic changes in coordination of Pt with support Ti and O atoms in the presence of (a) no adsorbate, (f) CO, and (d) OH. We find that the time averaged coordination number of Pt drops from 1.13 to 0.67 and 0.99 to 0.85 for Ti and O respectively for CO adsorption. The drop in coordination number is likely to induce migration of the Pt atom. The weaker the interaction between Pt and the surface, the more likely Pt is to migrate, which agrees with previous studies.[2,3] The time averaged coordination number of Pt to Ti and O also drops when comparing the bare Pt case to OH. The time averaged coordination number of the OH(Ti5c) simulation, Figure 1(e), of Pt to O remains high at 1.12 indicating high interaction with the support surface. No migration is observed during the 5 ps trajectory for this case.

Conclusion

Using AIMD, we observe that the adsorption of CO leads to Pt migration while the adsorption of OH seems to inhibit migration of the Pt atom agreeing with previous experimental papers. The simulation of OH adsorbed to a nearby Pt atom with CO attached is the scope of future work to determine any competing factors between CO and OH.

References

1. Parkinson, G. S. Atomic Scale Insights into Single-Atom Catalysis. *Vak. Forsch. und Prax.* **30**, 45–49 (2018).
2. Bliem, R. *et al.* Dual role of CO in the stability of subnano Pt clusters at the Fe3 O4(001) surface. *Proc. Natl. Acad. Sci. U. S. A.* **113**, 8921–8926 (2016).
3. Parkinson, G. S. *et al.* Carbon monoxide-induced adatom sintering in a Pd–Fe3O4 model catalyst. *Nat. Mater.* **12**, 724 (2013).
4. Xu, L., Henkelman, G., Campbell, C. T. & Jónsson, H. Small Pd clusters, up to the tetramer at least, are highly mobile on the MgO(100) surface. *Phys. Rev. Lett.* **95**, 1–4 (2005).

5. Campbell, C. T., Parker, S. C. & Starr, D. E. The effect of size-dependent nanoparticle energetics on catalyst sintering. *Science (80-.).* **298**, 811–814 (2002).
6. Bartholomew, C. H. Mechanisms of catalyst deactivation. *Appl. Catal. A Gen.* **212**, 17–60 (2001).
7. Bliem, R. *et al.* Cluster nucleation and growth from a highly supersaturated adatom phase: Silver on magnetite. *ACS Nano* **8**, 7531–7537 (2014).
8. Sun, J., Ruzsinszky, A. & Perdew, J. Strongly Constrained and Appropriately Normed Semilocal Density Functional. *Phys. Rev. Lett.* **115**, 1–6 (2015).
9. Hjorth Larsen, A. *et al.* The atomic simulation environment - A Python library for working with atoms. *J. Phys. Condens. Matter* **29**, (2017).
10. Chodera, J. D. A Simple Method for Automated Equilibration Detection in Molecular Simulations. *J. Chem. Theory Comput.* **12**, 1799–1805 (2016).
11. William G. Hoover. Canonical dynamics: Equilibrium phase-space distributions William. *Phys. Rev. A* **31**, 1695–1697 (1985).
12. Sprik, M. Coordination numbers as reaction coordinates in constrained molecular dynamics. *Faraday Discuss.* **110**, 437–445 (1998).
13. Humphrey, N., Bac, S. & Mallikarjun Sharada, S. Ab Initio Molecular Dynamics Reveals New Metal-Binding Sites in Atomically Dispersed Pt 1 /TiO 2 Catalysts. *J. Phys. Chem. C* (2020). doi:10.1021/acs.jpcc.0c06771

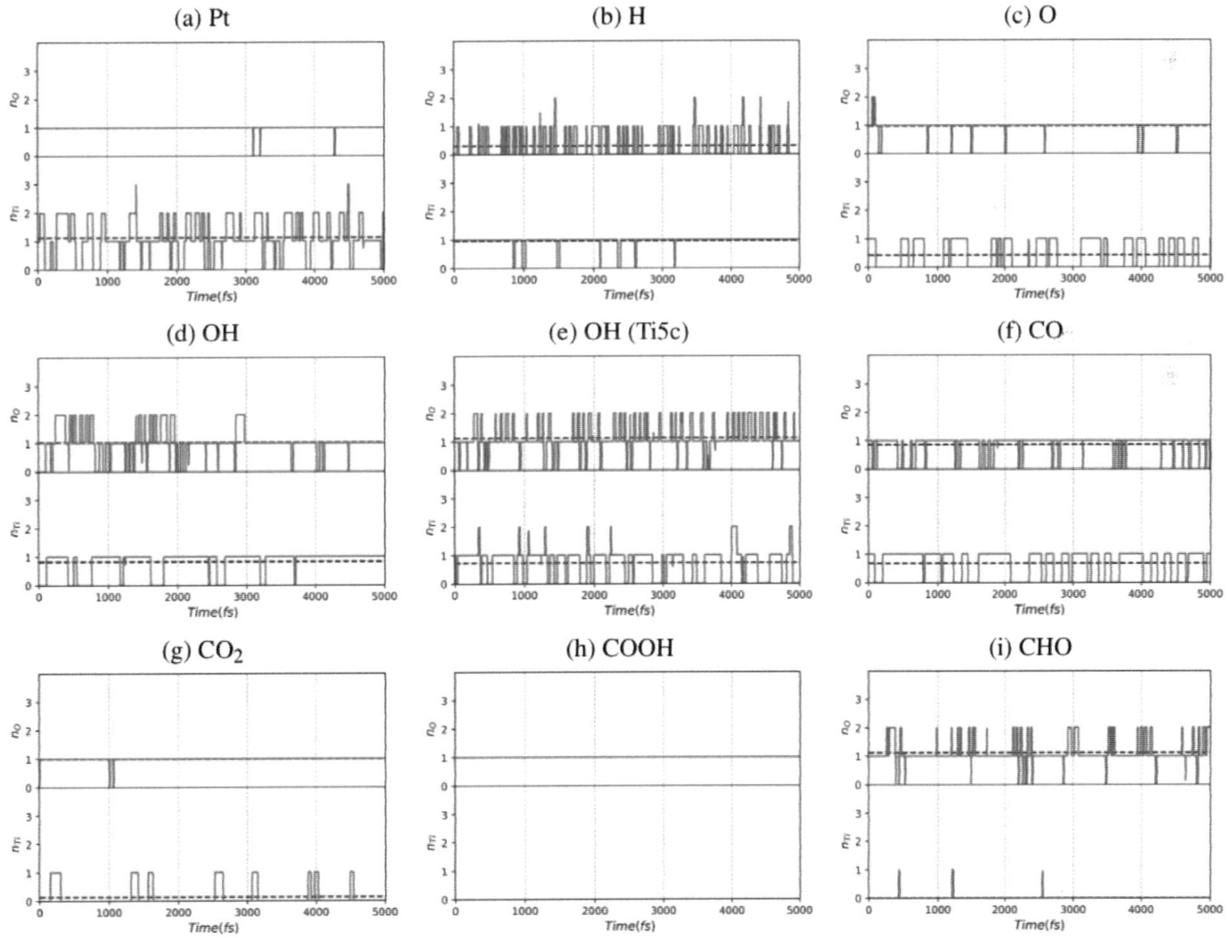

628234 Uncovering a Universal Molecular Mechanism of Salt Ion Adsorption at Solid/Water Interfaces

Abstract:

Understanding the molecular mechanism of salt ion adsorption at solid/water interfaces is essential for all energy and membrane-based applications involving electrolytes. Because water is a polar solvent and salt ions are charged, a long-standing puzzle involving solid/water interfaces is how do the electric fields exerted by the salt ions and the interfacial water molecules polarize the charge distribution in the solid and how does this polarization, in turn, influence ion adsorption at any solid/water interface. Here, using state-of-the-art polarizable force fields derived from quantum chemical simulations, we perform all-atomistic molecular dynamics simulations to investigate the adsorption of various ions comprising the well-known Hofmeister series at the graphene/water interface, including comparing with available experimental data. Our findings reveal that, in vacuum, the ionic electric field - induced polarization of graphene results in a significantly large graphene - ion polarization energy, which drives all salt ions to adsorb to graphene. On the contrary, in the presence of water molecules, we show that the water molecules and the salt ions exert waves of molecular electric fields on graphene which destructively interfere with each other. This remarkable phenomenon is shown to cause a water-mediated screening of more than 85% of the graphene–ion polarization energy – a finding which remains consistent for all the salt ions of different types (e.g., kosmotropic, chaotropic) and valency (e.g., monovalent, divalent) considered in our study. Further, by investigating model superhydrophobic and superhydrophilic solid/water interfaces, which form asymptotic limits on the water density and structuring possible at any solid/water interface, we demonstrate that this phenomenon occurs universally at all solid/water interfaces. We also show that as a result of the significant water – mediated weakening of the ion – solid interactions, ion specific effects are governed primarily by a molecular mechanism involving an interplay between the ion – water and water – water interactions. Finally, we compare our simulation results to the classical theory of image charge interactions, including the Onsager – Samaras theory, and highlight the importance of discrete effects beyond the mean-field level on the energetics of salt ion adsorption at solid/water interfaces.

624609 Predicting the Equilibrium Adsorption Morphologies of Surfactant Molecules at Metal-Water Interfaces Via Advanced Molecular Dynamics Simulations

Abstract:

Self-assembly of surfactant molecules at the metal-water interfaces is understood to play a key role in many useful applications, including corrosion inhibition, improving selectivity in heterogeneous catalysis, modulating electrochemical reactions and synthesis of anisotropic metal nanoparticles. Surfactants are understood to adsorb at metal-water interfaces in high-density adsorbed morphologies. Current molecular simulation techniques are unable to effectively study the formation of these morphologies. Therefore, determining the most thermodynamically stable adsorbed morphology of surfactants has remained an unsolved problem. In this work, we report a new free energy sampling methodology that enables the study of formation of high-density adsorption morphologies. The methodology is generally applicable for studying other adsorption systems. Using our methodology, we have studied adsorption of cationic and uncharged surfactant molecules on metal-water interfaces. Our free energy calculations show that the equilibrium adsorption morphology of cationic molecules vary with their alkyl tail lengths. Cationic molecules with small tail lengths *(C-4)* adsorb in a monolayer of molecules lying flat onto the surface, whereas the molecules with large tail lengths *(C-12)* adsorb in a mixed morphology such that some molecules lie-flat on the surface and rest of the molecules are aggregated as hemi-micelles on the surface. As opposed to cationic molecules, uncharged molecules adsorb in much higher densely packed morphologies.

622291 Understanding the Role of Charge Distribution and Pore Size for Water Vapor Adsorption in Idealized Nanoporous Materials

Abstract:

Since water vapor is present in major gaseous streams such as industrial flue gas, air, etc.,separating it efficiently can lead to augmenting water supply, increasing efficiency for proton con-duction, better CO_2 capture performance in presence of water and many others [1, 2]. In this pursuit, nanoporous materials such as metal-organic frameworks (MOFs), having a high pore volume and surface area, can act as a media for separating water vapor from these mixtures. Recent studies have shown that MOFs are capable of separating water vapor from air in various capacities including the case of atmospheric water harvesting (AWH) units [3, 4]. Computational studies have revealed that water vapor adsorption on nanopores are highly dependent on pore geometry, functional groups and topology [5]. We seek to provide further understanding on the role of electrostatic interactions in water vapor adsorption in nanoporous materials.

To do so, we designed idealized carbon-based cylinders (ICC) and performed Monte Carlo simulations to understand the role of electric potential and pore size in the water vapor adsorption. We found that the maximum water vapor uptake was independent of the charge distribution but was proportional to pore size. Our study also confirmed that charge arrangement on pore surface plays a significant role in influencing the hydrophilicity of ICCs. We observed that a surface was found to be more hydrophilic when charges are alternated (from positive and negative) along the circumference than when they were alternated along the length of the cylinder. Also, through wa-ter vapor adsorption density profiles, we found the water adsorption sites were also dependent on charge distribution even at the highest relative humidity. We also formed the electric potential maps for these ICCs and found a strong relationship between the electric potential inside a pore and the water vapor adsorption sites. We hypothesize that MOFs having similar electrostatic characteristics to the ICCs would also exhibit identical water vapor adsorption behavior and they can be potentially used for AWH technology and other water vapor separation processes. Further, novel MOFs may be designed by adding chemical moieties to existing MOFs to induce a desired water vapor adsorption and separation behavior as observed in the ICCs.

References

[1] Song Li, Yongchul G. Chung, and Randall Q. Snurr. "High-Throughput Screening of Metal–Organic Frameworks for CO2 Capture in the Presence of Water". In:Langmuir32.40 (2016). PMID: 27627635, pp. 10368–10376.doi: 10.1021/acs.langmuir.6b02803.
eprint:https://doi.org/10.1021/acs.langmuir.6b02803.url:https://doi.org/10.1021/acs.langmuir.6b02803.

[2] Stefan K. Henninger, Hesham A. Habib, and Christoph Janiak. "MOFs as Adsorbents for Low Temperature Heating and Cooling Applications". In: Journal of the American ChemicalSociety131.8 (2009). PMID: 19206233, pp. 2776–2777.doi:10.1021/ja808444z.
eprint:https://doi.org/10.1021/ja808444z.url:https://doi.org/10.1021/ja808444z.

[3] Hyunho Kim et al. "Water harvesting from air with metal-organic frameworks powered by natural sunlight". In:Science356.6336 (2017), pp. 430–434.issn: 0036-8075.doi:10.1126/science.aam8743. eprint: https://science.sciencemag.org/content/356/6336/430.full.pdf.url:https://science.sciencemag.org/content/356/6336/430.

[4] Xingyi Zhou et al. "Atmospheric Water Harvesting: A Review of Material and Structural De-signs". In: ACS Materials Letters2.7 (2020), pp. 671–684.doi:10.1021/acsmaterialslett.0c00130. eprint: https://doi.org/10.1021/acsmaterialslett.0c00130.url:https://doi.org/10.1021/acsmaterialslett.0c00130.

[5] Aurélie U. Ortiz et al. "What makes zeolitic imidazolate frameworks hydrophobic or hydrophilic? The impact of geometry and functionalization on water adsorption". In: Phys.Chem. Chem. Phys.16 (21 2014), pp. 9940–9949.doi:10.1039/C3CP54292K.url:http://dx.doi.org/10.1039/C3CP54292K.

630679 Phase Behavior of Confined Multiple Sites Associating Lj Fluids in Functionalized Slit Pore: A Monte Carlo Study

Abstract:

Fluids within nonporous materials have very common occurrence in many natural, biological and industrial processes such as shale gas extraction, microfluidics devices, fabrication of nanomaterials, nanotribology etc. The fluid phase behavior at nanoscale confined in different pores is significantly different than those in a bulk phase. The competitive interaction between fluids and surface atoms lead to various kind of phase transitions including layering, prewetting, vapor-liquid, melting/freezing, capillary condensation, along with a shift in critical properties of the system. Molecular insight of those phenomena is very essential for efficient design of micro or nano scale devices such as micro-electromechanical systems (MEMS), nanoelectromechanical systems (NEMS), Lab-on-a-chip (LOC) etc.

Various properties of coexistence phases and critical properties are assessed for multiple sites associating fluids at different slit widths as well as different functionalized surfaces using Grand-Canonical Transition Matrix Monte Carlo (GC-TMMC) method. In this work, one-site, two-sites (opposite to each other) and four-sites(tetrahedral) associating fluid model is considered in which Lenard-Jones potential is used for isotropic interaction between fluid molecules and a short-range directional square-well potential is used for site-site interaction. All interaction parameters are normalized with respect to parameters of fluid particle which is correspond the argon-graphite system. The structural features of coexistence phases are examined through orientation profile, monomer fraction and density profile. The effect of surface site density on the phase behavior under confinement is critically investigated. The presence of functionalized surface significantly alters the orientation of neighboring fluid particles, leading to rich coexistence phase behavior. For example, the orientation of two sites associating fluid molecules is normal to surface, which promote bonding with surface and propagate the bond formation to the bulk of the system.

630573 Molecular Dynamics Simulations Uncover Mechanisms of Increased Stability in Binary Colloidal Suspensions of Microparticles and Nanoparticles

Abstract:

This study focuses on the stability of neutral colloidal silica microparticles in aqueous suspension with low concentrations of highly charged zirconia nanoparticles. DLVO theory modified with additional depletion interactions was applied to describe particle interaction potentials and forces. Atomic force microscopy measurements identified and quantified an effective repulsion between microparticles that increased as nanoparticle concentration increased, suggesting that charged nanoparticles can be used to modulate the stability of microparticle suspensions. To better understand the mechanisms by which the effective repulsion arose, molecular dynamics (MD) simulations were performed. Systems containing 10^{-6} to 10^{-3} volume fraction of zirconia nanoparticles were simulated over timescales up to ten seconds. In the absence of nanoparticles, microparticles quickly aggregated during simulations. However, even in the smallest concentration of nanoparticles studied, an apparent kinetic barrier arose that slowed microparticle aggregation significantly. To quantify the apparent kinetic barrier and the effective pair interactions between microparticles in the presence of nanoparticle halos, enhanced sampling was employed. Umbrella sampling was used to ensure sampling of rare states and to construct a continuous potential of mean force as a function of microparticle-microparticle distances. Simulation results indicate a kinetic barrier ranging from about 2 to 6 k_BT exists for microparticle-microparticle interactions, in agreement with previous theories. This barrier grows as nanoparticle concentrations increase, suggesting that nanoparticle concentration is a driving factor in stabilizing microparticle suspensions as was observed in atomic force microscopy experiments. Ultimately, stabilization and aggregation were studied for systems of many microparticles using MD by applying potential of mean force curves from the umbrella sampling. Thus, the stabilizing effects of nanoparticles were preserved while their physical presence in the simulation box was no longer required. Simulations indicate that the kinetic barriers found during umbrella sampling simulations affect the rate of microparticle aggregation, matching experimental observations.

624868 Multidimensional Free Energy Landscapes for the Binding of Functionalized Nanoparticles to Lipid Bilayers

Abstract:

Nanoparticles (NPs) functionalized with organic ligands offer a potential nanotechnology platform for drug/gene delivery and phototherapy. These applications require specific interactions with cell membranes to facilitate efficient cell uptake or membrane adsorption while minimizing cytotoxicity. These interactions can be tuned by varying the properties of the ligand coating. However, nanoparticle interactions with the cell membrane remain difficult to predict *a priori*, hindering nanoparticle design. To address this challenge, previous experiments have correlated NP binding to model lipid bilayers with the octanol/water partition coefficient of the grafted ligand, yet the mechanism for this binding process is still largely unexplored. Molecular dynamics (MD) simulations offer the capability to understand these processes, explore the impact of new design motifs, and predict binding behavior. A challenge for molecular simulations is modeling the free energy landscapes underlying the binding of large NPs. Past simulation studies of NP-bilayer binding have utilized the center-of-mass (COM) distance between the NP and the bilayer to calculate free energy differences for the binding of NPs with short (< 1nm) ligands. Recently, atomistic MD simulations of NPs grafted with longer functionalized ligands have shown hysteresis in the COM-based measurement and suggest that the number of contacts between the ligands and the bilayer must contribute to the pathway. Sampling contacts in simulations that only bias the COM distance is challenging, indicating that multiple collective variables must be simultaneously biased to obtain the correct reversible path and accurate binding predictions. In this work, we parameterize coarse-grained (CG) simulation models of functionalized NPs based on prior atomistic calculations. We utilize the CG models to determine the free energy for binding NPs to model lipid bilayers as a function of multiple collective variables using multi-dimensional free energy calculation techniques including 2D umbrella sampling and metadynamics. Using the string method, we then efficiently find minimum free energy pathways for binding and compare the binding properties of several functionalized NPs. These predictions offer new mechanistic insight into the binding pathway for these NPs, allowing us to better understand which NP features contribute to enhanced binding to guide further NP design.

628420 Molecular Dynamics Investigation of Crystal-Melt Interface Kinetics and Its Stability Limits during Horizontal Ribbon Growth of Silicon

Abstract:

Horizontal ribbon growth (HRG) is an emerging method currently being developed for the fabrication of high-quality and low-cost single-crystal silicon wafers for solar cells. This work intends to improve the fundamental knowledge of HRG processes by modeling a continuous silicon crystal growth from molecular dynamics (MD) using the forced velocity solidification (FVS) approach. We studied growth modes of Si(100) and steps vicinal to Si(111) at high pulling speeds (~0.005 – 0.05 Å/ps), determined kinetics from interface temperature free of thermostatic effects, and investigated the stability of growth kinetics at high growth rates. We predicted kinetic coefficients from precise interface temperature facilitated by the FVS method, which increased kinetic coefficients for rough and stepped growths by ~40%. Also, we report an increase in step kinetic coefficient as step-size increased and a break-down of step propagation at a large step size of 23.10 Å due to the onset of interface roughening. The thermal gradients analysis across the melt and crystal parts of the interface suggests that interface roughening develops at higher pull speeds when the liquid's heat gradient is significantly low. Finally, we show that crystal-melt interface instabilities appear in the form of grain boundary grooves that served as nucleation sites for the formation of stacking faults and dislocations at undercooling as low as ~4 K—a value lower than previously observed from molecular dynamics.

630920 Effect of Structural Variations of Sugar-Based Surfactants (alkyl polyglucosides) on Interfacial and Surface Tension

Abstract:

Sugar-based surfactants produced from renewable raw materials such as starch and fats and their derivatives have attracted considerable interest due to their sustainable nature and biocompatibility. Amongst these "green surfactants" we have studied the particular family of Alkyl glucosides (AG) and alkyl polyglucosides (APG). APG surfactants are non-ionic surfactants characterised by the length of the alkyl chain tail (hydrophobic region) and the average number of glycose units linked to itforming part of the head group (hydrophilic region). While the general structure of an APG consists of only one sugar-based head group and a linear alkyl chain tail group, the overarching interest is to explore other possible morphologies, the incorporation of side-chains, other chemical groups (e.g. aromatic rings) to enhance the "tunability" of the overall surfactant [1].

In order to assess the effect of the different morphologies on the interfacial and surface tension, large scale atomistic canonical Molecular Dynamics (MD) simulations were performed for systems consisting of amphiphiles at the flat interfaces of water + *n*-decane and water + air. APG surfactants were modelled using the polymer compatible force field (*pcff+*). The values of interfacial tension were calculated using the Irving-Kirkwood method [2], through the integration of the difference between the normal and tangential pressure profiles against the coordinate perpendicular to the interface.

The surfactant with a single-glucoside head and a linear tail with 12 carbon atoms (APG_{12}) is used as a reference and benchmark to assess the difference with the other structural variations. This APG can reduce the interfacial tension for the water/oil system to close-to-zero values, while for the water/air system the tension is reduced to around 30 mN/m. Simulations have been run for over 30 different combinatorial variants of APG_{12} changing either the structure of the tail group, the head or both. Two of the more promising alternatives are shown in Figure 1, corresponding to a single-glucoside head and a branched tail with 12 carbon atoms (isomer), where one branch has 8 carbon atoms and the other one has 4 carbons. This structure can reach a lower critical micelle concentration, decreasing the interfacial tension faster than the regular APG_{12} configuration. Another example of an improved morphology consists of using a double-glucoside head and a linear tail with 12 carbon atoms. The surfactant with two glucoside rings shows a good performance at low surface coverages, decreasing the tension even faster than the branched structure. However, with a wider hydrophilic head, the system reaches the point of surface saturation before the other two structures.

One of our key findings is that branching of the tails provides a more stable interface with a lower critical micelle concentration. Amongst the options for branching tails, the more symmetrical the branches, the lower the surface coverage needed to reach the saturation point. Simulations provide key evidence on how the structure of the surfactant plays an important role in the stability of the interface. In particular, for APG's, given the bulkier nature of the headgroup, the surfactant benefits from tail groups that fill the space more evenly. This space-filling is typically carried out by adding co-surfactants to the formulations but can be easily engineered within the main surfactant.

This work encourages the validation of these findings through experiments and showcases the potential that molecular modelling has for the optimization of molecular fluids towards the ultimate deployment of bio-based surfactants in industrial, pharma and commercial applications.

[1] T. O. Gaudin et al., "Impact of the chemical structure on amphiphilic properties of sugar-based surfactants: A literature overview," *Advances in Colloid and Interface Science*, 270, 87–100 (2019).

[2] J. H. Irving and J. G. Kirkwood. *Journal of Chemical Physics*, 18:817–829, 1950.

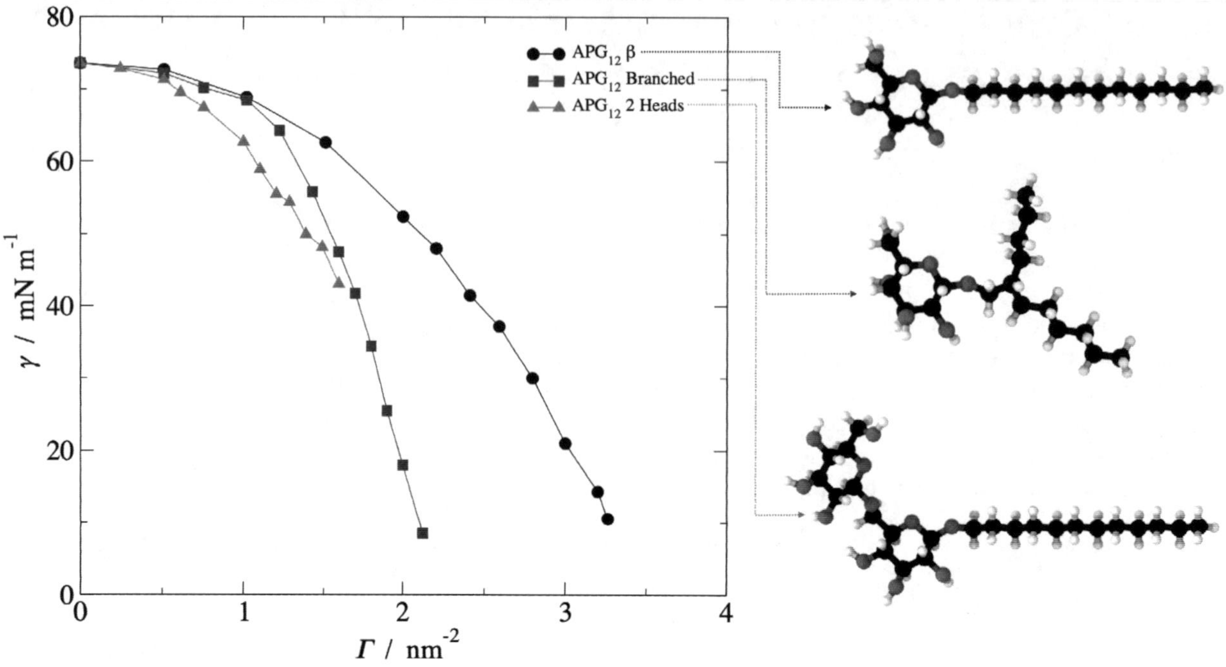

Figure 1. Surface tension of APG surfactants at the water/air interface at 298 K as a function of the surface coverage, Γ, or number of surfactants at the interface per area obtained from molecular dynamics simulations. Black symbols correspond to the APG with a single sugar head and a linear tail of 12 carbon atoms. Blue symbols represent the APG with a single sugar head and a branched tail with 12 carbon atoms in total (a branch with 8 carbons and another one with 4 carbons). Red symbols correspond to the APG with a double sugar head and a linear tail of 12 carbon atoms.

630659 Facet-Selective Roles of Chloride and Polyvinylpyrrolidone (PVP) in Facilitating the Growth of Ag Nanowires and Nanocubes

Abstract:

Ag nanocrystals are used in various applications in which the performance relies on their morphology. Though Ag nanocrystals with different shapes and sizes have been synthesized, control over the yield and selectivity is still difficult to achieve due to a lack of understanding of the growth mechanisms. Chloride and the capping agent PVP are usually hypothesized to affect Ag nanocrystal growth, but their exact roles remain controversial.

In this work, we applied density-functional theory (DFT) calculations to investigate the co-adsorption of Cl^- and PVP on Ag(100) and Ag(111) at different Cl^- coverages, with the aim to understand their synergistic roles in facilitating the growth of Ag nanocrystals. A weak facet-selectivity of PVP binding is found at low (or zero) Cl^- coverages and Ag truncated octahedra are thus predicted. A selective desorption of the PVP adlayer on Ag(111) [but not on Ag(100)] at 0.50 monolayer coverage of Cl^- is observed, which may facilitate anisotropic growth kinetically. No facet-selectivity presents at higher Cl^- coverages, as PVP leaves both surfaces due to stronger repulsion when more chloride is introduced.

Using *ab initio* thermodynamics calculations, a phase diagram, in which regions are classified based on various coverages of Cl^- and PVP, was constructed as a function of the chemical potentials of Cl^- and PVP. We see growth under thermodynamic and kinetic controls at different conditions. One pathway with increasing Cl^- concentration is predicted to thermodynamically produce truncated octahedra, cuboctahedra, and cubes, while it also indicates the switch to kinetic control to grow nanowires with decahedral seeds, which agrees well with experimental observations.

629176 Computational Study of Coarse-Grained Ionic Liquids at Electrified Interfaces

Abstract:

In the last couple of decades ionic liquids (ILs) have gained significant attention due to their unique physical and chemical properties. ILs are good candidates for battery electrolytes application due to low volatility, moderate reactivity, low flammability, and a wide liquid range than most organic solvents, thus imposing safer and higher-energy batteries. Systems of ILs have been extensively studied experimentally and computationally. For instance, the most commonly used all-atom molecular dynamics (MD) method delivers atomic-level details of IL systems for use in vast technical applications. MD method can handle systems of small length and time scale and coarse-grained (CG) simulation techniques may be used to extend all-atom computational simulations to mesoscale level. The present work is focused on studying CG ILs behavior at electrified interfaces. Selected ILs were confined between ideal graphite electrodes and density profiles for the cations and anions were calculated for various potentials. Obtained results are in close agreement with the reference all-atom MD simulations. CG systems considered in this work were at least eight times bigger than the reference all-atom systems allowing more realistic representation of the ILs at electrified interfaces. Research results represented in this work open opportunities to advance our knowledge in liquid-state batteries with modified electrodes structure for high power capacity battery development.

625368 Revisiting the Trappe Force Field for Organophosphorus Compounds: Sarin, DMMP, and Dimp

Abstract:

Even three decades after signing the Chemical Weapons Convention, organophosphorus chemical warfare agents (CWAs), such as sarin, remain a threat. There is therefore a critical need to develop novel methods for detection, protection, and destruction of these chemicals which motivates research on their physicochemical properties. Sarin is a highly toxic chemical compound, therefore an experimental investigation of its properties is usually conducted with simulants. To validate the performance of a simulant, molecular modeling can be employed for a direct comparison of physical-chemical properties of the compounds, such as vapor-liquid equilibrium properties.

The results of classical molecular modeling of sarin and simulants and their agreement with experimental data rely on force fields. In recent years there have been several force fields proposed for sarin and its most common simulant dimethyl methylphosphonate (DMMP). However, other simulants frequently used in experiments received less attention from the molecular simulation perspective, e.g. to date, there has not been any force field and no simulation data for diisopropyl methylphosphonate (DIMP).

We compare the literature force fields for sarin and DMMP, focusing specifically on the vapor-liquid equilibrium for the pure species. We performed Monte Carlo and molecular dynamics simulations for the calculation of liquid densities and vapor pressures developing the entire binodal curves. The TraPPE force field was shown to outperform other force fields. Therefore, it was extended to the DIMP molecule and employed for the calculation of its phase equilibrium properties. We demonstrated the ability of the TraPPE force field to give excellent predictions of the experimental equilibrium liquid densities and vapor pressures of DIMP in a wide range of temperatures. Moreover, since the critical points of sarin and the simulants have not been measured experimentally (they decompose at temperatures lower than critical), the simulation results were in particular use for estimation of the critical parameters of the compounds for further usage with the Peng-Robinson equation of state [1]. The results are expected to find applications in simulations of phase behavior of mixtures containing sarin or its simulants, and studies of adsorption of these species.

References:

[1] Emelianova, A.; Basharova, E.; Kolesnikov, A; Villaseco Arribas, E.; Ivanova, E. V.; Gor, G. Y. "Force Fields for Molecular Modeling of Sarin and its Simulants: DMMP and DIMP" J. Phys. Chem. B, 2021, DOI: 10.1021/acs.jpcb.0c10505

626447 Development of a Coarse Grained Discontinuous Molecular Dynamics Forcefield for Peptoids

Abstract:

Peptoids, or poly N-substituted glycines, are synthetic biocompatible peptidomimetics which have been used to develop antimicrobial agents, lung surfactants and drug delivery vehicles. They are protease resistant and have enhanced cellular uptake, which further make them attractive candidates for biological applications. Since peptoids lack native backbone hydrogens connected to an electronegative atom and hence lack backbone hydrogen bonding, their secondary structure is governed primarily by steric interactions (in the absence of specific interactions between side chains). Furthermore, unlike peptides, which have trans configuration as the prevalent amide bond isomer, peptoid amide bonds can have both cis- and trans- configurations. This allows for peptoids to exhibit a variety of secondary structures that are not observed in peptides.

Due to these differences and the varied accessible secondary structures between peptoids and peptides, force fields fitted for peptides have very limited applicability for peptoids. Weiser and Santiso [1] recently developed an atomistic force field for peptoids based on CGenFF, NTOID, which has the capability to model peptoids with amide bonds in the cis- and trans- conformation. In this work, we present an extension of this approach to develop coarse-grained models for use in discontinuous molecular dynamics (DMD) simulations.

We focus on discontinuous molecular dynamics because it is inherently faster than molecular dynamics and has been previously used for combinatorial screening of peptides. In order to develop our coarse-grained models, we use the relative entropy minimization method [2] to obtain bonded interaction parameters, using NTOID as the reference potential for which we implement a square well potential module in the software package VOTCA. For the intermolecular interaction parameters, we use a top-down coarse-graining approach using the square-well SAFT-γ equation of state [3] as implemented in the software package DESPASITO.

References:

[1] L.J. Weiser and E.E. Santiso, J. Comput. Chem. 40, 1946 (2019)

[2] M.S. Shell, Adv. Chem. Phys. 161, 395 (2016)

[3] A. Lymperiadis, C.S. Adjiman, G. Jackson, A. Galindo, Fluid Phase Eq. 274, 85 (2008)

628090 Data-Driven Many-Body Models for Molecular Fluids

Abstract:

Two of the most challenging problems at the intersection of electronic structure theory and molecular dynamics simulations are the accurate representation of intermolecular interactions and the development of reduced-scaling algorithms applicable to large systems. To some extent, these two problems are antithetical, since the accurate calculation of non-covalent interactions typically requires correlated, post-Hartree-Fock methods whose computational scaling with respect to system size precludes the application of these methods to large systems. I will describe recent advances in the development and application of our data-driven many-body models for molecular fluids which exhibit chemical accuracy from the gas to the condensed phase.

628695 Development of All-Atomistic Polarizable Force Fields for Molecular Simulations of Electronic Polarization Effects in Interfacial Phenomena

Abstract:

Molecular simulations along with the tools of statistical mechanics can provide vital mechanistic insights on the behavior of pure fluids and electrolytes at solid/water interfaces, which has broad applications in several scientific disciplines, including electrochemistry, membrane science, colloids, biophysics, and catalysis. At any solid/water interface, because water is a polar solvent and salt ions are charged species, they can exert strong electric fields which can in turn result in a significant electronic polarization of the solid. However, previous molecular dynamics (MD) simulation studies have typically neglected electronic polarization effects due to challenges associated with self-consistently modeling the ion–solid and water–solid polarization interactions at the solid/water interface. Here, we present a theoretical framework to self-consistently model electronic polarization effects at solid/water interfaces, and then apply it to investigate the graphene/water interface, which has shown tremendous promise for energy- and membrane-based applications. Using quantum chemical simulations carried out using the symmetry-adapted perturbation theory (SAPT), we first developed all-atomistic polarizable force fields which can accurately model the interactions of water molecules and salt ions with graphitic surfaces. Next, we applied our theoretical framework to carry out a comprehensive investigation of the role of electronic polarization effects on the contact angle of pure water on graphite (multilayer graphene). We demonstrate that the graphene-water polarization interactions have a significant impact on the interfacial entropy of water compared to the graphene–water dispersion interactions. This results in a significant energy–entropy compensation in the calculated solid–liquid (graphite –water) work of adhesion, which directly determines the water contact angle on graphite. We show that the simulated contact angle of water on graphite obtained using our all-atomistic polarizable force field is in quantitative agreement with the available experimental data, whereas an implicit modeling of the graphene–water polarization energy using a Lennard–Jones (LJ) potential results in a significant underestimation of the water contact angle by as much as 23°. Finally, in addition to modeling pure water, we also model electrolytes containing salt ions at the graphene/water interface. In this case, our study shows how electronic polarization effects contribute to a coupling between the graphene–ion and graphene–water interactions via electric fields, which is completely disregarded when the intermolecular interactions are instead modeled using pair-wise additive potentials, such as the LJ potential. In summary, we present a multiscale methodology to model electronic polarization effects using a combination of quantum chemical and classical MD simulations, and highlight the role of polarization interactions on the chemical physics of wetting and salt ion adsorption at the graphene/water interface.

628846 Benchmark of Topology Automated Force-Field Interactions (TAFFI)

Abstract:

Recent trends in the development of atomic force-fields include machine learning based potentials (MLP) and quantum chemistry derived force-fields. An important issue for these methods is the trade-off between the accuracy and complexity of the force field, where the high accuracy is achieved at the expense of transferability compared to classical force fields. Here we present an all-atom version of Topology Automated Force-Field Interactions (TAFFI), TAFFI-gen, which provide the development of force fields with the desired degrees of complexity. This is achieved by building an arbitrary extendable and transferable force field using self-consistent atom typing and DFT-based parameterization schemes. The performance of the derived parameters will be presented in terms of liquid property predictions revealing the representability limitation.

628948 Developing Phase-Transferable Machine Learning Force Fields of Molten Salts

Abstract:

A fundamental understanding of the behavior of molten salts is important for their applications in molten salts reactors, which are considered a leading candidate for the next generation of nuclear reactors. In order to obtain accurate physical properties in molecular dynamics (MD) simulations, accurate and computationally efficient force fields are required. In recent years, machine learning (ML) techniques have been widely applied in force field development. ML force fields are able to capture many-body interactions with accuracy close to *ab initio* calculations, and they are not much more computationally expensive than classical force fields. ML force fields are particularly appealing for simulations of molten salts, where polarizability (*i.e.*, many-body interactions) are often important.

ML force fields are commonly trained on *ab initio* MD (AIMD) simulations, and evaluated by their ability to reproduce *ab initio* energies and forces. However, this training procedure is necessarily performed with very small simulation boxes and short time scales, due to the large computational cost of AIMD simulations. The ML force field is then generally used to simulate much larger systems over longer timescales and calculate properties that would be difficult-to-impossible to evaluate with AIMD. Therefore, an important question to consider is whether a model trained on small simulation boxes can predict properties that can only be observed in larger systems. To rigorously answer this question, it is necessary to know the properties of the model used to generate the training data, which is not possible with AIMD.

In order to investigate the effect of small training system sizes on ML force fields, we trained an ML force field for $MgCl_2$ to reproduce an induced dipole model, named the polarizable ion model (PIM). $MgCl_2$ was selected because the liquid structure has long range ordering that PIM simulations capture and which is not observed in small box sizes (e.g., AIMD) due to system size effects. The PIM also captures complicated many-body interactions with induced dipoles, and is therefore more sophisticated than simple fixed charge models. Moreover, most physical properties of the PIM for $MgCl_2$ can be, or have already been, accurately calculated. This makes PIM $MgCl_2$ a great candidate to examine how the accuracy of a ML force field depends upon the time- and length-scales of the simulations used to train the model.

We develop a training scheme to make the ML force field phase-transferrable. Our training scheme efficiently generates a large amount of uncorrelated training data from both the crystalline and liquid phases by a two-step procedure. We evaluated a wide range of physical properties, including density, self-diffusivity, crystal and liquid structure, and melting point. We find excellent agreement between the ML force field and the PIM used for training. The ML force field of $MgCl_2$ is phase-transferrable, as it successfully reproduces accurate physical properties in both liquid and crystalline phases, compared to the original PIM. The melting point of the ML force field is also within 10 K of the PIM. Of particular note, we find that when performing the simulations in larger boxes, the ML force field reproduces the long range ordering of $MgCl_2$ liquid, even though the force field is entirely trained from PIM simulations in small simulation boxes where such structures cannot form. Our work provides strong evidence that ML force fields of molten salts can correctly capture properties that appear on time and length scales beyond those observed in the training simulations. By training an

ML force field on a computationally tractable model, we were able to rigorously test this hypothesis and validate our training procedure for developing ML phase-transferable force fields of molten salts.

630379 Interface Force Field for sp^3 Diamond Crystal Structures: Regression-Driven Development

Abstract:

The Interface Force Field (IFF) project centers around the construction of an all-atom force field and surface model database for simulation of nanostructures of 1 to 1000 nm size, including minerals, metals, oxides, and (bio)polymers.[1] To achieve this aim, the IFF project develops model systems that mimic experimentally observed behavior, which is validated through benchmarks such as density, surface energy, and melting point.

Via an ancillary electron method, we constructed a silicon (Si) model which reproduces a density of 2.312 g/cm^3 at 298 K, compared to 2.329 g/cm^3 as experimentally measured.[2] Our model is representative of experimentally observed Si melting temperature near 1687 K, largely owing to accurate description of surface energy which we demonstrate through the calculation of the Si(111) surface energy within 5% error of experimental estimates.[3] A similar model was constructed for carbon. Si and C were assembled to simulate the SiC layer that results from heterogeneous nucleation between C and molten Si during melt infiltration experiments. In the solid-state diffusion-controlled regime of the reaction occurring at the interface, the diffusivity of C through the initial nucleation layer of SiC governs further layer growth. We calculated the diffusivity of C through SiC layers of various thickness in order to investigate the effect of the time-dependent permeability on layer growth.

Where diffusivity is critical to reaction rate, as in the SiC system, a model that captures nanoscale system dynamics is necessary for micron-scale material simulation. Through accurate description of diffusivity in SiC, our Si and C ancillary electron models advance the capabilities of the Interface Force Field for ceramic materials.

References:

(1) Heinz, H., Lin, T.-J., Kishore Mishra, R., & Emami, F. S. (2013). Thermodynamically ConsistentForce Fields for the Assembly of Inorganic, Organic, and Biological Nanostructures: The INTERFACE Force Field. Langmuir, 29(6), 1754–1765. https://doi.org/10.1021/la3038846

(2) Assael, M. J., Armyra, I. J., Brillo, J., Stankus, S. V., Wu, J., & Wakeham, W. A. (2012). Reference Data for the Density and Viscosity of Liquid Cadmium, Cobalt, Gallium, Indium, Mercury, Silicon, Thallium, and Zinc. Journal of Physical and Chemical Reference Data, 41(3), 033101. https://doi.org/10.1063/1.4729873

(3) Tyson, W. R., & Miller, W. A. (1977). Surface free energies of solid metals: Estimation from liquid surface tension measurements. Surface Science, 62(1), 267–276. https://doi.org/10.1016/0039-6028(77)90442-3

630849 Integrated Strategy for Formulation of Realistic Potential Models for Simulation of Block Copolymers

Abstract:

Block copolymers (BCPs) have received much attention during the past few years because of their capability to self-assemble in various environments from high to low concentrations. Their ability to self-assemble makes them invaluable in producing different types of complex ordered structures for advanced materials. Therefore, BCPs are utilized extensively in nanostructured applications such as catalysts, organic photovoltaics, semiconductors, and optical coatings. Modeling BCPs is a challenging task as they have a vast variety of time and length scales important in their physical, structural, and dynamic characterization properties. Given the large time and length scales over which this self-assembly evolves, simplified mesoscale models are often employed because of their computational efficiency. However, these simplified mesoscale models possess some limitations that mediate their ability to realistically simulate BCPs. Chief among these are the use of parameters, such as the Flory-Huggins χ parameter, that do not account for the asymmetric interaction of the copolymer blocks and lack of a robust method to optimize the mesoscale energy parameters. While the commonly studied poly(styrene-b-methyl methacrylate) system reasonably approximates a block copolymer system in which the interactions of blocks with other blocks of their own type are relatively equal energetically (and thus this block copolymer is one we refer to as a relatively symmetric block interaction energy system), the vast majority of block copolymers deviate from this behavior. To address these limitations, in our work a coarse-grained mesoscale polymer model has been developed which utilizes interaction energy models and parameters that allow for the asymmetry of block interactions that occurs in many BCPs. Additionally, a detailed all-atom force field has been optimized for hydrocarbon polymers to allow atomistic simulations of block copolymers that reproduce with good accuracy the density, cohesive energy, and phase behavior of simple hydrocarbon block copolymers. This more detailed all-atom model has been used in conjunction with experimental data on block copolymer systems to parameterize more higher accuracy mesoscale homopolymer and block copolymer polymer models. This paper describes our approach to optimizing this all-atom force field to reproduce relevant bulk physiochemical and conformational properties and to properly account for long-range energy and pressure corrections. Specifically results of optimizing force fields for polystyrene and polyisoprene homopolymers will first be discussed. The block copolymer formed from these two homopolymers (i.e. poly(styrene-b-isoprene)) is known to possess a complex asymmetric phase diagram that has been experimentally well-characterized. Results of the ability of our models to reproduce this complex phase behavior will be discussed as a test case for our modeling approach and models.

632562 Development of Coarse-Grained Forcefields for the Self-Assembly of Ceramide-Based Lipid Membranes

Abstract:

Ceramides are a crucial component of the lipid lamella in the stratum corneum (SC) layer of mammalian skin, where barrier function is known to be localized [1]. Studying the SC lipid membrane can be challenging, as it is composed of at least 14 unique ceramides, along with cholesterol and free fatty acid of various lengths. Even when considering simplified mixtures that contain only a subset of these lipids, as are commonly studied via both experiment and simulation, challenges still exist. While most biological membranes form fluid-like bilayers with high lipid mobility, lipids in the SC instead form dense, gel-like multi-layer structures with very low lipid mobility [2]. As such, accurate modeling of membranes formed by SC lipids requires large simulation sizes to capture multilayer behavior and very long simulation times to allow the lipids to reach their equilibrium morphologies. This is further complicated by the fact that it is generally agreed upon that at least some ceramide lipids span multiple leaflets, making the use of pre-assembled multilayer configurations unlikely to be representative of experiment. In order to capture the larger length and timescales needed to represent the SC lamella, coarse-grained (CG) forcefields that allow for the self-assembly of large multilayer structures are necessary.

Here, we report our work to develop an accurate, transferable CG forcefield for studying the key lipids in the SC [3-7], optimized to match experimental data and the atomistic CHARMM forcefield [8,9]. This work utilizes the multistate iterative Boltzmann inversion (MS-IBI) method [10] combined with simulated wetting simulations to achieve the correct balance between hydrophobic and hydrophilic interactions [3]. MS-IBI is an extension of the original, single-state IBI methodology [11] commonly used to self-consistently optimize CG forcefields by matching the radial distribution function (RDF) of a CG model with that of the target system (typically a system simulated using atomistic models). MS-IBI builds upon the IBI approach by optimizing a single forcefield to match the RDFs from several thermodynamic states simultaneously; this methodology also allows a single forcefield to be optimized using different molecules with shared chemical topologies as target states. By using multiple states and molecules, MS-IBI has been shown to reduce structural artifacts and increase transferability between phases and molecules; this methodology also allows using multiple ensembles simultaneously, e.g., NVT and NPT, enabling the density-pressure relationship to be accurately captured.

Using the MS-IBI approach, forcefields for ceramide NS, free-fatty acid and cholesterol that provide close agreement with key structural properties from atomistic simulations and experiment, when available, have been developed [3-6]. The models are also able to reproducibly self-assemble multilayer structures. We further demonstrate the transferability of the SC forcefield developed using this approach by applying the models optimized for ceramide NS and FFA to other ceramides, namely NP, AP, and AS, finding close agreement with atomistic CHARMM simulations and experiment, without the need to reoptimize the non-bonded parameters [7]. To improve the reproducibility and ability to disseminate the forcefield, forcefield files are prepared that are compatible with the general purpose Foyer atom-typing library [12] as part of the Molecular Simulation and Design Framework (MoSDeF) [13] .

[1] Bouwstra, J.A. and Ponec, M., 2006. The skin barrier in healthy and diseased state. Biochimica et Biophysica Acta (BBA)-Biomembranes, 1758(12), pp.2080-2095.

[2] Moore, T.C., Hartkamp, R., Iacovella, C.R., Bunge, A.L. and McCabe, C., 2018. Effect of ceramide tail length on the structure of model stratum corneum lipid bilayers. Biophysical journal, 114(1), pp.113-125.

[3] Moore, T.C., Iacovella, C.R., Hartkamp, R., Bunge, A.L. and McCabe, C., 2016. A coarse-grained model of stratum corneum lipids: free fatty acids and ceramide NS. The Journal of Physical Chemistry B, 120(37), pp.9944-9958.

[4] Moore, T.C., Iacovella, C.R., Leonhard, A.C., Bunge, A.L. and McCabe, C., 2018. Molecular dynamics simulations of stratum corneum lipid mixtures: A multiscale perspective. Biochemical and biophysical research communications, 498(2), pp.313-318.

[5] Moore, T.C., Iacovella, C.R. and McCabe, C., 2016. Development of a coarse-grained water forcefield via multistate iterative Boltzmann inversion. In Foundations of Molecular Modeling and Simulation (pp. 37-52). Springer, Singapore.

[6] Moore, T.C. Shamaprasad P, Iacovella C.R., Bunge A.L., McCabe C., Multiscale Simulations of Stratum Corneum Lipid Mixtures, in preparation

[7] Frame, C.O., Shamaprasad P, Iacovella C.R., Bunge A.L., McCabe C., Transferrable models for the Simulation of Stratum Corneum lipids, in preparation

[8] Klauda, J.B., Venable, R.M., Freites, J.A., O'Connor, J.W., Tobias, D.J., Mondragon-Ramirez, C., Vorobyov, I., MacKerell Jr, A.D. and Pastor, R.W., 2010. Update of the CHARMM all-atom additive force field for lipids: validation on six lipid types. The journal of physical chemistry B, 114(23), pp.7830-7843.

[9] Guo, S., Moore, T.C., Iacovella, C.R., Strickland, L.A. and McCabe, C., 2013. Simulation study of the structure and phase behavior of ceramide bilayers and the role of lipid headgroup chemistry. Journal of chemical theory and computation, 9(11), pp.5116-5126.

[10] Moore, T.C., Iacovella, C.R. and McCabe, C., 2014. Derivation of coarse-grained potentials via multistate iterative Boltzmann inversion. The Journal of chemical physics, 140(22), p.06B606_1.

[11] Reith, D., Pütz, M. and Müller-Plathe, F., 2003. Deriving effective mesoscale potentials from atomistic simulations. Journal of computational chemistry, 24(13), pp.1624-1636.

[12] Klein, C., Summers, A.Z., Thompson, M.W., Gilmer, J.B., McCabe, C., Cummings, P.T., Sallai, J. and Iacovella, C.R., 2019. Formalizing atom-typing and the dissemination of force fields with foyer. Computational Materials Science, 167, pp.215-227.

[13] Summers, A.Z., Gilmer, J.B., Iacovella, C.R., Cummings, P.T. and McCabe, C., 2020. MoSDeF, a python framework enabling large-scale computational screening of soft matter: Application to chemistry-property relationships in lubricating monolayer films. Journal of chemical theory and computation, 16(3), pp.1779-1793. https://mosdef.org

632828 Development and Benchmarking of Small Molecule Force Fields in the Open Force Field Initiative

Abstract:

The Open Force Field Initiative (OpenFF) is academic/industrial collaboration focused on the development of 1) extensible, open source toolkits for constructing, applying, and evaluating force fields; 2) the curation of public datasets necessary to build high-accuracy biomolecular force fields; and 3) the generation of improved molecular force fields for biomolecular and other soft matter applications. In this talk, I will discuss recent efforts by OpenFF to provide automated optimization and benchmarking of small molecule force fields using both extensive quantum mechanical data and substantial condensed phase properties. I will describe the recent generations of general organic molecule force fields developed by OpenFF and their performance on large-scale benchmarks developed in cooperation with pharmaceutical industrial partners, including protein-ligand binding free energies. I will also talk about current efforts to capture greater physical detail including new charging schemes, data-based choices of interaction types, and incorporation of Wiberg bond order information to capture conjugated bond behavior without expansion of duplicative atom types. I will also discuss the tools we have developed for easily parameterizing molecules with Open FF tools and using them across a range of simulation platforms as well as how the framework is being applied in a fully transferable way to biopolymers.

623880 Integrative Modeling of the Smc-5/6 Protein Complex

Abstract:

Structural Maintenance of Chromosomes (SMC) complexes are important chromatin modulators. The Smc5/6 complex is directly involved in the regulation and repair of DNA in eukaryotes. Despite overarching topological similarities with other members of the Smc family like cohesin and condensin, Smc5/6 has distinct functions whose molecular bases remain poorly elucidated. Here, we develop an integrative model of the budding yeast Smc5/6 in complex with the SUMO ligase Nse2, and the Nse5 and Nse6 proteins. Specifically, we use crystal structures of Nse2 available from the PDB, our cryo-electron-microscopy derived 3 Å resolution structure of the Nse5/6 subcomplex, comparative models of the globular head and hinge regions of Smc5/6 and molecular mechanics models of their coiled-coil arm regions. A multiscale model representation consisting of the aforementioned structurally covered components as rigid bodies, and all other regions as (coarse-grained) flexible beads is scored using spatial restraints designed from our dataset of ~337 unique inter-residue chemical crosslinks, in addition to fundamental physical interactions like chain connectivity and excluded volume. Structural degrees of freedom of alternate models that satisfy the input information are sampled, and a representative ensemble of good scoring models are filtered out to describe the pentamer at a precision of 30 Å. The integrative model qualitatively and quantitatively corroborates the shape of the Smc5/6 arm region from negative staining 2D electron microscopy class averages, which were not used as input data for modeling. Unlike several examples of other SMC structures in the past literature, as well as contemporary structural investigations of the Smc5/6 complex, we provide a rigorous 3D model of the coiled-coil arm region that is structurally consistent with the input data. We find that in contrast to cohesin and condensin, the Smc5/6 arms do not fold back onto themselves; rather they interact uniquely with the Nse-s 2, 5 and 6, such that the Nse5/6 subcomplex organizes spatially as a linchpin that connects distal parts of the pentameric holocomplex.

629125 Molecular Dynamics Modeling Based Investigation of Protein Instability during "Non-Isothermal" Freezing Process

Abstract:

Purpose:

Bulk protein therapeutics are generally frozen for long term storage to avoid product degradation and facilitate product transportation. As a result, freezing of water induces solute rejection, creating regions of high solute concentration, which may impact protein to be destabilized and lose its activity. However, the underlying mechanism and molecular details of cryostabilization of proteins at freezing microenvironment are still poorly understood. Molecular dynamics (MD) simulations have been developed to investigate how protein conformation and stability be impacted during the bulk freezing process. The simulation results from this current work will be further validated with previous experimental data on freezing bulk protein therapeutics.

Methods:

Lysozyme (LYZ) protein solution in 10 mM phosphate buffer at pH 7.0 was chosen as initial conditions applied into this current MD simulations. The MD simulation trajectories and their analyses were carried out using GROMACS, VMD and Pymol software. All atom MD simulations were applied using CHARMM36 force field with temperatures gradually decreased from -3˚C to -26 ˚C with 5 – 6 ˚C temperature intervals. The initial protein crystal structure of LYZ was taken from the protein data bank entries (pdb: 1AKI). The protein was initially modeled using the psfgen tool in VMD, and then further solvated in a rectangular simulation box with the charge neutralized with $NaPO_4$. According to experimental work, an ice seeding approach was adopted to initiate ice formation in the MD simulation work. The simulation system was equilibrated in the NPT ensemble by initially applying 1000 cycles of a conjugate-gradient minimization scheme followed by a short 500-ps MD run in NPT assemble. The temperature was controlled using the Langevin thermostat, and the pressure was controlled by the Nose-Hoover barostat. The simulations were carried out using periodic boundary conditions. This briefly equilibrated system of LYZ protein buffer solution was further subjected to MD simulation sets in the NVT ensemble. Computations were performed on HPC, a supercomputer cluster at the University of Connecticut.

Results:

Ice formation was successfully initiated by ice seeding approach in the model protein buffer solution under freezing process. The convergence of the potential energy was monitored to reach the system equilibration along with decreased temperature, and then Root-mean-square-deviation (RMSD) analysis was applied to evaluate protein conformational changes. Significant RMSD fluctuations were captured at initial stage of freezing for the LYZ which is close to ice-water interface, whereas RMSD levels off to be stable at final freezing process. Moreover, the most flexible and unstable residues in LYZ were also identified during freezing process with the assessment of counting hydrogen bond formation within protein itself and with surrounding solvent water molecules.

Conclusions:

Therapeutic protein buffer solution under "non-isothermal" freezing process have been successfully modeled in MD simulations. Analyses of the resulted trajectories reveal the effect of temperature, ionic strength, solute concentration on the protein folding and unfolding states, biomolecular interactions, and protein aggregations.

Keywords:

Protein therapeutics, Lysozyme, Freezing, Molecular Dynamics Simulations, Non-isothermal

624764 Insight in the Glass Transition of the TIP4P/Ice Model

Abstract:

Over the past decades, research has shown us the capability of complex liquids to exhibit various polymorphic or polyamorphic forms when undergoing solidification processes (crystallization of glass transition). Water is the most famous of these complex liquids and a prime choice in the investigation of structural changes occurring during the glass transition. Therefore, we investigate the glass transition of water using Molecular Dynamics simulations and the TIP4P/Ice model. We navigate the phase diagram of this model in order to determine the temperature of glass transition at various pressure ranging from 1 to 8000 bar. In addition, we analyze the resulting structure for each of the thermodynamic conditions using the tetrahedral order parameter, the radial distribution function, and the coordination numbers. This allows the determination of the structure of our final system (LDA, HDA, or VHDA). Lastly, we monitor the evolution of local changes such as the local density fluctuations often due to the large hydrogen bond network.

624613 Obtaining the Equilibrium Adsorption Morphologies of Surfactant Molecules at Metal-Water Interfaces through Advanced Molecular Dynamics Simulations

Abstract:

Adsorption of surfactants on metal-water interfaces is useful for many applications, such as, in corrosion inhibition, heterogeneous catalysis, electrochemical reactions, and synthesis of anisotropic metal nanoparticles. At these interfaces, surfactant molecules are understood to adsorb in high-density adsorbed morphologies. These morphologies, however, have not been studied effectively from the current simulation techniques.

In this work, we report a new free energy sampling methodology that enables studying the formation of high-density adsorption morphologies. Using this methodology, we have studied adsorption of cationic and uncharged surfactant molecules of different alkyl tail lengths on metal-water interfaces. Our free energy calculations show that cationic molecules with small tail lengths (C-4) adsorb in a monolayer of molecules lying flat onto the surface, whereas the molecules with large tail lengths (C-12) adsorb in a mixed morphology such that some molecules lie-flat on the surface and rest of the molecules are aggregated as hemi-micelles on the surface. The uncharged molecules adsorb in much higher densely packed morphologies than cationic molecules.

627655 Accelerated Design of Flame Retardant Polymeric Nanocomposites Via Machine Learning

Abstract:

Improving the flame retardancy of polymeric materials is an increasingly important strategy to limit the exposure of life safety to more fire hazards, especially in this era when those polymeric materials are widely used almost everywhere in housing, vehicles, and commercial products. However, the wide variety of nanocomposite designs prevent rapid identification of the optimal composition for a given application. In this study, we applied machine learning methods, including *k*-nearest neighbor (*k*-NN), support vector regression (SVR), random forest (RF), and gradient boosting regression (GBR), to predict the flame retardancy index (FRI) and related features of cone calorimetry for different types of flame retardant polymeric nanocomposites. The quality of machine learning predictions was evaluated based on the statistical values, such as the coefficient of determination R^2, root mean square error (RMSE), average relative deviation (ARD), and absolute average relative deviation (AARD). We studied how physical features of polymeric nanocomposites affected flame retardancy using the correlation matrix of each feature, which in turn was used to guide the design of polymeric nanocomposites for flame retardant application. Following the guidelines deduced from these models, a high-efficiency flame retardant polymeric nanocomposites was designed and synthesized, of which the experimental results were compared with the machine learning predictions.

625254 Full Atomistic Polymer Structure Generation Using Reverse-Mapping Technique, Application to Poly(methy methacrylate)

Abstract:

Polymer materials are now widely used in our society. The weight reduction, flexibility, and toughness of the materials are the important features. Molecular simulation is a useful tool to examine materials properties from the microscopic viewpoint. However, polymer structure has not only micro-scale structural feature but also meso-scale structural ones such as entanglements, phase separation, etc. It is important to reproduce such structures to investigate polymer materials properties by molecular simulation. We have been developing full-atomistic entangled polymer structure generation scheme using reverse-mapping technique from the beads-spring models. We use a mapping relation proposed by Everaers et al. [1]. In the previous studies, we reported polymer structure generation for amorphous bulk models and for the adsorbed chains on a solid surface [2,3]. We analyzed generated structures in detail and found their entanglement molecular weight were in good agreement with references [4]. In this study, we are going to focus on Poly(methyl methacrylate), PMMA. PMMA is known as a polymer that has characteristic profile of mean square internal distance depending on its tacticity. As a result of reverse-mapping, the profile for the polymer chains has same character with that for beads-spring model basically. To generate PMMA structure accurately using reverse-mapping, it is considered that coarse-grained model which is finer than beads-spring model is necessary. Here we are going to utilize chemically accurate coarse-grained models for PMMA in reverse-mapping process to reproduce detail structural feature of polymer chains.

[1] R. Everaers, H. A. Karimi-Varzaneh, N. Hojdis, F. Fleck, and C. Svaneborg,, "Kremer-Grest models for commodity polymer melts: Linking theory, experiment and simulation at the Kuhn scale," *Macromolecules*, 53, 1901, 2020.

[2] H. Nitta and T. Ozawa, "Efficient Generation of Polymer Amorphous Structure By Reverse-Mapping from Beads-Spring to Full-Atomistic Model", AIChE 2018 Annual Meeting

[3] H. Nitta and T. Ozawa, "Efficient Generation of Full-Atomistic Polymer Amorphous Structure Adsorbed onto Solid Surfaces by Reverse-Mapping technique", AIChE 2019 Annual Meeting

[4] H. Nitta and T. Ozawa, "Full-atomistic entangled polymer structure generation using reverse-mapping from breads-spring models." 2020 Annual Meeting

630694 Uncertainty Quantification in Transition-Matrix Monte Carlo Simulations

Abstract:

Flat histogram methods vastly improve the efficiency of Monte Carlo methods used to simulate a wide variety of phenomena including phase equilibrium, self-assembly and aggregation in biological materials, colloids, polymers, ionic liquids and other complex fluids. However, the iterative convergence of these methods complicates quantification of the uncertainty. We develop a methodology for on-the-fly quantification of uncertainty in transition-matrix Monte Carlo simulations of fluid phase separation and self-assembly with Van der Waals and charged interactions. This new methodology will be compared with existing methods by considering both accuracy and efficiency.

632674 Comparative Role of Geometric and Chemical Parameters for the Formation of Pair-Wise Nanomolecular Complexes

Abstract:

The design and the prediction of the formation of pair-wise assemblies between proteins and other biomolecules in nanoscale dimensions are essential for understanding numerous biological processes. Despite the abundant protein topology data and advances in computations of self-assembled nano-complexes, the lack of generalized descriptors for the pairwise interactions of nano molecules impedes the understanding and prediction of their role in the biosystem.

Here, to understand the comprehensive parameters for the formation of nano-biomolecular complexes, the protein complex information is investigated. The protein structural data from protein databanks allow us to analyze the various descriptor sets to establish the probability for two proteins to form a complex. In addition to the chemical (CH) descriptors at the molecular level, the geometric (GE) and graph-theoretical (GT) are newly introduced to provide a universal description of the local properties of any nanostructures. GT-descriptors were calculated using graph theory and included parameters, such as Ricci curvature, multi-fractal nodal dimension, Gaussian network model were combined with GE (e.g. local chirality and pocketness), and CH (e.g. hydrophobicity and charge) features. The feature correlation dynamics validate the significant contribution of graph network features in the interaction prediction algorithms. As the proof-of-concept applications of nanomolecular complexes' interfacing sites prediction, the SARS-Cov-2 nucleocapsid protein dimer and experimentally proven protein and nanoparticle pairs are analyzed. The rapid and straightforward prediction of interface sites in pairwise molecular complexes will provide simplified design rules, new descriptors for the engineering of a wide range of pair-wise complexes of biomolecules with nanoscale dimensions.

626570 Drugsea: A Bioinformatic Tool to Identify Classes of Drugs with Differential Sensitivity

Abstract:

There is a pressing need for targeted therapies in diseases such as cancer. While targeted therapies have the potential for efficacy with minimal effects on normal tissues, it is estimated that only seven to eight percent of cancer patients are candidates for targeted therapies. To facilitate the identification of drugs which may bear therapeutic benefit for patients with particular genotypes or phenotypes, we present Drug Set Enrichment Analysis (drugSEA). drugSEA is an algorithm which analyzes drug sensitivity data from 1,351 drugs profiled on 327 cancer cell lines to identify drug targets that are differentially sensitive for individual genotypes or phenotypes. We first tested and validated our approach using synthetic data. Next, we further validated our method using published datasets to recapitulate known drug sensitivities towards EGFR, PI3K, and HDAC inhibitors. Finally, we used drugSEA to identify novel drugs which exhibit sensitivity in cancer cell lines based on their RNA expression of cystine/glutamate antiporter, xCT. Collectively, this study presents drugSEA as a bioinformatic tool which can identify drug targets with differential sensitivity towards individual genotypes or phenotypes.

627783 Hybrid Monte Carlo – Molecular Dynamics Simulations

Abstract:

Inherent to all computer simulations to understand biological structure and function, or perform drug design/discovery, is the need to efficiently sample the necessary phase space. However, there are a number of situations where the standard molecular dynamics simulations will fail to produce converged results [1, 2, 3, 4]. For example, ligand-protein binding energies are strongly affected by protein hydration state [2], and slow water exchange with the surroundings can lead to incorrect results [2, 3].

To address this issue, we present a hybrid Monte Carlo-molecular dynamics approach, which utilizes the strengths of each method to optimize sampling efficiency, and enable previously intractable calculations to be performed. This work builds on the software GPU Optimized Monte Carlo (GOMC) [5, 6], which is linked to the molecular dynamics software NAMD via a Python manager code that oversees information transfer and schedules the execution of each code [7, 8]. A number of illustrative examples are provided to highlight the potential of this approach, including prediction of the phase behavior of water, adsorption in a graphite slit pore [4], and hydration of protein binding pockets [3].

References

[1] Y. Deng and B. Roux, "Computation of binding free energy with molecular dynamics and grand canonical Monte Carlo simulations," J. Chem. Phys., vol. 128, p. 115103, 2008.

[2] G. Ross, E. Russell, Y. Deng, C. Lu, E. Harder, R. Abel and L. Wang, "Enhancing Water Sampling in Free Energy Calculations with Grand Canonical Monte Carlo," Journal of Chemical Theory and Computation, vol. 16, p. 6061–6076, 2020.

[3] M. Samways, H. Bruce Macdonald and J. Essex, "grand: A Python Module for Grand Canonical Water Sampling in OpenMM," Journal of Chemical Information and Modeling, vol. 60, p. 4436–4441, 2020.

[4] I. Ben-Shalom, C. Lin, T. Kurtzman, R. Walker and M. Gilson, "Simulating Water Exchange to Buried Binding Sites," Journal of Chemical Theory and Computation, vol. 15, p. 2684–2691, 2019.

[5] Y. Nejahi, M. Barhaghi, J. Mick, B. Jackman, K. Rushaidat, Y. Li, L. Schwiebert and J. Potoff, "GOMC: GPU Optimized Monte Carlo for the simulation of phase equilibria and physical properties of complex fluids," SoftwareX, vol. 9, p. 20–27, 2019.

[6] Y. Nejahi, M. Barhaghi, G. Schwing, L. Schwiebert and J. Potoff, "Update 2.70 to GOMC: GPU Optimized Monte Carlo for the simulation of phase equilibria and physical properties of complex fluids," SoftwareX , vol. 13, p. 100627, 2021.

[7] J. Phillips, R. Braun, W. Wang, J. Gumbart, E. Tajkhorshid, E. Villa, C. Chipot, R. Skeel, L. Kale and K. Schulten, "Scalable molecular dynamics with NAMD," Journal of Computational Chemistry, vol. 26, p. 1781–1802, 2005.

[8] J. Phillips, D. Hardy, J. Maia, J. Stone, J. Ribeiro, R. Bernardi, R. Buch, G. Fiorin, J. Hénin, W.

Jiang, R. McGreevy, M. Melo, B. Radak, R. Skeel, A. Singharoy, Y. Wang, B. Roux, A. Aksimentiev, Z. Luthey-Schulten, L. Kalé, K. Schulten, C. Chipot, E. Tajkhorshid, " Scalable molecular dynamics on CPU and GPU architectures with NAMD," Journal of Chemical Physics, vol. 153, p. 044130, 2020.

622947 Designing a Python Interface for Molecular Dynamics and Monte Carlo Particle Simulations: HOOMD-Blue v3.0

Abstract:

HOOMD-blue [1] is a general-purpose toolkit that performs molecular dynamics and Monte Carlo simulations of particles. Much has changed in the scientific Python ecosystem since HOOMD-blue's initial release in 2008. In this talk, we describe the design and implementation of the next major release of HOOMD-blue, v3.0. This release includes a new object-oriented Python interface, adds integrations with commonly used packages like NumPy, allows users to implement customizations to the simulation run loop, and offers a flexible system for accessing and logging computed quantities. As one example, HOOMD-blue v3.0 integrates with the MoSDeF [2] toolkit, allowing users to generate complex initial conditions with mBuild, atom-type the system with foyer, define force fields with gmso, and then run those simulations with HOOMD-blue – all in one Python script or notebook. I will also describe the software engineering practices we employ. We have made a major effort to increase the amount of testing we perform in v3.0 including new and rewritten unit and validation tests using the pytest framework that execute on continuous integration platforms. We provide binaries through Docker and Singularity images that support high performance computing clusters such as PSC Bridges2 and SDSC Expanse. We also provide generic container images for workstations as well as conda packages through conda-forge. To minimize developer effort, we use continuous integration systems to automate the release process.

[1]: Anderson, J. A., Glaser, J., & Glotzer, S. C. (2020). HOOMD-blue: A Python package for high-performance molecular dynamics and hard particle Monte Carlo simulations. *Computational Materials Science*, *173*, 109363. http://glotzerlab.engin.umich.edu/hoomd-blue/

[2]: Thompson, M. W., Gilmer, J. B., Matsumoto, R. A., Quach, C. D., Shamaprasad, P., Yang, A. H., Iacovella, C. R., McCabe, C., & Cummings, P. T. (2020). Towards molecular simulations that are transparent, reproducible, usable by others, and extensible (TRUE). *Molecular Physics*, *118*(9–10), e1742938. https://mosdef.org/

627683 Understanding Water Slip on Defective Boron Nitride Surfaces: Combined Quantum and Classical Modeling of Solid–Liquid Friction

Abstract:

Defects such as atomic vacancies are commonly encountered in real two-dimensional (2D) materials, yet their effect on interfacial phenomena is seldom considered. In this work, we realistically model the frictional properties of hexagonal boron nitride (hBN), an up-and-coming 2D material, by considering five different types of vacancy defects present in it. In recent years, hBN has been employed in various applications, such as membrane separations and energy harvesting, where hBN–water interfaces are ubiquitous. Thus, understanding solid–liquid friction on realistic hBN surfaces is important to design and develop such applications. Here, we present a combined quantum-classical approach to achieve this objective. To this end, we use quantum-mechanical density functional theory (DFT) to obtain point charges to accurately represent the unique electric field produced by each of the five defects considered. Subsequently, using classical molecular dynamics (MD) simulations, we study the role of the vacancy size and composition in modulating the hBN–water friction coefficient, and the resultant slip length for water flow on hBN. We also quantify the effect of vacancy concentration on the hBN–water friction coefficient. Our work provides realistic limits for the friction coefficient of hBN, by considering the effect of the most-probable defects in that material.

627157 Molecular Simulations Study of Corrosion Inhibitors Performance on Different Surface Morphologies in CO_2-Saline Aqueous Phase

Abstract:

Corrosion inhibitors (CIs) well-rooted their presence in the oil and gas industry to combat internal metal deterioration. The global market of corrosion inhibitors has reported 7.2 billion US$ in 2017 and forecasting to reach US$10.1 by the end of 2025 [1]. Despite the extensive application of CIs in industrial applications, their successful implementation is based in most cases on a trial-and-error testing approach. A deep understanding of the atomistic interactions near the aqueous-solid interface is the key answer to achieving improved CIs performance on metallic substrates that are in continuous contact with aqueous harsh media [2]. Computational quantum mechanics (QM) and molecular simulations can be implemented as powerful tools to screen the effectiveness of organic inhibitors prior to experimental validation. Several studies in the literature have focused on extracting CIs adsorption insights using quantum mechanics and molecular simulations with vacuum or pure water conditions of single inhibitor adsorption [3] -[5]. However, inhibitor adsorption in CO_2-saline solution conditions have not been reported nor considerably examined the multiple CIs aggregation behavior [6]. Besides, the adsorption of such organic compounds not only depends on the competitive adsorbate environment, but is significantly influenced by the metallic surface morphological properties [7] -[8].

This work belongs to a long-term project on using computational tools to design *ad-hoc* corrosion inhibitors for their practical implementation, providing a relationship between their structural properties and their surface protectiveness ability in different environments. We have recently published results concerning all-atom molecular simulations and chemical quantum calculations to understand the effect of the environment on the surface adsorption of corrosion inhibitor (CI) molecules on iron surfaces [9]. Three CIs abbreviated: TEPA, iTEPA and HC-iTEPA studied were selected to systematically investigate the influence of the alkyl tail, N-pendant group, imidazoline and benzene rings, on the CIs adsorption behavior. The quantum molecular parameters in water solvation anticipated higher electron transfer ability of iTEPA in aqueous conditions compared to TEPA and iTEPA, thus, leading to stronger adsorption on the iron surface, corroborated by the molecular dynamics (MD) classical simulations. MD simulations showed nearly 53%, 39%, 59% reduction in adsorption energies for single inhibitor molecule of TEPA, iTEPA, and HC-iTEPA when shifting from water to CO_2-saline media, respectively. The formation of water double adsorption layer contributed to decreasing the CIs adsorption energies. Nevertheless, the multi-inhibitors study revealed strong adsorption of TEPA and iTEPA on the iron surface, while HC-iTEPA neglected cooperative adsorption and aggregated as a spherical-like micelle with lower surface coverage propensity.

In the present contribution, building on previous work, we will present and discuss for the first time predictions concerning the adsorption behavior of CI molecules with different amine and imidazoline-based groups on different surface morphologies in dense CO_2-saline aqueous phase. MD and DFT calculations revealed that defected wide-groove surfaces significantly enhanced the multi-inhibitors cooperative adsorption behavior. Results of multi-inhibitors adsorption energy (E_{ads}) followed the order of *rough2>rough1>smooth* substrates. Furthermore, iTEPA/HC-iTEPA binary multi-inhibitor synergistic effect supported the multi-inhibitor surface interactions. DFT calculations complemented the MD simulation findings, revealing the most preferential sites responsible for the inhibitors electron donation\acceptance ability towards/from the iron surface.

We acknowledge financial support for this work from Khalifa University of Science and Technology (project RC2-2019-007)

KEY WORDS: corrosion inhibitors, adsorption, molecular simulations, surface roughness.

REFERENCES

[1] P.R. Newswire, Global Corrosion Inhibitors Market is Expected to Reach USD 7.14 Billion in 2019: Transparency Market Research, Bc-Transparency-Market-R. (2013).

[2] A.A. Olajire, Corrosion inhibition of offshore oil and gas production facilities using organic compound inhibitors - A review, J. Mol. Liq. 248 (2017) 775–808. https://doi.org/10.1016/j.molliq.2017.10.097

[3] S.K. Saha, M. Murmu, N.C. Murmu, I.B. Obot, P. Banerjee, Molecular level insights for the corrosion inhibition effectiveness of three amine derivatives on the carbon steel surface in the adverse medium: A combined density functional theory and molecular dynamics simulation study, Surfaces and Interfaces. 10 (2018) 65–73. https://doi.org/10.1016/j.surfin.2017.11.007

[4] H. Lgaz, I.M. Chung, R. Salghi, I.H. Ali, A. Chaouiki, Y. El Aoufir, M.I. Khan, On the understanding of the adsorption of Fenugreek gum on mild steel in an acidic medium: Insights from experimental and computational studies, Appl. Surf. Sci. 463 (2019) 647–658. https://doi.org/10.1016/j.apsusc.2018.09.001

[5] E. Aquino-Torres, R.L. Camacho-Mendoza, E. Gutierrez, J.A. Rodriguez, L. Feria, P. Thangarasu, J. Cruz-Borbolla, The influence of iodide in corrosion inhibition by organic compounds on carbon steel: Theoretical and experimental studies, Appl. Surf. Sci. 514 (2020) 145928. https://doi.org/10.1016/j.apsusc.2020.145928

[6] A. Mishra, C. Verma, H. Lgaz, V. Srivastava, M.A. Quraishi, E.E. Ebenso, Synthesis, characterization and corrosion inhibition studies of n-phenyl-benzamides on the acidic corrosion of mild steel: experimental and computational studies, J. Mol. Liq. 251 (2018) 317–332. https://doi.org/10.1016/j.molliq.2017.12.011

[7] N.M. El Basiony, E.E. Badr, S.A. Baker, A.S. El-Tabei, Experimental and theoretical (DFT&MC) studies for the adsorption of the synthesized Gemini cationic surfactant based on hydrazide moiety as X-65 steel acid corrosion inhibitor, Appl. Surf. Sci. 539 (2021) 148246. https://doi.org/10.1016/j.apsusc.2020.148246

[8] C. Verma, L.O. Olasunkanmi, E.E. Ebenso, M.A. Quraishi, I.B. Obot, Adsorption Behavior of Glucosamine-Based, Pyrimidine-Fused Heterocycles as Green Corrosion Inhibitors for Mild Steel: Experimental and Theoretical Studies, J. Phys. Chem. C. (2016). https://doi.org/10.1021/acs.jpcc.6b04429

[9] S. Alareeqi, D. Bahamon, R. P. Nogueira, L. F. Vega. Understanding the relationship between the structural properties of three corrosion inhibitors and their surface protectiveness ability in different environments. Appl. Surf. Sci. 542 (2021) 148600. https://doi.org/10.1016/j.apsusc.2020.148600

624648 What Makes Paint Stick? Insights from Computer Simulations of Epoxy Resin Binding on Iron Oxide / Oxide-Hydroxide Surfaces

Abstract:

The infrastructure of our modern society relies on large scale metal structures, which have a very large embedded carbon dioxide footprint. Protective organic coatings are key to sustainability as they extend the life of assets in aerospace, automotive, and freight by providing protection against corrosion. Epoxy resins are widely used due to their good heat and chemical resistance, favourable mechanical properties and good adhesion to a range of substrates. In all cases, the performance of the final solid-polymer system is dependent on the physicochemical properties of the interface and the interaction between the polymer and the solid substrate. However, experimental methods to characterize this interaction are limited and mostly are destructive of the interface. Computer modelling provides an excellent tool to investigate the surface-polymer interface at an atomistic level.

Here we perform atomistic molecular dynamics simulations to investigate the binding of common epoxy resin components to different iron oxide/ oxide-hydroxide surfaces. We calculate the binding energy, a measure of the energy difference between the bound adsorbate and the adsorbate at infinite distance from the surface, alongside free energy calculations which are computationally more demanding but provide valuable insight into the underlying free energy landscape that drives adsorption.

In epoxy resin applications the composition of the solid substrate is highly varied, with pre-treatments and production processes leading to a non-uniform surface chemistry and roughness. To reflect this, we investigate three different iron oxide/ oxide-hydroxide surfaces; hematite (Fe_2O_3) and magnetite (Fe_3O_4) and goethite (FeOOH) and consider different surface hydroxylation levels. We compare our findings with experimental results and find that the binding of epoxy components to each surface is quantitatively different. From these results, we suggest reasons for this trend based on the surface termination and gain fundamental understanding on how epoxy resins bind to metal substrates.

628710 Free Energy Calculation with Non-Uniform Windows for Improved Computational Efficiency: A Proof of Concept

Abstract:

Umbrella sampling combined with WHAM (weighted histogram analysis method) is widely used for computing free energy profiles along specified reaction coordinates. Success of the approach relies on the adequate sampling in each sampling window. For larger molecules with long relaxation times, it becomes costly to achieve this criterion. Using oligomer adsorption on a crystalline cellulose surface as a case study, we show that the actual sampling time required to reach a robust free energy profile, in the conventional approach with uniform windows, is much larger than that seen in typical literature. Lesser sampling time can result in significant errors. Inspection of sampling quality in each window shows that it is often a small number of windows being insufficiently sampled that limits the overall performance. By allowing non-uniform window sizes and biasing potentials, optimization can be applied to obtain numerical parameters that enable most efficient sampling. We show that accurate free energy profiles can be obtained with much less overall simulation time if computational resources are redistributed and optimized between different windows.

621894 Quick CH_4 Hydrate Nucleation By Ice-Catalytic Effect through Hydrate like Order Propagation: A Molecular Dynamics Study

Abstract:

In the present work, heterogeneous nucleation of methane hydrate in the presence of ice (hexagonal ice-Ih) was studied by molecular dynamics simulations using different initial methane concentrations (from 3 to 10 mole %) at 250 K and 50 MPa. The ice growth rate was observed to decrease with CH_4 concentration. The newly grown ice shows stacking faults with a combination of layers of hexagonal and cubic ice. The growing ice produced an accumulation of methane molecules near the advancing interface and promoted formation of partial hydrate cages at the ice surface. These partial cages have a short lifetime, detached from the ice interface, and migrated into the bulk that works as a seed for methane hydrate nucleation. A growing ice front thus works as a catalyst-promoting agent and initiates a faster rate of hydrate nucleation compared to the homogeneous system. The nucleation time of methane hydrate was significantly reduced in the presence of ice and catalytic effect of ice has been seen in the form of hydrate order fluctuations from the enhanced ice front. The space-time correlation highlights the propagation of hydrate like order parameters and insights the actual mechanism that promotes the hydrate nucleation. The statistical data of nuclei location indicates that sI methane hydrate prefer to nucleate near the ice interface or into the bulk, while surface nucleation was observed only in the case (4/40), where growing front is cubic ice.

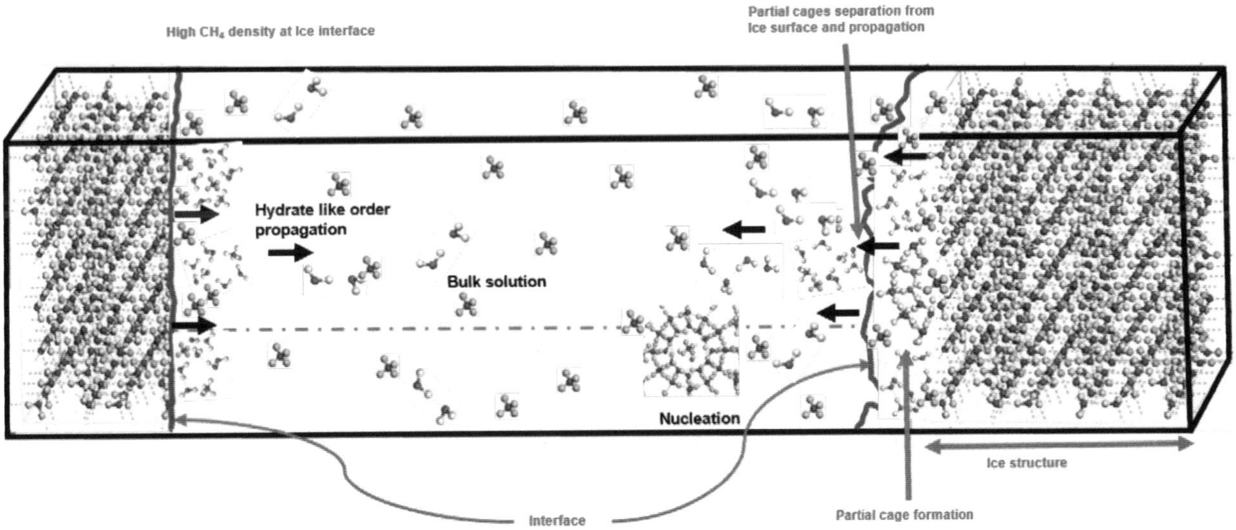

629083 Leaching Rate of Silver Ions from Engineered Nanosilver

Abstract:

Nanosilver is the third largest nanomaterial today and is used extensively into consumer products, including socks, food containers, bandages, and cosmetics. In addition, it exhibits remarkable antibacterial activity and strong plasmonic performance. Yet, there are growing concerns about the adverse effects of nanosilver on human health and the environment. Even though silver metal is not soluble in water, nanosilver particles of less than 10 nm release silver ions from the particle (leaching) into aquatic environments. This silver ion leaching is associated with nanosilver toxicity. However, the precise effect of nanosilver characteristics on silver leaching and the underlying mechanisms for toxicity are poorly understood.

Here, the mechanisms of silver ion release from nanosilver particles and their leaching rate is studied by reactive Molecular Dynamics simulations. Specifically, the effect of particle characteristics (size, crystallinity, specific surface area) and environmental conditions (temperature, concentration) on the leaching rate of nanosilver is quantified. Figure 1 shows the fraction of silver ions in nanosilver particles as function of particle size obtained by reactive molecular dynamics simulations (circles). The results are in excellent agreement with experiments of flame- and wet-made nanosilver (triangles) in aqueous suspensions. For comparison, the estimated silver ion fractions assuming one (squares) and two (diamonds) silver oxide layers are shown.Small nanosilver particles (with diameter of 3 – 4 nm) exhibit a large fraction of silver ions, present in the form of silver oxide at the particle surface. This fraction corresponds to a double layer of silver oxide. In contrast, larger nanosilver (7 – 8 nm) exhibit a smaller fraction of silver ions (corresponding to a single layer), consistent with experiments (Sotiriou et al., 2012).

The employed reactive molecular dynamics simulations are used, for the first time, to quantify the nanosilver oxidation rate and ion release rate from the particle surface in aqueous solutions that are responsible for the nanosilver toxic manifestations. Understanding the mechanism of silver ion release rate from nanosilver particles will determine their environmental fate and adverse effects on human health. Minimising these release rates will also enable the expanded use of silver nanoparticles and their nanocomposites as bioimaging materials or plasmonic biosensors.

References

Sotiriou, G. A., Meyer, A., Knijnenburg, J. T., Panke, S., & Pratsinis, S. E. (2012). Quantifying the origin of released Ag+ ions from nanosilver. Langmuir, 28(45), 15929-15936.

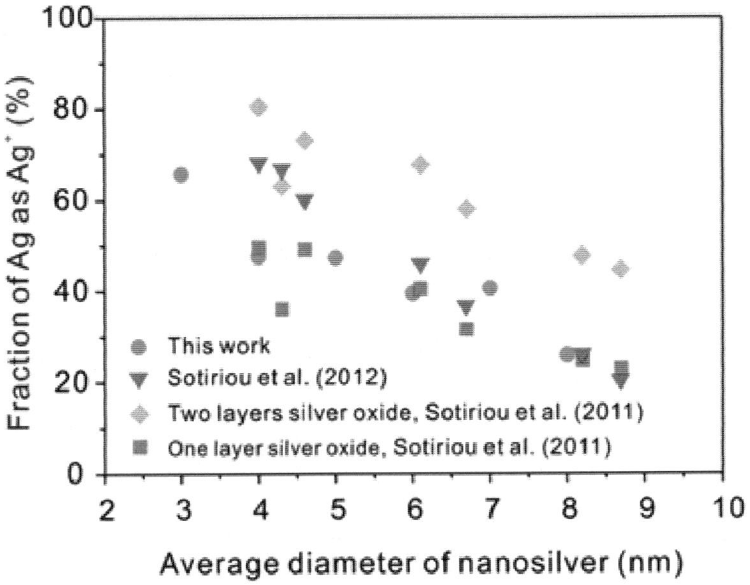
100
80
60
40
20
0
Fraction of Ag as Ag⁺ (%)
This work
Sotiriou et al. (2012)
Two layers silver oxide, Sotiriou et al. (2011)
One layer silver oxide, Sotiriou et al. (2011)
2
3
4
5
6
7
8
9
10
Average diameter of nanosilver (nm)

630778 Predicting Catalytic Activity of Heteropoly Acid Using Machine Learning Model with Small Dataset

Abstract:

Deprotonation Enthalpy (DPE) is a decisive probe to theoretically analyze the ease of proton removal from acid catalysts, facilitating quick catalyst screening. Reducing heavy computational dependence has been popular with Machine Learning (ML) models in material science. The availability of such models can learn the trend in the catalyst properties and further fast-track their selection. The DPE accurately explains the activity of the Keggin-type ($H_nM_{12}XO_{40}$) Heteropoly acids (HPAs) class of acid catalysts. Thus, the dataset for the model is developed on HPAs with the periodic table properties of their central (X) and addenda (M) atoms as feature variables, and their DPE is obtained from Density Functional Theory (DFT) calculations. The model evaluation metrics used are root mean squared (RMS) error and variance score to develop a reliable ML model with the least bias and high accuracy. Accordingly, Gradient Boosting Regression (GBR) Algorithm returned exceptional fitting with a high variance score of 0.88 for the test set and 0.97 for the train set. This supervised learning model showed the least error among the popular linear and ensemble models. Thus, the GBR model predicted DPE with an RMS error of just 5.6 kJ/mol (~0.05 eV) in test data for an average of 100 random train-test splits. It is important to note that the GBR algorithm was trained on small datasets, considering that obtaining massive data on a specific class of catalysts is an arduous task. Herein, the electronic and elemental periodic table properties selected as feature inputs for the model suggest an intrinsic trend with the target descriptor (i.e., DPE)—ensuring good predictions on a small dataset. Further, this algorithm efficiently understands the residual error between true and predicted values while developing the model. This yields an ensemble of these models; therefore, providing an excellent fit to the dataset.

First-principles insights into thermal degradation mechanisms of aqueous 2-amino-2-methyl-1-propanol (AMP) for CO_2 capture

Bohak Yoon and Gyeong S. Hwang

McKetta Department of Chemical Engineering, University of Texas at Austin, Austin, Texas 78712, USA

2-Amino-2-methyl-1-propanol (AMP) has been considered an attractive amine-based solvent for CO_2-capture, but its thermal degradation still remains a challenge. In this talk, based on first-principles simulations, we will present the underlying reaction mechanisms of AMP degradation to cyclic 4,4-dimethyl-1,3-oxazolidin-2-one (DMOZD), a major product, in CO_2-loaded aqueous solution. We have evaluated possible degradation pathways including the relative stabilities of intermediates and products based on implicit solvent calculations, and then predicted the free-energy barriers for key elementary steps using metadynamics-biased *ab initio* molecular dynamics. Our calculations show that DMOZD may be formed from carbamic acid of AMP primarily following a direct ring-closure mechanism. This is in contrast to the case of monoethanolamine (MEA) in which the formation of cyclic 2-oxazolidinone (OZD), the primary initial product, is more likely to occur through an isocyanate intermediate. Our work also demonstrates that the reaction of DMOZD with AMP can be markedly suppressed compared to the reaction of OZD with MEA which leads to the formation of more stable products in aqueous MEA. Our further analyses clearly illustrate that the presence of methyl groups in AMP derivatives makes their solvation structure and dynamics noticeably different from those of MEA derivatives. The solvation effect turns out to be largely responsible for the distinct difference in degradation behavior between AMP and MEA in aqueous solution.

625583 Theoretical Investigations of Air Separation Using Cobalt-Containing Complexes and Solvents

Abstract:

The most common commercial process for O_2 separation from N_2 is cryogenic distillation, but the process is expensive and energy intensive. More efficient and effective O_2 separation from the air is needed. One possibility is to use solvent-based absorption coupled with vacuum or thermal regeneration, but solvents with improved O_2 working capacity need to be developed. Cobalt (Co) containing solvents are of particular interest because of their high affinity for O_2. In this presentation, we show screening and theoretical calculations of solvents to solve a Co metal containing complex, salcomine (Co(salen)), and O_2 and N_2 interactions with the systems from quantum mechanical calculations in the gas phase. All possible spin states for the O_2-Co(salen)/solvent systems were considered. The experimental O_2 solubilities in Co(salen) solved in different solvents at room temperature and the theoretical O_2 interactions with the Co(salen)/solvent system were compared. It was found that when the interaction is below -23 kJ/mol, O_2 solubility in the system will be low and is determined by physical absorption. When O_2 interaction with the system is stronger than -30 kJ/mol, at low Co(salen) concentration in solvent, the O_2:Co ratio is 1:1. At higher Co(salen) concentrations in the solvent, the O_2:Co ratio is 0.5. When the O_2 interaction is around -25 kJ/mol, the O_2:Co ratio is about 0.5 independent of the Co(salen) concentration in solvents. In contrast, N_2 interactions with the same systems were found to be much lower, about -16 kJ/mol. Furthermore, the calculations show that by adding ethyl groups on Co(salen), the O_2 interaction could be increased.

The NIST database was screened by specifying some specific sub-structures. Three promising solvents were identified with low vapor pressures. Our molecular modeling calculations further predict that the viscosities for those solvents are low, which was experimentally confirmed. In addition, the quantum mechanical calculations suggest that O_2 solubilities in Co(salen) solved in the solvents would be high. We will also discuss the calculation results in the condensed phase.

626211 Investigating the Role of Lewis Acid-Base Interactions in Halide Perovskite Solutions

Abstract:

When processed in solution, metal halide perovskite solar cells, provide a unique cost advantage over existing solar cell technologies. However, the morphology and solar cell efficiency of the final crystal film is known to be heavily influenced by the perovskite material's many components of (A, B, and X-site atoms) and solvent selection. As these effects remain poorly understood, studying the nucleation and early-stage growth of these materials in solution would provide much needed insight into the effect of solvents on final film morphology. Molecular simulation approaches have a significant role to play in this endeavour by uncovering the solvent-solute interactions (described by Lewis acid-base chemistry) which influence the formation and aggregation of moieties that will ultimately nucleate and grow into thin films. In this study, we leverage first principles density functional theory (DFT) and molecular dynamics (MD) to elucidate the role of these interactions in the pre-nucleation and early-stage growth of metal halide perovskites. Specifically, we:

- **Quantify the binding affinity** of Lewis bases (solvents) towards Lewis acids (perovskite salts and cations) – **over 600 combinations were explored.**
- **Identify metrics** towards selecting ideal solvent candidates for perovskite processing without the need for additional computation (**rules of thumb and simple linear models**).
- **Uncover their impact** on the solution precursors and solid-state intermediate structures preceding the final perovskite film.

A key result from this study is that no single metric can describe solvent interactions towards A-site cations (e.g., Cs^+) and B-site cation species (PbI_2). For example, the Gutmann donor number (DN), a Lewis affinity scale metric that has gained popularity in recent years, adequately describes solvent interactions with isolated B-cation salt complexes (BX_n). Here B represents B-site cations of group 14 and 15 elements (Pb, Sn, Ge, Bi, and Sb) and X represents the perovskite halide ions (Cl-, Br- and I-). On the other hand, the lithium cation affinity (LCA) scale best describes solvent interactions towards group 1 A-site cations (Cs^+ and Rb^+), while the methyl ammonium cation affinity (MACA) approximates the affinity of solvents to the methylammonium cation (MA^+) in perovskite solutions. Using computation, we elucidate how the affinity of solvents towards these Lewis acids can **impact the formation of higher order iodoplumbate complexes and the intercalation of A-site cation species into the perovskite framework** – two key reactions occurring in solution that are essential to the nucleation and growth of the final perovskite film. Finally, through leveraging our generated metrics, one could design processing protocols that optimize the quality of the final perovskite film via judicious solvent selection.

630451 Predicting Interactions of Aβ and the NMDA Receptor Using Molecular Docking and Simulations

Abstract:

Despite decades of research, the fundamental molecular mechanism of Alzheimer's disease (AD) is unknown. Aβ, a 40-42 amino acid long peptide, has been proposed as a central molecule in pathogenesis. Two mutations of the second residue in the N-terminus of Aβ have opposite effects in developing the disease. A2T is a neuroprotective variant and A2V is a causative mutation. Another protein of interest, when studying Alzheimer's pathology, is the NMDA receptor (NMDAR), a post-synaptic receptor family critical to neuroplasticity and consequently learning and memory. The NMDAR protein is composed of four subunits and is activated by the coincidence of glutamate and glycine together with a membrane depolarization. Several treatments for AD focus on regulating the activity of NMDAR through inhibition, e.g., memantine. While there is evidence that suggests a link between Alzheimer's via direct or indirect interactions between the NMDA receptor with Aβ, the structural details and binding modes remains largely unknown. The goal of this study is to use *in silico* methods (molecular docking and simulations), to provide insight into NMDA - Aβ interactions and guide experiments that can elucidate their role in Alzheimer's pathogenesis. Our methodology includes an *ab-initio* rigid body docking step, to generate unbiased interaction models of the NMDAR- Aβ complex and discover preferred Aβ binding regions on NMDAR. This is then followed by a flexible docking step to predict a more representative structure of the Aβ peptide:NMDAR complex, and a molecular dynamics simulation step to study the strength and stability of the complexes in an environment that closely represents *in vivo* conditions. We find that a single amino acid substitution with known changes to developing AD results in distinct interaction profiles in our rigid body docking experiments. We also find overlaps between active molecule binding locations and predicted docking models with all Aβ isoforms.

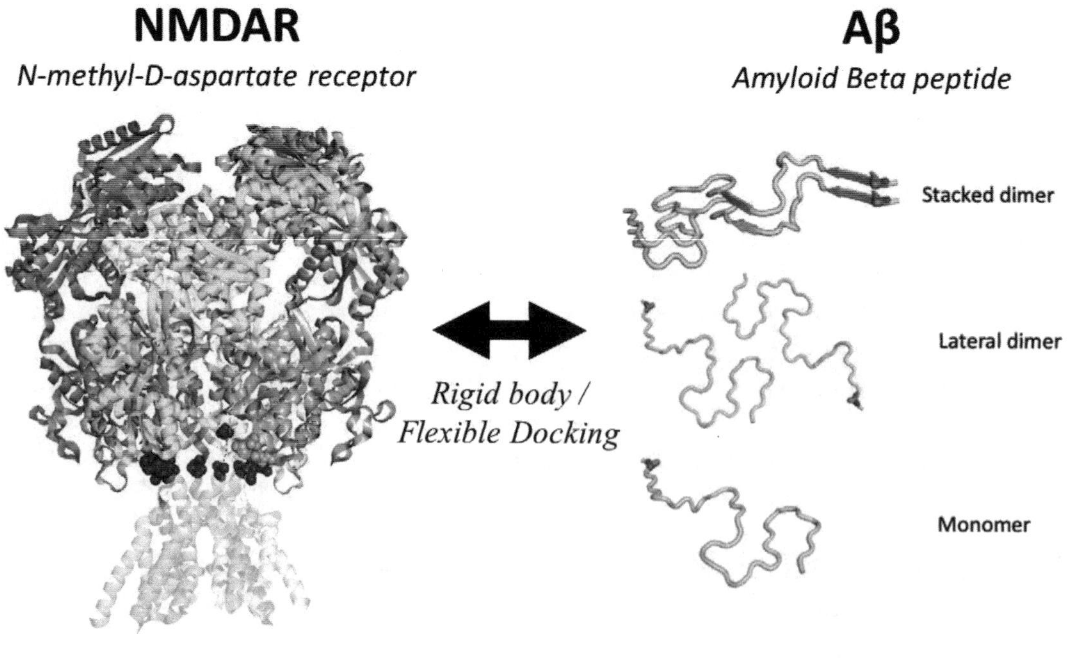

630858 Mixed-Surfactant Adsorption on Single-Walled Carbon Nanotubes: Insights from Molecular Simulations

Abstract:

Single-walled carbon nanotubes (SWCNTs) present unique mechanical, electrical, and optical properties but many difficulties exist in the practical application of this material. The main barrier to more widespread use is the need to separate and purify the heterogeneous mixtures (in terms of size and chirality) produced during synthesis. Numerous attempts have been made to design methods to efficiently separate particular SWCNT chiralities, but such techniques have yet to demonstrate success beyond the bench scale. These approaches usually involve a two-step process – first dispersion of the SWCNT material in aqueous solution via the use of either small-molecule or polymeric surfactants followed by manipulation of the dispersed mixture via some external action (centrifugation, phase separation, etc.). Recently, via near-infrared fluorescence emission measurements, it has been shown that varying the relative amounts of deoxycholate (DOC) and dodecyl sulfate (SDS) results in a sharp transition in the two-phase partitioning behavior of the dispersed SWCNTs. Furthermore, the location of this transition appears to be different for each SWCNT chirality. We shed light on the nature of this transition by using molecular simulations to investigate competitive adsorption in multi-surfactant systems. Resulting packing motifs are then compared across a range of relative surfactant concentrations. Free energy perturbation is also used to determine the relative solvation energies of these complexes. This work will be used to help design separation protocols that allow for the sequential extraction of specific SWCNT chiralities.

632441 Engineering Novel Proteins for the Sustainable Extraction of Rare-Earth Metals

Abstract:

Rare-earth elements (REEs) have applications in many consumer electronics such as motors and batteries. However, the current state-of-the-art extraction process is environmentally damaging due to the heavy use of acidic and organic solvents. The physiochemical similarity of the multiple REE species also complicates purification and no current extraction technique presents selectivity with the group. Recently, a protein was discovered that demonstrates preferential binding with REE ions as opposed to more biologically common species such as calcium. Named 'lanmodulin' due to its preference for lanthanides and its similarity to calmodulin, this protein has served as a starting point for our investigation into the design of metal-binding proteins that can selectively remove specific REEs of choice. Using both computational molecular modeling and experiment, we have begun to investigate which peptide characteristics lead to the observed REE affinity. Binding to specific protein domains has been characterized using UV/Vis spectroscopy. The next generation of engineered protein designs have been expressed via various synthetic biology techniques and it is hoped that this approach will yield a sustainable method to catch-and-release specific REE ions from new sources such as electronic and industrial wastes.

625243 Uncovering Magnetic Properties Composition Dependence of Substituted Ferrites for Magnetically Mediated Energy Delivery Applications: DFT and ML Study

Abstract:

We use computational methods to drive the design of substituted ferrites. Substituted ferrites are magnetic materials based off of the bulk structure of magnetite, i.e., Fe_3O_4 (**Figure 1a**). In this structure, two of the Fe ions are octahedrally coordinated with formal charges of +2 or +3, and one is tetrahedrally coordinated with a formal charge of +3 (**Figure 1b**). Substituted ferrites have the formula $M_xFe_{(3-x)}O_4$,where M is the substituent atom(s), and can be Mn, Co, Ni, Cu, or Zn; whereas x corresponds to the number of substitutions for the substituent atom(s). Capitalizing on their strong and tunable magnetic properties (e.g., magnetic moment, magnetic saturation and/or magnetic anisotropy), substituted ferrites can be used for magnetically mediated energy deliver (MagMED) applications. For biomedical applications, a goal is to optimize the magnetic saturation and anisotropy so that energy can be delivered (e.g., to cancer cells) effectively and safely. Recent literature suggests that the magnetic properties of substituted ferrites—such as magnetic moment—can be tuned via their compositions, as illustrate **Figure 1c**. In this work, we use a combination of Density Functional Theory (DFT) and machine learning (ML) to learn the relationship between composition and magnetic performance of substituted ferrites. Magnetic moment and magnetic saturation are calculated using DFT for ~666 composition—with normal and inverse spinel bulk structures —with x ranging from 0.0625 to 1. Substitutions are made at the octahedrally and tetrahedrally coordinated Fe^{2+} and Fe^{3+} sites respectively, which are highlighted in gray in **Figure 1b**. DFT values, for the magnetic moment and saturation, are input to machine-learning (ML) model to correlate it to composition. Specifically, an XGBoost algorithm is used to correlate the magnetic moment and saturation to tabulated properties of the substituent atom(s), such as electronic properties, nuclear properties, atomic sizes, and redox energies. All of these "features" of the substituent atom(s) are taken from established databases and do not require further calculations. Our first ML model which predicts if an specific composition results in normal or inverse spinel structures, **Figure 1d** shows parity plot comparing the difference in electronic energy (ΔE) for normal vs. inverse spinel structures as computed with DFT and predicted with our ML model. The developed ML model enables a rapid screening of vast chemical spaces of ferrites composition with surrogate ML models, which can be used to identify design rules and guide materials selection for substituted ferrites.

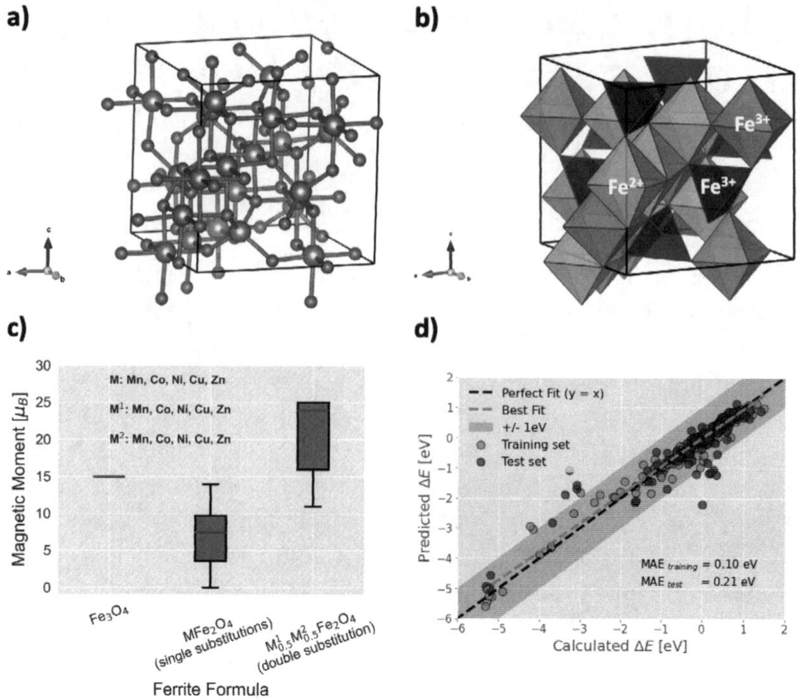

Figure 1. (a) Ball-and-stick style rendering (Fe and O atoms are colored in tan a red), b) Polyhedral style representation—purple, orange and gray polyhedral represent tetrahedral (Fe^{2+}), octahedral (Fe^{3+}), and octahedral (Fe^{2+}) sites respectively, and c) Magnetic moment (in Bohr Magneton (μ_B) for 3d transition metal single and double, substituted ferrites. M, M^1, M^2 correspond to the 3d transition metal substituent atom(s) (Data accessed from Materials Project dataset), and d) parity plot comparing the saturation magnetization computed from DFT and ML model.

627039 Combining Data-Driven and Physics-Based Simulations for Predicting Geometries and Energies of Biomass Intermediates on Rh Surfaces

Abstract:

Biomass compounds adsorbed on surfaces are challenging to study due to the large number of possible species and adsorption geometries. In this work, elementary steps and intermediate species of linearly-structured biomass compounds are studied in the gas-phase and on the Rh(111) surface. Possible intermediates and elementary reactions are generated by a recursive bond-breaking algorithm. These structures are used as the input of a unsupervised Mol2Vec algorithm to generate vector descriptors. We first develop a data-driven scheme to classify the reactions and predict the reaction energies in gas-phase. The data-driven model can be further extended to Rh surface species as well. The lowest mean absolute error (MAE) of our prediction on adsorption energies is 0.39 eV, and the relative ordering of different surface adsorption geometries is relatively accurate. We show that combining geometries from density functional tight-binding (DFTB) calculations with energies from machine-learning predictions provides a novel workflow for rapidly assessing the geometry stability on the Rh(111) surface.

632054 Computational Algorithms for Molecular Profiling of Activating Mutations in Kinases

Abstract:

Computational algorithms to assess the mutational landscapes of protein kinases mutated in cancer are desirable since the prevalence of mutations (some activating, some passenger) make clinical decisions for patient enrollment in drug trials challenging. While data-driven techniques provide a scalable approach to assess the mutational effect on protein structure and function, they often overlook the mechanism underlying the effect thereby resulting in high false positive rates.

Firstly, we implement a data driven technique that is a significant improvement over SIFT and PolyPhen2 in predicting driver and passenger mutations in kinases based on machine learning algorithm that implements features from the biochemical space. Moreover, through implementing combined computational studies of molecular dynamics (MD) as well as enhanced sampling simulations, we obtain mechanistic understanding of conformational transition in kinase and effectively predict the experimentally determined activation status of mutation in kinase by identifying hydrogen bond fingerprint in the activation loop and the αC-helix regions and this outperforms existing standard methods with drastically reduced false positives when validated against experiments while also leading to the unraveling of convergent mechanisms in the oncogenic activation of mutations in ALK kinase which is known to have implications in pediatric cancer. Finally, through this technique we propose to append explainability and interpretability to the supervised-ML algorithm through engineering features from the structural transition of the kinase from the inactive to the active state, with the structural transition being obtained through enhanced sampling technique of metadynamics.

We acknowledge financial support from the NIH and computational support from XSEDE.

625074 An Integrated Computational and Experimental Approach to Identifying Inhibitors for Sars-Cov-2 3CL Protease

Abstract:

The newly evolved SARS-CoV-2 has caused the COVID-19 pandemic. Although vaccines are the most efficient way to end the COVID-19 pandemic and a couple of vaccines have been authorized for emergency use to control the pandemic, safety issues for certain people, *e.g.*, those with allergies and those with immune disorders, are still concerned. Furthermore, we have observed SARS-COV-2 variants spread (CDC, 2021). SARS-CoV-2 main protease 3CLpro, cleaving the transcript polyprotein, is essential for the rapid replication of the virus. Inhibiting this protease may open an alternative avenue towards therapeutic intervention. (Anand et al., 2003; Ratia et al., 2008). Many peptide inhibitors were designed to mimic natural viral polypeptides and covalently bind to the active site Cys_{145} of $3CLpro^{SARS\text{-}CoV\text{-}1}$. The other category of $3CLpro^{SARS\text{-}CoV\text{-}1}$ inhibitors includes non-covalent or reversible covalent inhibitors, which have advantages regarding side effects and toxicity. In this work, we aim to identify non-covalent inhibitors against $3CLpro^{SARS\text{-}CoV\text{-}2}$. In particular, we first implemented the molecular docking approach to quickly identify potential 3CLpro inhibitors against SARS-CoV-2 in previous research (Jin et al., 2020). To narrow down $3CLpro^{SARS\text{-}CoV\text{-}2}$ inhibitor candidates, an integrated docking-based virtual screening and quantitative structure–activity relationship (QSAR) method was then conducted, on the basis of the aforementioned $3CLpro^{SARS\text{-}CoV\text{-}1}$ inhibitors and half-inhibition concentration IC_{50} data. A 3D QSAR model attempts to correlate 3D molecular structure to biological activity, often using a variety of molecular descriptors such as physicochemical, topological, electronic and steric properties (Nantasenamat et al., 2009). In particular, 3D Atomic Property Fields (APF) QSAR methods developed by ICM calculate physico-chemical properties of superimposed chemicals and utilize their half-inhibition data to weight contributions for each property through Partial-Least-Squares (PLS) regression modeling (Totrov, 2008). The developed QSAR model gave a quantitative ligand-based virtual screening approach to further evaluate the physico-chemical properties of compounds and estimate their IC_{50} values against $3CLpro^{SARS\text{-}CoV\text{-}2}$. Those top candidates were further evaluated in enzyme activity assay and IC_{50} experiment.

After screening half million compounds, 288 hits in total were identified from the FDA-approved compound library and the IBScreen database. The compounds were all predicted to be bound within the active site of 3CLpro in a position similar to the crystallographic ligands. QSAR model assesses the physicochemical properties of identified compounds and estimates their inhibitory effects on $3CLpro^{SARS\text{-}CoV\text{-}2}$. The training dataset for the QSAR model had a good quality fit ($R^2 = 0.8967$), while the testing dataset suggested the predicted IC_{50} was still correlated to the actual IC_{50} ($R^2 = 0.7257$). The 288 identified hits were input into the developed QASR model to estimate half-inhibition values. The predicted IC_{50} for each compound were ranged from 0.35 μM to 46.7 μM. Top 71 compounds with predicted IC_{50}'s ranging from 0.35 μM to 19.86 μM, were selected for further evaluation in an enzyme activity assay. The top two compounds were confirmed by experiments to effectively inhibit the activity of $3CLpro^{SARS\text{-}CoV\text{-}2}$, with IC_{50} values of 19+/-3 μM and 38+/-3 μM, respectively. The functional groups pyrimidinetrione and quinoxaline were newly found in 3CLpro inhibitors, thus they are of high interest for lead optimization. In future studies, cellular infection and animal testing could be conducted to validate the efficacy and safety of the two newly identified compounds.

References:

Jin, Z., Du, X., Xu, Y., Deng, Y., Liu, M., Zhao, Y., Zhang, B., Li, X., Zhang, L., Peng, C., Duan, Y., Yu, J., Wang, L., Yang, K., Liu, F., Jiang, R., Yang, X., You, T., Liu, X., Yang, X., Bai, F., Liu, H., Liu, X.,

Guddat, L.W., Xu, W., Xiao, G., Qin, C., Shi, Z., Jiang, H., Rao, Z., and Yang, H. (2020). Structure of Mpro from SARS-CoV-2 and discovery of its inhibitors. *Nature* 582, 289-293.

Kitchen, D.B., Decornez, H., Furr, J.R., and Bajorath, J. (2004). Docking and scoring in virtual screening for drug discovery: methods and applications. *Nat Rev Drug Discov* 3, 935-949.

Liu, W., Zhu, H.M., Niu, G.J., Shi, E.Z., Chen, J., Sun, B., Chen, W.Q., Zhou, H.G., and Yang, C. (2014). Synthesis, modification and docking studies of 5-sulfonyl isatin derivatives as SARS-CoV 3C-like protease inhibitors. *Bioorg Med Chem* 22, 292-302.

Liu, Y., Liang, C., Xin, L., Ren, X., Tian, L., Ju, X., Li, H., Wang, Y., Zhao, Q., Liu, H., Cao, W., Xie, X., Zhang, D., Wang, Y., and Jian, Y. (2020). The development of Coronavirus 3C-Like protease (3CL(pro)) inhibitors from 2010 to 2020. *Eur J Med Chem* 206, 112711.

Nantasenamat, C., Isarankura-Na-Ayudhya, C., Naenna, T., and Prachayasittikul, V. (2009). A practical overview of quantitative structure-activity relationship. *EXCLI Journal* 8, 74-88.

Ramajayam, R., Tan, K.P., Liu, H.G., and Liang, P.H. (2010). Synthesis, docking studies, and evaluation of pyrimidines as inhibitors of SARS-CoV 3CL protease. *Bioorg Med Chem Lett* 20, 3569-3572.

Totrov, M. (2008). Atomic Property Fields: Generalized 3D Pharmacophoric Potential for Automated Ligand Superposition, Pharmacophore Elucidation and 3D QSAR. *Chemical Biology & Drug Design* 71, 15-27.

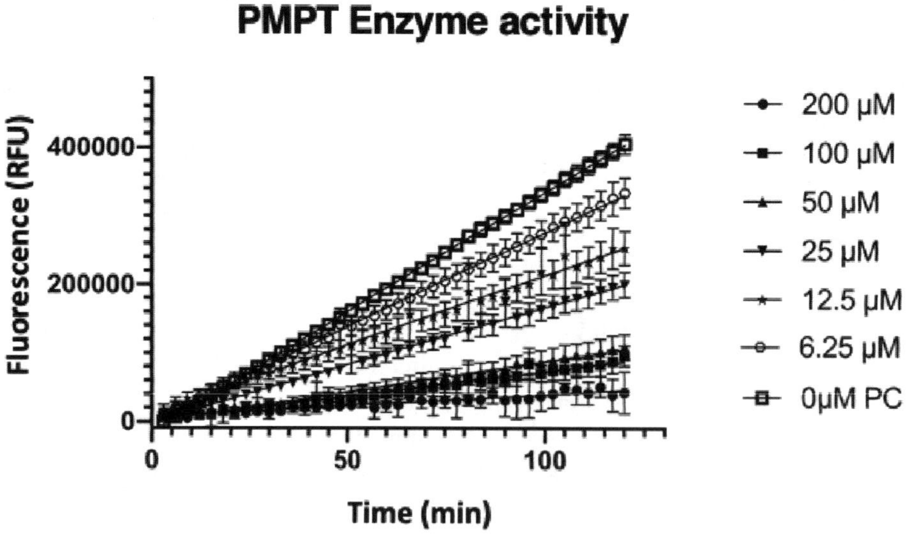

(A)

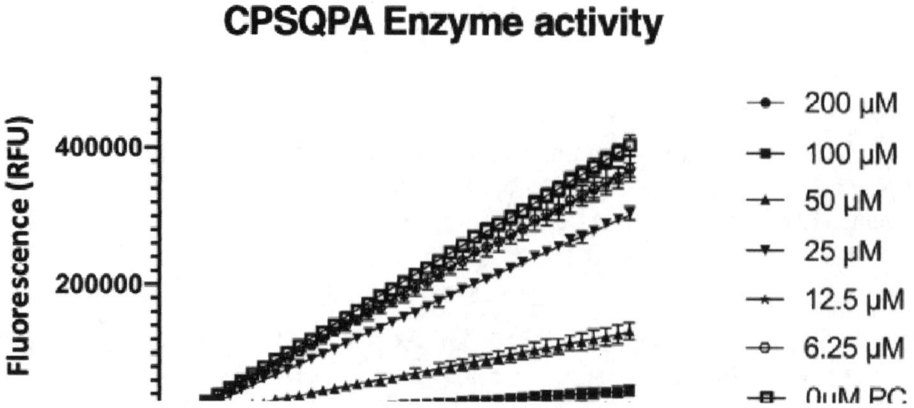

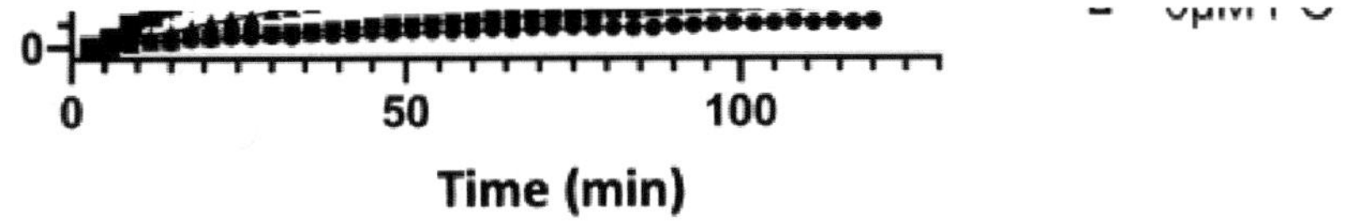
0
0
50
100
Time (min)

(B)

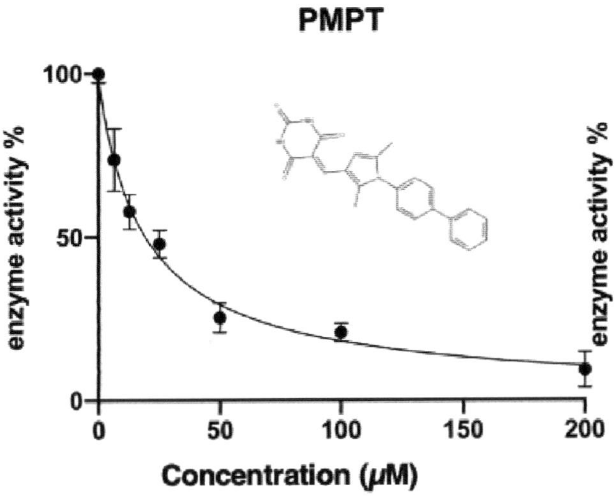
PMPT
100
50
0
enzyme activity %
0
50
100
150
200
Concentration (μM)

(C)

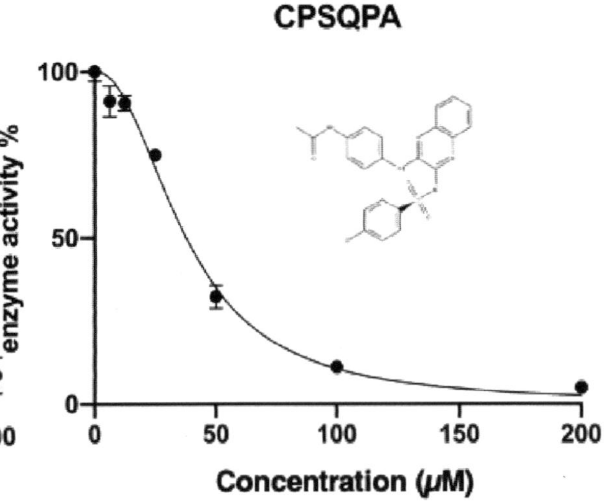
CPSQPA
100
50
0
enzyme activity %
0
50
100
150
200
Concentration (μM)

(D)

625286 Multi-Fidelity Deep Neural Network for the Prediction of Mechanical Properties of Polymer Matrix Composites for AI-Assisted Materials Design

Abstract:

Automotive material design is a traditional inverse problem of optimizing mechanical properties, and it requires a lot of predictions of mechanical properties and behavior depending on constructed materials. One of the efforts to minimize trial and errors from the design of engineering plastic materials to commercialization is the establishment of a Computer-Aided Engineering (CAE) system for automated material design. Even with an automated workflow, the mechanical properties and behavior of polymeric materials must be predicted to operate the CAE system. Engineering plastics used in automobile materials require excellent long-term heat resistance, unlike general-purpose plastics, and mechanical properties with minimum tensile strength of 500 kg/cm^2 and impact strength of 6 $kgf{\cdot}cm/cm^2$ or more over a wide temperature range. In order to consider the aforementioned conditions, the information on mechanical properties and behaviors required when designing polymer matrix composites includes tensile strength, elastic modulus, maximum load, maximum stress, and breaking point stress. Many of them can be easily derived from stress-strain curves (S-S curve) and creep curves obtained from tensile tests, impact tests, and creep tests, or can be obtained from models with sufficient accuracy. Therefore, the prediction of the S-S curve and creep curve is the key to the mechanical property behavior, and if this part is automated, the entire CAE can be automated.

In this study, the S-S curve and creep curve prediction model considering microstructure for the combination of 2~5 components of matrix types such as amorphous polymer and semi-crystalline polymer and fillers such as ceramic powder, glass fiber (GF), carbon fiber(CF), carbon nano tube, etc., which are often used as materials for automobiles. This time, version 1 developed a model using point data for measuring the mechanical properties of pure materials, and version 2 was developed to predict the mechanical properties of pure materials without information on the mechanical properties of the pure material. In this presentation, as the first achievement of the study, a data-driven model was created based on the mechanical property behavior of the combination that the research team has already accumulated.

In this paper, a system for predicting mechanical property behavior of binary polymer matrix composites is developed using deep neural network methodology by learning tensile test data for two different compositions in a binary combination of matrices and fillers. Numerical analysis methods such as FEM, which are conventionally used to predict the behavior of S-S curve, take a long time to calculate, and for complex models, calculation performance is poor. Therefore, studies are being attempted to overcome this through (large-scale) data-based prediction. The polymer matrix composites for which mechanical properties are predicted are a combination of matrices such as PA6, PP, PC and PA6,6 and fillers such as Al_2O_3, Al_2O_5Si, Si_3N_4, and Bn. This material combination is currently actively used as a main material for transportation equipment such as electric vehicles, and there are many tensile tests and physical property data, which is an advantage for prediction using big data-based machine learning. Tensile test data is removed because there are many unnecessary string information, and all strings are converted to integers and used for prediction. Test data are stress (MPa), stroke elongation (%), stroke (mm), and load (N) data over time, measured at 0.001 sec intervals. Since it is data at a time interval of 0.001sec, about 1.5 million data had to be learned in the tensile test, so the time interval is adjusted to improve performance.

This prediction model predicts the stress value according to the strain at a given combination and composition by using the following tensile test data of the polymer composite resin. The strain at the given combination and composition and the elastic modulus, density, molecular weight, and structure (as SMILES) information of each matrix/filler pure material were entered. Increased performance. Tensile test data was randomly divided into train, validation, and test data set to proceed with learning. The deep neural network structure used 5 layers, and the hyperparameter values were obtained through bayesian optimizer.

Finally, when the verification was conducted through cross-validation, compared with the existing experimental data, the performance of R^2=~0.83 value was shown, and the mechanical properties under conditions that were not used for training were predictable. As a result, this is after verification and performance verification of functional matrix test data mainly used in automobiles and electronic products, and the proposed hybrid prediction model showed improved performance compared to the existing first principle model or data-based model. The proposed model is predicted by reflecting not only data but also physical properties, and it is expected that it will be possible to predict only mechanical and physical properties without relying on experimental data in the future.

Although accuracy results are obtained with single fidelity previously, there is not always sufficient accuracy data for the matrix and filler, so studies based on multi-fidelity that use less expensive experimental data sets are also in progress. By adding one more single-fidelity model to the single-fidelity model above, we complete the Multi-fidelity Deep Neural Network in which two Single-fidelity NNs using different fidelity data are layered. Multi-fidelity DNN uses tensile test data (high-fidelity, HF) and CAE simulation data (low-fidelity, LF) universally, so it has an average of 5% performance with fewer data sets compared to models using only HF data. It is expected to show improvement. With the addition of the LF data-driven single-fidelity model, human errors and experimental errors in the test can be compensated for, thereby improving the predictive performance in situations where the mechanical property data set is insufficient.

The prediction of mechanical properties of polymer matrix composites has been less studied than inorganic or metals, and the information and properties of pure materials and data-driven machine learning without microstructures or microscopic images are the first attempts by our researchers. Ultimately, for materials mainly used in automobiles and electronic products, we want to support the function of automatically designing engineering plastics that meet the target properties with only pure material information without tensile test data in matrix/filler combinations in 2 to n components.

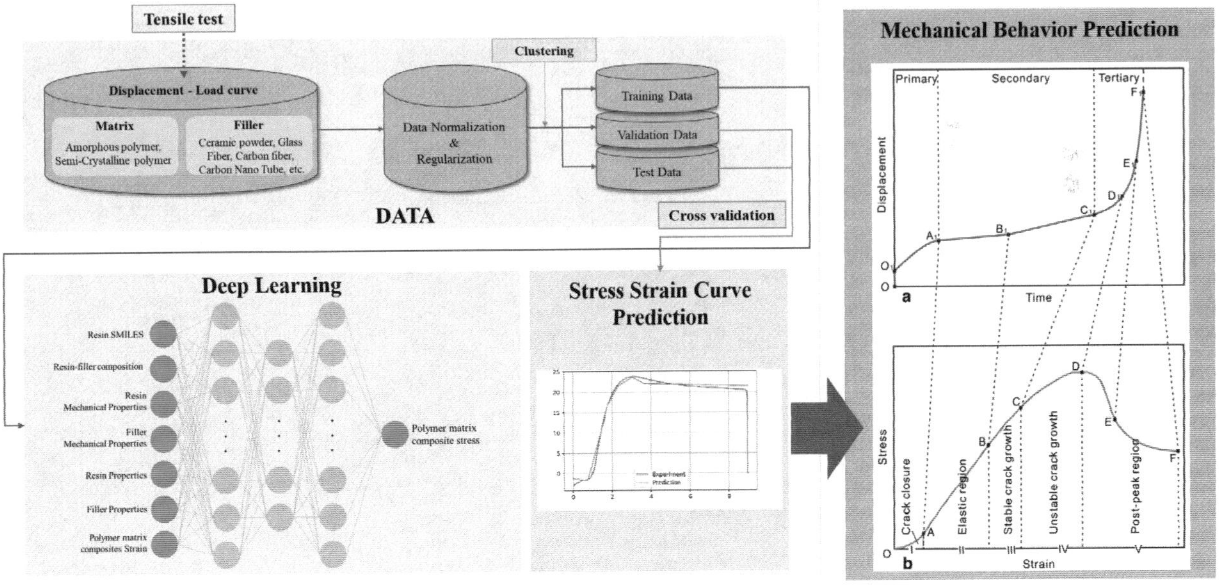

626222 Assessment of Commercially Available Packaging Films for Hot Sauce Using Contact Angle Analysis Combined with Principal Component Analysis

Abstract:

The aggressive nature of hot sauces prevents its long-term storage in many commercially-available packaging films. The goal of this work is to monitor changes in surface properties, focused on surface energy and wettability, of packaging films during exposure to hot sauce. Two monolayer and six multilayer packaging films were assessed during immersion in hot sauce at elevated temperature to accelerate aging for up to 12 weeks. Changes in contact angles and surface energies were measures of surface degradation of the films. Principal component analysis (PCA) was carried out based on these data to reduce the dimensionality and discern the most influential variables in the dataset. For PCA, contact angle measurements as well as calculated surface energies were inputs. Cluster analysis of the results of PCA led to separate the packaging films into two groups (Cluster 1 & Cluster 2), with different extents/types of surface degradative phenomena. Cluster 1 films have higher total surface energy compared to Cluster 2 films. Cluster 1 films exhibited greater changes in the values of principal components 1 and 2 during long-term exposure to hot sauce, which is representative of the more stable surface properties of Cluster 2 films over time. Combined, these results suggest that Cluster 2 films may perform better as hot sauce packaging materials. These types of approaches may also be appropriate for screening of films for other packaging applications.

626815 Combined Experimental, Molecular Simulations and Machine Learning Approach to Optimize Aqueous Amine Solvents for CO_2 Capture

Abstract:

Tertiary amines are attractive for industrial CO_2 capture because they tend to be more stable, more energy-efficient and show a higher CO_2 absorption capacity than other amines. One of the current challenges in the CO_2 capture industry is identifying tertiary amines with a higher CO_2 absorption rate, which is their main weakness. Understanding the relationship between amine structure and properties and CO_2 absorption rate is required for efficient amine screening. Despite numerous attempts, this goal has been elusive.

First, we have developed a computational approach based on molecular modeling to predict the CO_2 absorption rates and energies in aqueous tertiary amines. The predictions are validated using experimental data. The key to the new model's excellent accuracy is to include the solvation effects in the computation of the free energy barrier for the reaction of CO_2 absorption (https://doi.org/10.1021/acs.jcim.0c01386).

In a second step we have generated a diverse library of 100 tertiary amines which represent a large portion of chemotypes relevant for industrial amine-based chemical CO_2 absorption. Using the molecular simulations-based model, we have calculated the CO_2 absorption rates and energies for those 100 aqueous amine solvents. As a check, some points have been verified experimentally.

In a third step we have used the dataset of the CO_2 absorption rates and energies for the 100 aqueous amines to build quantitative structure-property relationship models. These models allowed us to perform a large-scale virtual screening of commercially available amines and to identify the amines with optimal CO_2 absorption rate and energy. For several amines the predictions have been verified experimentally. The approach and the results will be presented.

623631 Maintaining Atomic Precision in Thiolate-Protected Alloy Nanoclusters

Abstract:

Thiolate-protected metal nanoclusters (TPNCs) have attracted great interest as a unique class of atomically precise nanomaterials with a wide range of applications, which includes sensing, imaging, and catalysis. Due to their high stability at magic size compositions ($M_n(SR)_m$), TPNCs are often regarded as inorganic-organic hybrid molecules, thus differentiating them from larger metal nanoparticles. With a growing catalogue of structures, TPNCs make up the ideal "playground" to distill fundamental nanomaterials properties from data-driven analyses[1]. Recent efforts in the field have sought to modulate TPNC properties through heterometal doping, achieving enhanced catalytic activity and realization of new magic sizes. However, with the increase in structural complexity, new challenges arise in structural determination with atomic precision. The problem is only magnified at larger TPNC sizes (> 100 metal atoms), which offer a massive configuration space of potential alloy chemical orderings (i.e. how the metals position themselves in a TPNC). Herein, we introduce a computational framework to predict thermodynamically favorable dopant positions and concentrations in alloy TPNCs. We first performed Density Functional Theory calculations on Ag-doped Au TPNCs that range in size (36- to 279-metal atoms), metal composition, and chemical ordering. Our efforts yielded a rich dataset of 360 optimized alloy TPNC structures. Next, we extended the Bond-Centric Model[2] (which determines metal nanoparticle stability) to capture alloy LPNC stability. Importantly, we coupled data science with our domain TPNC knowledge to develop physics-based tuning parameters that were trained and validated with our extensive dataset. The resulting model accurately predicts thermodynamic stability across LPNC size, shape, metal composition, and ligand type. Finally, we incorporated our model into our previously developed nanoparticle genetic algorithm[2] to conduct a high-throughput screening across the alloy TPNC space. The analysis produced a vast population of hypothetical alloy TPNCs, enabling us to determine structure-based metal mixing behavior of ligated nanostructures.

1. Cowan, M. J.; Mpourmpakis, G., Towards elucidating structure of ligand-protected nanoclusters. *Dalton Trans.* **2020,** *49* (27), 9191-9202.

2. Dean, J.; Cowan, M. J.; Estes, J.; Ramadan, M.; Mpourmpakis, G., Rapid Prediction of Bimetallic Mixing Behavior at the Nanoscale. *ACS Nano* **2020,** *14* (7), 8171–8180.

630847 Addressing Well-Posedness in a Data-Driven Manner: Model Problems and Physics-Informed Neural Networks

Abstract:

In recent years, the (re)emergence of physics-informed neural networks (PINNs) [1,2] has been proposed as an alternative to direct numerical simulations (DNS) for producing solutions to differential equations (DEs). The key idea is that the easy access to derivatives of the approximant can be used to create a loss function based on the residuals of the DE at all discretization locations in the (space-)time domain. This is in addition to another important loss term--that the solution should match provided values or derivative(s) at particular supervised points--conditions analogous to traditional initial and boundary conditions for DNS. A major strength of the PINN approach is that the supervisory data need not be prescribed at the domain locations typically used for initial and boundary conditions, but can instead be opportunistically sampled throughout the domain of interest (``internal" as opposed to ``boundary" conditions).

Another apparent advantage is that the variational solution of the PINN problem will return results even if the original problem is not well posed (whether due to overspecification of underspecification of constraints). However, though the method will produce an approximant at all domain locations, and a well-trained PINN will extrapolate reasonably to all locations where the DE-residual term is enforced, this does not mean that the usual issues of well-posedness, critical for DNS, are absent for PINNs. Here, we explore (for both the under-constrained and the over-constrained case) how the lack of well-posedness affects the approximant produced by the PINN training. We use this variational approach for PINNs, but also for more traditional numerical discretizations, and conclude with some recommendations on metrics for diagnosing over- and under-determinedness in PINN training.

[1]: Raissi, Maziar, Paris Perdikaris, and George E. Karniadakis. "Physics-informed neural networks: A deep learning framework for solving forward and inverse problems involving nonlinear partial differential equations." Journal of Computational Physics 378 (2019): 686-707.
[2]: Lagaris, Likas, Fotiadis. "Artificial neural networks for solving ordinary and partial differential equations." IEEE Transactions on Neural Networks. 9(5) (1998): 987-1000.

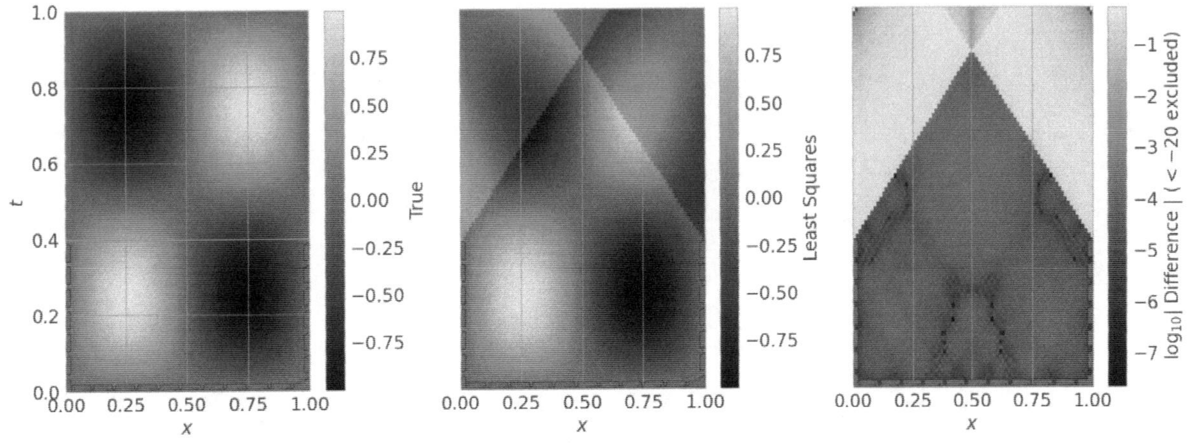

An underconstrained finite difference solution of the wave equation (left), what is systematically recoverable (middle) and the difference between the two (right). We solve $u_{tt} - c^2 u_{xx} = 0$ at all grid points, and match the true solution within red dashed regions (one grid column at left and one at right, and two grid rows across the bottom).

626780 Prediction of Atoms in Molecules with Matlab Graph Convolutional Network

Abstract:

Data-driven modeling methods is becoming more popular in chemical engineering and chemistry applications. Machine learning techniques are used for applications such as grouping similar equipment to detect outliers, fault detection and diagnosis, process inferential and optimization in chemical engineering space. Deep learning helps engineers with tasks such as identifying and tagging equipment, detecting defects, and developing predictive models to sort molecules based on their mass spectra data. Graph convolutional networks can be used directly on graphs and they utilize structural information. Molecules and social networks can be represented as graphs which makes them great inputs for graph convolutional networks. In a graph structure of a molecule, atoms in the molecules correspond to the nodes and chemical bonds correspond to the edges. With a graph convolutional neural network, types of unlabeled atoms in the molecule can be predicted from the molecules' graph representations.

In this talk, we will share user stories on applications of data science to molecules and we will share a MATLAB workflow example of how graph convolutional networks can be used to predict the atoms in a molecule using QM7 data set.

629782 Deployment of Machine Learning Models in Pharmaceutical Development

Abstract:

The advantage of model driven decisions in research organizations has long been recognized and applied in the pharmaceutical research and development. From the conceptualization of the Quality by Design framework as a series of mathematical constructs [1,2] to the application of digital twins [3]. Machine Learning methodologies are widely applicable and efficient tools to construct mathematical models of relevant pharmaceutical processes enabling process optimization, robustness characterization, and failure rate estimation. However, the widespread adoption of these approaches requires a combination of streamlined availability, and institutional know-how to interpret and incorporate the results in the decision-making process.

This talk addresses different approaches to leverage emerging and stablished tools in Machine Learning to pharmaceutical development applications by comparing the adoption of specific algorithms and computational platforms. Specifically, a generic workflow for data exploration analysis, design of experiments, feature selection, and model exploration has been developed and deployed with a combination of cloud computing and Jupyter notebooks. In our experience this approach has been more successful at integrating a model-based decision making culture in process development than previously attempted alternatives.

[1] Peterson, J.J., 2008. A Bayesian approach to the ICH Q8 definition of design space. *Journal of biopharmaceutical statistics*, *18*(5), pp.959-975.

[2] Garcia-Munoz, S., Luciani, C.V., Vaidyaraman, S. and Seibert, K.D., 2015. Definition of design spaces using mechanistic models and geometric projections of probability maps. *Organic Process Research & Development*, 19(8), pp.1012-1023

[3] Chen, Y., Yang, O., Sampat, C., Bhalode, P., Ramachandran, R. and Ierapetritou, M., 2020. Digital Twins in Pharmaceutical and Biopharmaceutical Manufacturing: A Literature Review. *Processes*, *8*(9), p.1088.

626098 Unsupervised Learning from Sets of Data Using Contrastive Latent Variable Models

Abstract:

In unsupervised learning, dimensionality reduction is an important tool for data exploration and visualization. Because these aims are typically open-ended, it can be useful to frame the problem as looking for patterns that are enriched in one dataset relative to another. These pairs of datasets occur commonly, for instance a population of interest vs. control or signal vs. signal free recordings.However, there are few methods that work on sets of data as opposed to data points or sequences. Here, we present a probabilistic model for dimensionality reduction to discover signal that is enriched in the target dataset relative to the background dataset. The data in these sets do not need to be paired or grouped beyond set membership. By using a probabilistic model where some structure is shared amongst the two datasets and some is unique to the target dataset, we are able to recover interesting structure in the latent space of the target dataset. The method also has the advantages of a probabilistic model, namely that it allows for the incorporation of prior information, handles missing data, and can be generalized to different distributional assumptions. We describe several possible variations of the model and demonstrate the application of the technique to de-noising, feature selection, and subgroup discovery settings.

627704 Energy Dispersive X-Ray Hyperspectral Image Analysis and Chemometrics for Catalyst Characterization

Abstract:

Multi-dimensional images with chemical information embedded at pixel-level (so-called Hyperspectral Imaging) is the next generation of Image Analysis and Machine Vision. This advanced technology has attracted the attention of different industry sectors in the last two decades, ranging from metalurgic, mining and pharma, to high-value-added manufacturing industries such as biotechnology and biomedical imaging. Recently, the combination of Scanning Transmission Electron Microscopy (STEM) and Energy Dispersive X-ray spectroscopy (EDX) has been developed to characterize the surface area of catalysts in the chemical industry. The measurements are done by scanning the electron beam over the sample and at the same time measuring the X-ray spectrum – so-called spectral imaging [1]. The spectral dimension of the hyperspectral images consists of X-ray counts acquired at different energy channels. Chemical data is corrupted with noise, which comes from the time-dependent arrival of discrete particles on the sensor. This noise typically follows a Poisson distribution, which represents the probability of occurrence or events (X-ray counts) during a given period of time.

The application of Multivariate Image Analysis (MIA) techniques and Multivariate Curve Resolution (MCR) models becomes essential for the analysis of EDX hyperspectral images. This contribution proposes a modeling framework that permits segregating hyperspectral X-Ray images into simpler images (so-called Distribution Maps, DM's), which can be directly related to each of the chemical compounds present in the mixture [2]. From these DM's, chemical-textural score images (SI's) are further obtained. Finally, from all these DM's and SI's, different types of features, such as quantitative, morphological or textural can be extracted and combined into new data structures [3]. This new source of information is used to build multivariate statistical models for process understanding and prediction purposes at a MIA image-based level. This approach allows us to study similarities and differences between and within types of catalysts -i.e. different samples of the same catalyst across batches, and across locations, and the potential effects on quality properties of interest. To illustrate the chemometric framework for catalyst characterization, STEM-EDX images of real industrial catalyst will be used.

[1] R. Haswell, D. W. McComb and W. Smith, Preparation of site-specific cross-sections of heterogeneous catalysts prepared by focused ion beam milling, Journal of Microscopy (2003) Vol. 211, Pt 2, pp. 161-166
[2] J.M. Prats-Montalbán, A. de Juan, A. Ferrer, Multivariate image analysis: a review with application, Chemometrics and Intelligent Laboratory Systems, 107: 1-23, 2011.
[3] C Duchesne, JJ Liu, JF MacGregor, Multivariate image analysis in the process industries: A review, Chemometrics and Intelligent Laboratory Systems 117, 116-128, 2012.

634389 Applications of machine learning in production enhancement and field monitoring in Oil & Gas industry

Abstract:

Due to the fast advancement of machine learning technologies and availability of diverse types of data collection platforms, we have been seeing the growing machine learning applications in a wide spectrum of industries, spanning from emerging industries, e.g., autonomous driving, to traditional ones, such as Oil & Gas (O&G).

In this talk, we will discuss three different machine learning applications in the O&G industry. In specific, we will discuss a machine learning-based approach for understanding the multiphase flow system studied in the experimental fluid mechanics; which shows the potential of developing a complimentary methodology to assist the first principal physics; secondly, we will describe a deep learning-based object detection system that can be used to monitor the operation field or the natural environment, with the goal to improve operation safety and meet the regulation requirement ; Thirdly, we will introduce the content understanding of Process & Instrument Diagrams (P&IDs) using computer vision techniques.

Each of the above studies represents a unique usage scenario in the value chain of O&G industry. We would like to demonstrate the diverse types of machine learning applications and discuss their respective technical challenges that O&G industry presents.

AUTHOR INDEX